Adaptive Filtering

I0988286

Adaptive Filtering

Fundamentals of Least Mean Squares with MATLAB®

Alexander D. Poularikas
University of Alabama, Huntsville, AL

CRC Press
Taylor & Francis Group
Boca Raton London New York

CRC Press is an imprint of the
Taylor & Francis Group, an **informa** business

MATLAB® is a trademark of The MathWorks, Inc. and is used with permission. The MathWorks does not warrant the accuracy of the text or exercises in this book. This book's use or discussion of MATLAB® software or related products does not constitute endorsement or sponsorship by The MathWorks of a particular pedagogical approach or particular use of the MATLAB® software.

CRC Press
Taylor & Francis Group
6000 Broken Sound Parkway NW, Suite 300
Boca Raton, FL 33487-2742

© 2015 by Taylor & Francis Group, LLC
CRC Press is an imprint of Taylor & Francis Group, an Informa business

No claim to original U.S. Government works

Printed on acid-free paper
Version Date: 20140814

International Standard Book Number-13: 978-1-4822-5335-1 (Paperback)

This book contains information obtained from authentic and highly regarded sources. Reasonable efforts have been made to publish reliable data and information, but the author and publisher cannot assume responsibility for the validity of all materials or the consequences of their use. The authors and publishers have attempted to trace the copyright holders of all material reproduced in this publication and apologize to copyright holders if permission to publish in this form has not been obtained. If any copyright material has not been acknowledged please write and let us know so we may rectify in any future reprint.

Except as permitted under U.S. Copyright Law, no part of this book may be reprinted, reproduced, transmitted, or utilized in any form by any electronic, mechanical, or other means, now known or hereafter invented, including photocopying, microfilming, and recording, or in any information storage or retrieval system, without written permission from the publishers.

For permission to photocopy or use material electronically from this work, please access www.copyright.com (http://www.copyright.com/) or contact the Copyright Clearance Center, Inc. (CCC), 222 Rosewood Drive, Danvers, MA 01923, 978-750-8400. CCC is a not-for-profit organization that provides licenses and registration for a variety of users. For organizations that have been granted a photocopy license by the CCC, a separate system of payment has been arranged.

Trademark Notice: Product or corporate names may be trademarks or registered trademarks, and are used only for identification and explanation without intent to infringe.

Library of Congress Cataloging-in-Publication Data

Poularikas, Alexander D., 1933-
Adaptive filtering : fundamentals of least mean squares with MATLAB / author, Alexander D. Poularikas.
pages cm
Includes bibliographical references and index.
ISBN 978-1-4822-5335-1 (paperback)
1. Adaptive filters--Mathematical models. 2. Adaptive signal processing--Mathematics. 3. Least squares--Data processing. 4. MATLAB. I. Title.

TK7872.F5P678 2014
621.382'2--dc23 2014027999

Visit the Taylor & Francis Web site at
http://www.taylorandfrancis.com

and the CRC Press Web site at
http://www.crcpress.com

Contents

Preface

This book is written for the applied scientists and engineers who want or need to learn the fundamentals about this subject, but are not experts in this specific field. It is also written to accompany a first graduate course in digital signal processing. In this book, we have selected the most important part, the least mean-square (LMS) adaptive filter, which is part of the statistical signal processing area. The adaptive filters have found their use in many and diverse fields such as communications, control, radar, sonar, and seismology.

The aim of this book is to present in a fundamental way the area of adaptive filtering concentrating to a specific type of adaptive filters and their variations, known as the LMS adaptive filter. All the adaptive filters are of the finite duration or finite impulse response filters (FIRs). Since the signals are random or nonrandom with additive random signals, noise, an introduction to random variables and stochastic processes is also included. Because many of the formulas contain vectors and matrices, appropriate chapters were included.

The book contains all the material necessary for the reader to study the content without the need to refer to other sources. It includes appendices to provide the needed extra supporting material. It also includes a number of MATLAB® functions and m-files for practising and verifying the material provided in the text. In addition, it provides more than 110 Book MATLAB functions and script files, which specifically were developed to verify the theory and proposed examples. It also provides many computer experiments to illustrate theories such as the Wiener and adaptive filtering. At the end of each chapter, numerous problems are provided to help the reader develop a deeper understanding of the presented material. Detailed solutions or hints and suggestions for solving all the problems are provided.

Chapter 1 covers vectors and their properties; Chapter 2 covers matrices and determinants; Chapter 3 provides a synopsis of the deterministic discrete systems and signals; Chapter 4 covers the fundamentals of the discrete random signals, probability distributions, and their properties as well as their simulations; Chapter 5 presents the Wiener filter and its practical uses; Chapter 6 explains how to find the eigenvalues and eigenvectors of a matrix and the properties of the error surfaces; Chapter 7 presents the steepest descent method and develops the Newton's algorithm; Chapter 8 covers the basics of the celebrated LMS adaptive filter algorithm and presents numerous examples for different types of applications; and Chapter 9 covers the variants of the LMS adaptive filters. Appendix 1 explains the use of the MATLAB, Appendix 2 covers the fundamentals of the matrix theory, Appendix 3 presents some mathematical formulas, and Appendix 4 develops the method of Lagrange multiplier method.

MATLAB® is a registered trademark of The MathWorks, Inc. For product information, please contact:

The MathWorks, Inc.
3 Apple Hill Drive
Natick, MA 01760-2098, USA
Tel: +1 508 647 7000
Fax: +1 508 647 7001
E-mail: info@mathworks.com
Web: www.mathworks.com

Author

Alexander D. Poularikas received his PhD from the University of Arkansas and became a professor at the University of Rhode Island. He was chairman of the engineering department at the University of Denver, Colorado, and then became chairman of the electrical and computer engineering department at the University of Alabama in Huntsville. He has published, coauthored, and edited 14 books. Dr. Poularikas served as an editor-in-chief of the *Signal Processing Series* (1993–1997) with Artech House and is now an editor-in-chief of the *Electrical Engineering and Applied Signal Processing Series* as well as the *Engineering and Science Primers Series* (1998 till present) with Taylor & Francis Group. He was a Fulbright scholar and is a lifelong senior member of the Institute of Electrical and Electronics Engineers (IEEE) and a member of Tau Beta Pi, Sigma Nu, and Sigma Pi. In 1990 and 1996, he received the IEEE Outstanding Educators Award, Huntsville Section.

Abbreviations

AMWM	asymmetric modified Welch method
ARMA	autoregressive moving average
BIBO	bounded input bounded output
BT	Blackman–Tukey
cdf	cumulative distribution function
CRLB	Cramer–Rao lower bound
DFT	discrete Fourier transform
DTFT	discrete-time Fourier transform
FFT	fast Fourier transform
FIR	finite impulse response
FT	Fourier transform
ENSS	error normalized step size
IDFT	inverse discrete Fourier transform
IFFT	inverse fast Fourier transform
iid	independent and identically distributed
IIR	infinite impulse response
LMMN	least mean mixed norm LMS algorithm
LMF	least mean fourth algorithm
LMS	least mean-square algorithm
LS	least squares
LSI	linear shift invariance
LSE	least-squares estimator
LTI	linear time invariance
MMSE	minimum mean-square error
MSE	mean-square error
MVU	minimum variance unbiased
NLMS	normalized LMS algorithm
pdf	probability density function
PNLMS	power normalized LMS
pr	probability
PSD	power spectral density
QAM	quadrature amplitude modulation
rv	random variable
RVSS	robust variable step size
RW	random walk
SCLMSF	self-correcting LMS filter
SCNLMSF	self-correcting normalized LMS filter
SCWF	self-correcting Wiener filter
SMWM	symmetric modified Welch method
SNP	signal-to-noise power
SPMWM	symmetric predicted modified Welch method

TDLMS	transform domain LMS
VSLMS	variable step-size LMS
WN	white noise
WSS	wide-sense stationary process

MATLAB® Functions

`[m,z]=hist(x,a)`	m = counts in each bin; z = coordinates of the a bins; w = max(x)/length(z); $bp = m/(w*a)$ = probability per bin;
`[r,p,k]=residue(nu,de)`	Example: $F(z) = z^2/(z^2 - 0.04)$, $nu = [1 \quad 0 \quad 0]$, $de = [1 \quad 0 \quad -0.04]$, $r = [0.1000 \quad -0.1000]$, $p = [0.2000 \quad -0.2000]$, $k = 1$, hence $F(z) = 0.1(z/(z-0.2)) - 0.1(z/(z+0.2)) + 1$, the inverse is $f(n) = 0.1(0.2)^n - 0.1(-0.2)^n + \delta(n)$; r = residues, p = poles
`[r]=xcorr(x,y,'biased')`	Produces a biased cross-correlation; if $x = y$, r = autocorrelation; the r is an $2N-1$ symmetric function with respect to 0
`[r]=xcorr(x,y,'unbiased')`	Produces an unbiased cross-correlation; if $x = y$, r = autocorrelation
`\omega`	Produces the lowercase omega, the same for the rest of the Greek letters
`\Omega`	Produces the uppercase omega, the same for the rest of the Greek letters
`A(:,1)`	Gives $[1 \quad 4]^T$ for matrix $A = [1 \quad 2;4 \quad 3]$, all rows of the first column
`A(2,:)`	Gives $[2 \quad 4]$ for the matrix $A = [1 \quad 2;4 \quad 3]$, all columns for the second row
`a'*a`	$\boldsymbol{a}^T.\boldsymbol{b}^T$ is a row vector $\boldsymbol{c} = (a_1b_1, a_2b_2, \ldots, a_nb_n)$
`abs()`	Gives the absolute value of the complex expression in the parentheses
`acos()`	Produces the inverse of the cosine
`angle()`	Gives the angle of a complex number or vector
`asin()`	Produces the inverse of the sine
`atan()`	Produces the inverse of the tangent
`axis([0 2 -2 5])`	Creates axes with length 0 to 2 in the x-axis and −2 to 5 in the y-axis
`bar(z,p)`	Plots bars at the x-coordinates of z and heights p
`bar(z,pb)`	Plots the probability distribution with z bars
`color designation`	'y' = yellow, 'm' = magenta, 'c' = cyan, 'r' = red, 'g' = green, 'b' = blue, 'w' = white, 'k' = black
`colormap(gray)`	Produces black figure; see also `meshgrid()`
`conj()`	Gives the complex conjugate of a complex number or expression
`conj(z)`	Gives the conjugate value of the complex number z, $\text{conj}(3 - j*5) = 3.0000 + 5.0000i$
`contour(X,Y,Z,15)`	Produces 15 contours [see also `meshgrid()` below]; contour (Z, [2.5 3.1 5]) will produce 3 contours at heights 2.5, 3.1, and 5
`contour(X,Y,Z,V,'g')`	$\boldsymbol{V} = [a \quad b \quad c]$ is a vector with values a, b, and c which can be any number; the contour will have green color

(*Continued*)

`contour(Z,[1.6 3.1 5])`	Produces contours of the performance surface at heights 1.6, 3.1, and 5
`contour(Z,30)`	Creates 30 contours of the performance surface
`conv(x,y)`	Convolution between two vectors `x` and `y`; the output vector signal will have length:length (x) + length $(y) - 1$
`det(A)`	Gives the determinant of A; we can also write det([1 2;4 3]) for 2×2 matrix
`dimpulse(nu,de,5)`	Inverts the z-transform. Example: $z^2/(z^3+0.1)$, $nu = [0\ \ 1\ \ 0\ \ 0]$, $de = [1\ \ 0\ \ 0\ \ 0.1]$, gives 5 terms of the time function
`dimpulse(num,den,a)`	Gives a numbers of the time function $f(n)$, num = vector with the values of the numerator coefficients of the z-function in descending order; den = same as for the num but for the denominator
`dot`	$\boldsymbol{a}.*\boldsymbol{b}$ is a column vector with elements $\boldsymbol{c} = (a_1b_1, a_2b_2, \cdots, a_nb_n)'$
`eig(R)`	$[Q,D]$ = eig(R), R = a correlation matrix, Q = each one of its columns equal to an eigenvector, D = a diagonal matrix with the eigenvalues of the correlation matrix
`eye(4)`	Creates a 4×4 identity matrix, its main diagonal have unit value
`fft(x,N)`	Takes the DFT of the sequence (vector) $\boldsymbol{x}$ having N elements
`figure(2)`	Creates a second figure and the next plot will be introduced at this new figure
`filter(b,a,x)`	$\boldsymbol{b}$ = a vector of the input coefficients; $\boldsymbol{a}$ = a vector of the output coefficients; $\boldsymbol{x}$ = a vector of the input to the system; $y(n)$ = an output $y(n)+\sum_{m=1}^{p} a_m y(n-m) = \sum_{m=0}^{q} b_m x(n-m)$
`fliplr(x)`	Flips a vector from left to right, x = a vector
`floor(a)`	Gives the number without the decimal, floor(5.68) = 5, floor(− 2.6) = −3
`freqz();` `[H,W]=freqz(A,B,N)`	B = a vector of the b's coefficients of the numerator of the transfer function $H(e^{j\omega})$; A = a vector of the a's coefficients of the denominator of $H(e^{j\omega})$; N = number of bins in the frequency domain; W = a frequency vector in rad/sample of the filter $H(e^{j\omega})$ = transfer function
`grid on`	After `plot()`, we write `grid on` and a grid will appear with horizontal and vertical lines
`hist(a,α)`	Creates a histogram with α number of bins from the elements of the vector a
`hold on`	We make a plot first; then we write `hold on` and then we proceed for the second plot on the same figure
`ifft(X,N)`	Takes the inverse DFT (IDFT) of the frequency sequence (vector) $\{\boldsymbol{x}\}$ with N elements
`inv(A)`	Gives the inverse of matrix A, A^{-1} inv([1 2;4 3]) to invert a 2×2 matrix
`legend('FT')`	Produces a legend identifying with the type of plotting, here FT
`length(a)`	Gives the number of the elements of the vector a

(Continued)

`log(a₁)`	Gives the natural logarithm of the number a_1
`log10(a₁)`	Gives the logarithm of the number a_1 with base 10
`matrix: A(4:6,8:15)`	Produces a 3×8 matrix from the elements of ***A*** that contains the rows 4 to 6 and columns 8 to 15
`matrix: A.^n`	Raises every element in the power of n
`mean(a)`	Gives the mean value of the elements of the vector a
`meshgrid(x,y)`	Example: $x = 0:0.1:4$; $y = 0:0.1:4$; $[X,Y] =$ meshgrid(x, y); $Z = 4 * X.\char94 2 + Y.\char94 3$; % (the functions we want to produce); surf(X,Y,Z); colormap(gray); %black figure; see color above
`norm(x)`	Gives the norm of $\boldsymbol{x}$, the square root of the sum of the square of the vector $\boldsymbol{x}$ coefficients number; the contours will have color green
`plot(x,y)`	Plots a blue color graph x versus y
`plot(x,y,'k')`	Plots a black color graph, 'g'=green, 'r'=red $r = [0.1000 \quad -0.1000]$, $p = [0.2000 \quad -0.2000]$, $k = 1$, hence
`rand(1,n)`	Produces a vector of n elements that are white noise with 0.5 mean value; rand(1,n) – 0.5 is a vector of n elements which are uniformly distributed with zero mean
`randn(1,n)`	produces a vector of n elements that are normally distributed
`rank(A)`	Gives the rank of the matrix A, the number of independent columns of the matrix
`roots(a)`	Give the roots of the above polynomial; $\boldsymbol{a}$ is a vector of the coefficients of the polynomial, e.g. $a\lambda^3 + b\lambda^2 + c\lambda + d = 0$, then we write $a = [a \quad b \quad c \quad d]$
`sign()`	Examples: $\text{sign}(0.2) = 1$, $\text{sign}(-2) = -1$, $\text{sign}(0) = 0$, $\text{sign}([-2 \quad 0.1 \quad 0 \quad -0.3 \quad -5]) = [-1 \quad 1 \quad 0 \quad -1 \quad -1]$
`sqrt()`	Takes the square root of a number
`std(x)`	Standard deviation = square root of the variance, x = a data vector
`stem(x,y)`	Produces a set of lines of magnitude of the set $\{\boldsymbol{y}\}$
`stem(x,y,'k','filled')`	Produces a black color graph with the top circles filled with black color
`subplot(4,1,3)`	Produces a subplot on the third row (3) of a figure that is made of four rows (4) and one column (1);
`subplot(5,2,3)`	Produces a figure of 5 rows and 2 columns and plots the third subplot
`sum(a)`	Sums the elements of a vector a
`sum(X,1)`	Sums all rows of the matrix X
`sum(X,2)`	Sums all the columns of the matrix X
`surf(X,Y,Z)`	Produces a three-dimensional figure; see also meshgrid();
`title('Figure')`	Produces the title "Figure" on the top of the figure
`trace(A)`	Sums the main diagonal of the matrix A
`transpose()`	Transposes the matrix
`var(a)`	Gives the variance of the elements of vector a
`var(x)`	Gives the variance of the data vector x

(*Continued*)

`xcorr(x,y)`	Gives two-sided correlation of the two vectors; if $x = y$, the one-sided correlation starts from length(x) to length(x) + length(y) − 1; if $x < y$, MATLAB adds zeros to the smaller, the one-sided correlation starts from length(x) and ends at 2length(y) − 1
`xlabel('x_{cD}')`	On the x-axis will appear x_{cD}
`xlabel('x-axis')`	On the x-axis will appear x-axis
`ylabel('y-label')`	On the y-axis will appear y-axis

1 Vectors

1.1 INTRODUCTION

In this text we present vectors with boldfaced lowercase Latin letters, in columnar format. Matrices are presented with boldfaced capital Latin letters and numbers with lowercase Latin letters. For example, for three different types of three element vectors, we write

$$\boldsymbol{x} = \begin{bmatrix} 1 \\ 2.3 \\ -2 \end{bmatrix}; \quad \boldsymbol{y} = \begin{bmatrix} y_1 \\ y_2 \\ y_3 \end{bmatrix}; \quad \boldsymbol{z} = \begin{bmatrix} 2.33 \\ 4 + j5 \\ -j \end{bmatrix} \tag{1.1}$$

The elements y_1, y_2, and y_3 are constants. For any one of the above vectors, we could also write its equivalent form, for example, $\boldsymbol{x}^T = [1 \quad 2.3 \quad -2]$ or $\boldsymbol{x}^T = (1, 2.3, -2)$, where the letter T stands for the word **transpose**. The MATLAB® column vector format is written in the command window as follows:

```
>>x = [1 2.3 -2]';%column vector, accent indicates transpose;
>>x = [1; 2.3; -2];%column vector;
>>x = [1 2.3 -2];%row vector; () are not used in MATLAB for
>>     %vector manipulations;
```

We can consider a column vector as a special **matrix** with m rows, in this case 3, and n columns, in this case 1.

1.1.1 Multiplication by a Constant and Addition and Subtraction

1.1.1.1 Multiplication by a Constant

Using two vectors from (1.1), we write $a_1\boldsymbol{x}^T = [a_1 1 \quad a_1 2.3 \quad -a_1 2]$ and $a_1\boldsymbol{z}^T = [a_1 2.33 \quad a_1 4 + a_1 j5 \quad -a_1 j]$, where a_1 is a constant. Equivalent expressions in MATLAB are the vectors $a_1\boldsymbol{x}$ and $a_1\boldsymbol{z}$ (of course, in MATLAB all the elements must be numbers). We named the column vector $a_1\boldsymbol{x}$, though we could have given it any other name. Hence, in the MATLAB command window, we write

```
>>a1x = [a1*1  a1*2.3  -a1*2]T;
>>a1z = [a1*2.33  a1*4+a1*j*5  -a1*j]T;
```

1.1.1.2 Addition and Subtraction

We obtain summation and/or subtraction for the two vectors $\boldsymbol{a}$ and $\boldsymbol{b}$ as follows (each operation must be done separately):

$$z = \boldsymbol{a} \pm k_1 \boldsymbol{b} = \begin{bmatrix} a_1 \\ a_2 \\ a_3 \end{bmatrix} \pm \begin{bmatrix} k_1 b_1 \\ k_1 b_2 \\ k_1 b_3 \end{bmatrix} = \begin{bmatrix} a_1 \pm k_1 b_1 \\ a_2 \pm k_1 b_2 \\ a_3 \pm k_1 b_3 \end{bmatrix} \quad (1.2)$$

Having the vectors $\boldsymbol{a}$ and $\boldsymbol{b}$ in numerical forms, the MATLAB operations are given as follows:

```
>>a = [a1 a2 a3]';b = [b1 b2 b3]';%column vectors;
>>z = a+k1*b;%z is a column vector;
```

or

```
>>z = a-k1*b;%z is a column vector;
```

A vector $\boldsymbol{a}$ with two elements, for example, can be considered as a line in the x–y plane starting at the origin and ending at the point (a_1,a_2). If a vector has n elements, the vector starts from the origin of an n-dimensional space and ends at the point $(a_1, a_2, \ldots, a_n)$.

If $\boldsymbol{a}$, $\boldsymbol{b}$, and $\boldsymbol{c}$ are n-dimensional vectors, belonging to n-dimensional **vector space**, and v_1 and v_2 are constants, then the rules in Table 1.1 are applicable.

Figure 1.1 shows two vectors in two-dimensional vector space, their sum, and their difference.

TABLE 1.1
Properties of Vectors

1. $\boldsymbol{a}+\boldsymbol{b}$ is an n-dimensional vector
2. $\boldsymbol{a}+\boldsymbol{b} = \boldsymbol{b}+\boldsymbol{a}$
3. $(\boldsymbol{a}+\boldsymbol{b})+\boldsymbol{c} = \boldsymbol{a}+(\boldsymbol{b}+\boldsymbol{c})$
4. $\boldsymbol{a}+\boldsymbol{0} = \boldsymbol{a}$ ($\boldsymbol{0}$ is n-dimensional)
5. $\boldsymbol{a}+(-\boldsymbol{a}) = \boldsymbol{0}$
6. $v_1\boldsymbol{a}$ is an n-dimensional vector
7. $v_1(\boldsymbol{a}+\boldsymbol{b}) = v_1 a + v_2 b$
8. $(v_1+v_2)\boldsymbol{a} = v_1\boldsymbol{a}+v_2\boldsymbol{a}$
9. $v_1(v_2\boldsymbol{a}) = (v_1v_2)\boldsymbol{a}$
10. $1(\boldsymbol{a}) = \boldsymbol{a}$
11. $\boldsymbol{a}+\boldsymbol{b} = \boldsymbol{a}$ implies $\boldsymbol{b} = \boldsymbol{0}$ ($\boldsymbol{0}$ is n-dimensional)
12. $\boldsymbol{a}+\boldsymbol{b} = \boldsymbol{0}$ implies $\boldsymbol{b} = -\boldsymbol{a}$ ($\boldsymbol{0}$ is n-dimensional)
13. $\boldsymbol{0}\boldsymbol{a} = \boldsymbol{0}$
14. $\boldsymbol{a}\boldsymbol{0} = \boldsymbol{0}$ (both zeros are n-dimensional)
15. If $v_1\boldsymbol{a} = \boldsymbol{0}$, then $v_1 = \boldsymbol{0}$ or $\boldsymbol{a} = \boldsymbol{0}$
16. $-(-\boldsymbol{a}) = \boldsymbol{a}$

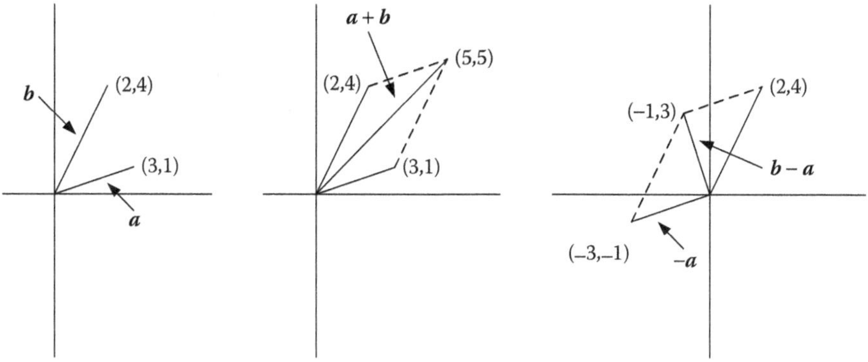

FIGURE 1.1

1.1.2 Unit Coordinate Vectors

For two-dimensional and three-dimensional vector spaces, for example, the unit vectors along the x-, y-, and z-axes are as follows:

$$\underbrace{e_x = [1 \;\; 0]^T, e_y = [0 \;\; 1]^T;}_{\text{Two-dimensional}} \quad \underbrace{e_x = [1 \;\; 0 \;\; 0]^T, e_y = [0 \;\; 1 \;\; 0]^T, e_z = [0 \;\; 0 \;\; 1]^T}_{\text{Three-dimensional}} \tag{1.3}$$

Therefore, the vector $\boldsymbol{a} = [a_1 \quad a_2 \quad a_3]$ can also be presented in the following form:

$$\boldsymbol{a} = a_1\boldsymbol{e}_x + a_2\boldsymbol{e}_y + a_3\boldsymbol{e}_z \text{ or } \begin{bmatrix} a_1 \\ a_2 \\ a_3 \end{bmatrix} = a_1\begin{bmatrix} 1 \\ 0 \\ 0 \end{bmatrix} + a_2\begin{bmatrix} 0 \\ 1 \\ 0 \end{bmatrix} + a_3\begin{bmatrix} 0 \\ 0 \\ 1 \end{bmatrix} = \begin{bmatrix} a_1 \\ 0 \\ 0 \end{bmatrix} + \begin{bmatrix} 0 \\ a_2 \\ 0 \end{bmatrix} + \begin{bmatrix} 0 \\ 0 \\ a_3 \end{bmatrix} = \begin{bmatrix} a_1 \\ a_2 \\ a_3 \end{bmatrix}$$

1.1.3 Inner Product

The **inner** product of two vectors with the **same** number of elements is given by

$$\alpha = \text{number} = \boldsymbol{a}^T\boldsymbol{b} = a_1b_1 + a_2b_2 + \cdots + a_nb_n = \sum_{i=1}^{n} a_ib_i$$

$$>> \alpha = \boldsymbol{a}'*\boldsymbol{b}; \%\text{MATLAB format}$$

$$>> \alpha = [a_1 \quad a_2 \cdots a_n] * \begin{bmatrix} b_1 \\ b_2 \\ \vdots \\ b_n \end{bmatrix}; \%\text{MATLAB format} \tag{1.4}$$

The MATLAB operation is as follows:

```
>>ab = a'*b;
```

where $\boldsymbol{a}$ and $\boldsymbol{b}$ are column vectors, as we have decided to present the vectors in this text, and $\boldsymbol{a}^T\boldsymbol{b}$ is a number and is the **inner** product of the two vectors. There are also other mathematical definitions of the inner product, for example, $\boldsymbol{a}\cdot\boldsymbol{b}$; $<\boldsymbol{a},\boldsymbol{b}>$; $(\boldsymbol{a},\boldsymbol{b})$. However, we use the above definition since it resembles the MATLAB format.

For example, for the vectors $\boldsymbol{a}^T = [2 \quad 1+j \quad 2*j]$ and $\boldsymbol{b}^T = [j \quad -2*j \quad 4]$, the inner product is given in MATLAB as follows:

```
>>ab = a'*b
>> 2.0000-8.0000i
```

Although the **semicolon** is written as a row vector in the presentation of the vector in MATLAB, it produces a column vector. For example, $\boldsymbol{x} = [1; \quad 3]$ is a column vector. The **prime** in MATLAB indicates a **transposition**, which means that it changes a row vector to a column vector and a column vector to a row vector.

Based on the inner product of vectors, we can find the square of a vector's magnitude using the inner product. The **magnitude** square and the magnitude of a vector (column) $\boldsymbol{a}$ are, respectively, given by

$$|\boldsymbol{a}|^2 = \boldsymbol{a}^T\boldsymbol{a} = (a_1^2 + a_2^2 + \cdots + a_n^2)$$

$$d(\boldsymbol{a},\boldsymbol{a}) \equiv \|\boldsymbol{a}\| = |\boldsymbol{a}| = (a_1^2 + a_2^2 + \cdots + a_n^2)^{1/2} = \left(\sum_{i=1}^{n} a_i^2\right)^{1/2}$$

The magnitude is also known as one of the definitions of the **norm** of a vector and it is given by the mathematical identification $\|\boldsymbol{a}\|$. In this text, we use as a norm the metric equivalent, the square root of the squares of the elements of vectors; thus, both are equivalent.

In MATLAB, the above operations are given as follows:

```
>>mas = a'*a;%mas is the sum of the squares of the individual
>>    %vector elements;
>>ma = sqrt(a'*a);%ma is the magnitude of the vector a; sqrt()is
>>    %a MATLAB function providing the square root of
>>    %the number in parentheses;
```

or

```
>>mas = sum(a.*a);%a.*a is a vector obtained by multiplying the
>>    %corresponding elements and the sum() adds them;
>>    %the dot,., indicates multiplication of corresponding
>>    %elements; the same can be done for two
>>    %different vectors of the same number of
>>    %elements;
>>ma = sqrt(sum(a.*a));%ma is the magnitude of the vector a;
```

1.1.4 Distance between Two Vectors

Definition 1.1.1: Let $\boldsymbol{a} = [a_1 \quad a_2 \cdots a_n]$ and $\boldsymbol{b} = [b_1 \quad b_2 \cdots b_n]$ be any two n-dimensional vectors of equal length. Then, the distance (nonnegative number) between these two vectors is given by the following expression:

$$d(\boldsymbol{a},\boldsymbol{b}) \equiv \left\|\boldsymbol{a}-\boldsymbol{b}\right\| = \left[(a_1-b_1)^2+(a_2-b_2)^2+\cdots+(a_n-b_n)^2\right]^{1/2} = \left[\sum_{i=1}^{n}(a_i-b_i)^2\right]^{1/2} \tag{1.5}$$

▲

Example 1.1.1

If the scalar $k>0$, find the magnitude of the vector $k\boldsymbol{a}$.

Solution:

$$\left|k\boldsymbol{a}\right| \equiv \left\|k\boldsymbol{a}\right\| = \left[(ka_1)^2 + (ka_2)^2 + \cdots + (ka_n)^2\right]^{1/2} = k\left(a_1^2+a_2^2+\cdots+a_n^2\right)^{1/2} = k\left|\boldsymbol{a}\right| \equiv k\left\|\boldsymbol{a}\right\|$$

The equivalent MATLAB format is as follows:

```
>>c = sqrt(sum(k*(a'*a)));%c = number;
```

■

1.1.5 Mean Value of a Vector

Definition 1.1.2 (mean vector): The mean value of an n-element vector is defined by

$$\bar{\boldsymbol{a}} = \frac{1}{n}(a_1+a_2+\cdots+a_n) = \frac{1}{n}\sum_{i=1}^{n}a_i \tag{1.6}$$

▲

If the values of the elements come out from a space of random numbers, then the mean value defined above gives an approximate value of the exact mean defined in the space. More will be discussed about random numbers and variables in Chapter 4.

Let us assume that a person picks up 9 resistors of 5 ohms each from a box containing a large number of resistors. A second person picks up another 9 resistors. Using MATLAB, we obtain the following:

```
>>a = 5+randn(1,9);%randn()is a MATLAB function creating a set of
>>    %numbers with mean value 0 and are normally (Gaussian)
>>    %distributed;
```

```
>>    %a is a row vector with 9 elements with mean
>>    %value close to 5 and normally distributed;

>> 4.5674    3.3344    5.1253    5.2877    3.8535    6.1909
>> 6.1892    4.9624    5.3273

>>b = 5+randn(1,9);

>>5.1746     4.8133    5.7258    4.4117    7.1832    4.8636
>>5.1139     6.0668    5.0593

>>ma = mean(a);%mean(a) gives the mean value of the elements of the
>>    %vector a, mean() is a MATLAB function;

>>4.9820

>>mb = mean(b);

>>5.3791
```

Another way to find the mean value of a vector is the following:

```
>>ma = sum(a)/length(a);%length(a)is a MATLAB function giving the
>>    %number of the elements of the vector;
```

Observe that the two means are not close to each other and to the mean value 5, which we would expect. This is due to the small number we picked from the space of normally distributed numbers. If we had considered a number greater than 1000, the mean value would have been closer to 5 and closer to each other. Figure 1.2 shows the distribution of 7000 numbers having a mean value close to 5. The mean value of these 7000 numbers is 5.0052. If we look at the mean value of another set of 7000 normally distributed numbers, we would find the mean value to be very close to 5 but not exactly the same as the one above. Therefore, if we repeat this experiment many times and plot the distribution of the means, we will observe a normal distribution around the value 5, but a very narrow normal distribution.

The normal distribution in Figure 1.2 may be obtained as follows:

```
>>x = 5+randn(1,7000);%we must add the semicolon because otherwise
>>    %all 7000 numbers will appear in the
>>    %command window;
>>hist(x,40);%hist(x,40) is a MATLAB function which produces the
>>    %distribution, here the normal, from the vector
>>    %x with 40 bins;
>>    %the color is always blue;
>>h = findobj(gca,'Type','path');
>>set(h,'FaceColor','w','EdgeColor','k');%creates a figure with
>>    %white, 'w', color
>>    %the bins and
```

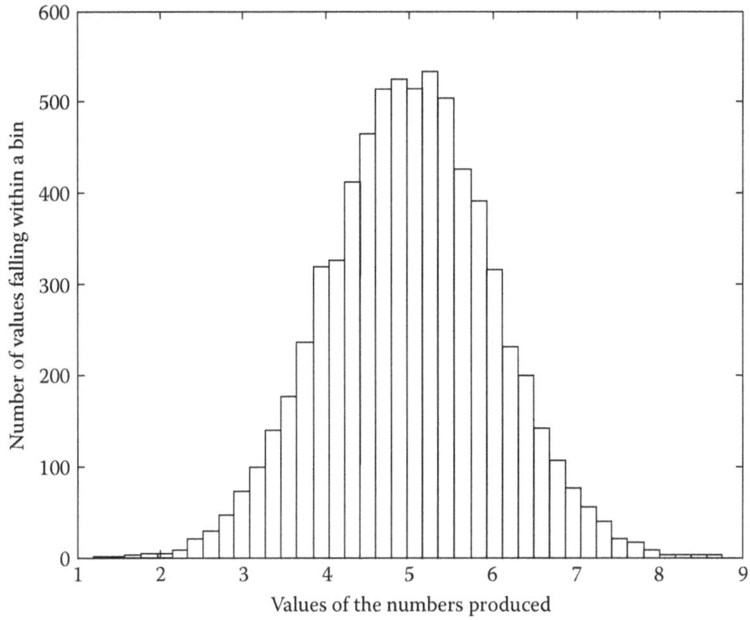

FIGURE 1.2

```
>>     %their lines black, 'k',
>>     %other colors are:
>>     %green, 'g'; red, 'r';
>>     %yellow, 'y';
>>     %cyan, 'c'; blue, 'b';
>>     %magenta, 'm';
>>xlabel('Values of the numbers produced');
>>ylabel('Number of values falling within a bin');
```

1.1.6 Direction Cosines

A three-dimensional vector $\boldsymbol{a}$, its components, and the three unit vectors along the three axes (x, y, and z) are shown in Figure 1.3. The cosines of the three angles created by the vector and the three axes are as follows:

$$\cos\theta_1 = \frac{a_1}{\|\boldsymbol{a}\|};\ \cos\theta_2 = \frac{a_2}{\|\boldsymbol{a}\|};\ \cos\theta_3 = \frac{a_3}{\|\boldsymbol{a}\|}; \qquad |\boldsymbol{a}| = \|\boldsymbol{a}\| \tag{1.7}$$

Example 1.1.2

We will find the cosines of the vector $\boldsymbol{a}^T = \begin{bmatrix} 1 & 1/2 \end{bmatrix}$; we will also write the vector in mathematical form: $\boldsymbol{a}^T = (1, 1/2)$.

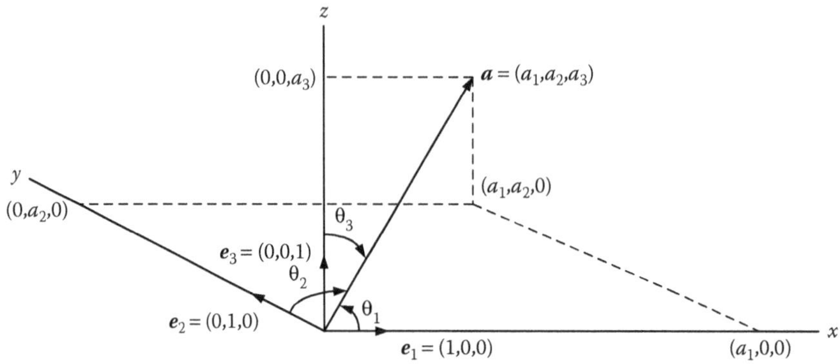

FIGURE 1.3

Solution: The direction cosines and their sum square are as follows:

$$\cos\theta_1 = \frac{a_1}{\|a\|} = \frac{1}{\sqrt{1^2+(1/2)^2}} = \frac{2}{\sqrt{5}},\ \theta_1 = 26.5°;\ \cos\theta_2 = \frac{a_2}{\|a\|} = \frac{1/2}{\sqrt{5}/2}$$

$$= \frac{1}{\sqrt{5}},\ \theta_2 = 63.5°;$$

$$\cos^2\theta_1 + \cos^2\theta_2 = \left(\frac{2}{\sqrt{5}}\right)^2 + \left(\frac{1}{\sqrt{5}}\right)^2 = 1$$

We can use the following MATLAB program:

```
>>a = [1 1/2];
>>cos1 = a(1)/sqrt(sum(a.*a));%cosine of theta one;
>>th1r = acos(cos1);%theta one in radians;
>>    %acos() is a MATLAB function that gives
>>    %the inverse of a
>>    %cosine; similar: asin(), atan();
>>th1 = th1r*180/pi;%theta one in degrees;
>>th1
>>26.5651;%theta one in degrees;
```

■

The cosine of the angle between two vectors $\boldsymbol{a}$ and $\boldsymbol{b}$ is

$$\cos\theta = \frac{\boldsymbol{a}^T\boldsymbol{b}}{\|\boldsymbol{a}\|\,\|\boldsymbol{b}\|} \tag{1.8}$$

Because $|\cos\theta| \le 1$, the inner product between the two vectors satisfies the Cauchy–Schwarz inequality:

$$|\boldsymbol{a}^T\boldsymbol{b}| \le \|\boldsymbol{a}\|\,\|\boldsymbol{b}\| \tag{1.9}$$

Another useful inequality is the following:

$$2|\boldsymbol{a}^T\boldsymbol{b}| \le \|\boldsymbol{a}\|^2 + \|\boldsymbol{b}\|^2 \tag{1.10}$$

1.1.7 The Projection of a Vector

To find the projection point $\boldsymbol{p}$ of the vector $\boldsymbol{b}$ on $\boldsymbol{a}$ (Figure 1.4), we observe that the magnitude of the vector $\boldsymbol{p} = \alpha\boldsymbol{a}$ is some multiple of the vector $\boldsymbol{a}$. The line from the end of vector $\boldsymbol{b}$ to vector $\boldsymbol{a}$, which ends at the end point of the vector $\boldsymbol{p}$, is perpendicular to vector $\boldsymbol{a}$. Hence, we have

$$\boldsymbol{p} = \alpha\boldsymbol{a}, \quad (\boldsymbol{b} - \alpha\boldsymbol{a}) \perp \boldsymbol{a} \quad \text{or} \quad \boldsymbol{a}^T(\boldsymbol{b} - \alpha\boldsymbol{a}) = 0 \quad \text{or} \quad \alpha = \frac{\boldsymbol{a}^T\boldsymbol{b}}{\boldsymbol{a}^T\boldsymbol{a}} \tag{1.11}$$

Therefore, the projection of $\boldsymbol{b}$ onto the line through 0 and the end of vector $\boldsymbol{a}$ is

$$\boldsymbol{p} = \alpha\boldsymbol{a} = \frac{\boldsymbol{a}^T\boldsymbol{b}}{\boldsymbol{a}^T\boldsymbol{a}}\boldsymbol{a} \tag{1.12}$$

Example 1.1.3

Find the orthogonal projection of $\boldsymbol{b}^T = [2 \quad 7 \quad 2]$ onto vector $\boldsymbol{a}^T = [5 \quad 7 \quad 4]$.

Solution: The magnitude of the projected vector is $\|\boldsymbol{b}_{(p)}\| = (\boldsymbol{a}^T\boldsymbol{b})/(\boldsymbol{a}^T\boldsymbol{a})^{1/2} = (10 + 49 + 8)/(25 + 49 + 16)^{1/2} = 7.0624$. The direction is given by the unit vector $\boldsymbol{a}/\|\boldsymbol{a}\| = (5,7,4)/(25 + 49 + 16)^{1/2} = [0.5270 \quad 0.7379 \quad 0.4216]$. Hence, $\boldsymbol{b}_{(p)} = \|\boldsymbol{b}_{(p)}\|[0.5270 \quad 0.7379 \quad 0.4216] = 7.0624[0.5270 \quad 0.7379 \quad 0.4216] = [3.7219 \quad 5.2113 \quad 2.9775]$ and also $\boldsymbol{b}_{(p)} = (\boldsymbol{a}^T\boldsymbol{b})/(\boldsymbol{a}^T\boldsymbol{a})\boldsymbol{a} = 0.7444\boldsymbol{a}$.

■

Definition 1.1.3 (projection theorem): Let $\mathcal{V}$ be a vector space, $\mathcal{S}$ be a subspace of $\mathcal{V}$, and $\boldsymbol{v}$ be any vector in $\mathcal{V}$. The vector $\boldsymbol{v}_p \in \mathcal{S}$ is a minimizing vector if and only if $\boldsymbol{e} = \boldsymbol{v} - \boldsymbol{v}_p$ is orthogonal to $\mathcal{S}$ ($\boldsymbol{e}$ is perpendicular to all vectors in $\mathcal{S}$).

▲

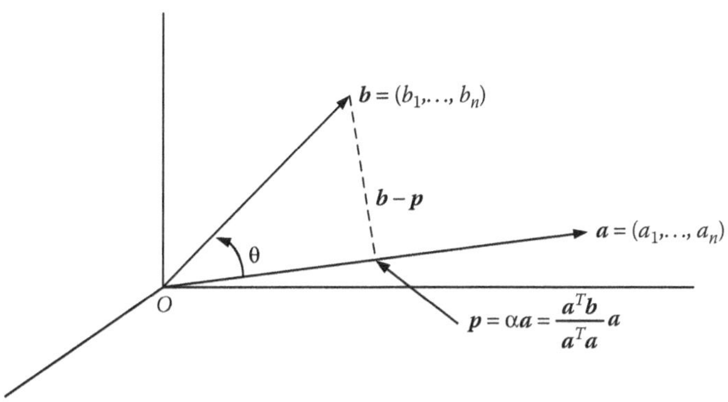

FIGURE 1.4

1.1.8 Linear Transformations

Functions from vector spaces to vector spaces are called transformations (or maps or operators).

Examples of transformations are as follows:

Shift

$$L\{\boldsymbol{x}\} = \boldsymbol{x} + \boldsymbol{a}$$

Let $\boldsymbol{a} = [2 \quad 1]^T$ and $\boldsymbol{x} = [1 \quad 1]^T$; the shift of $\boldsymbol{x}$ is shown in Figure 1.5 (shown by the circle). L is an operator, indicating a transformation.

Rotation and shift

Let the line l be described by $0.5[1 \quad 1]^T + (1-0.5)[-1 \quad 2]^T$ or $[0.5 \quad 0.5]^T + [-0.5 \quad 1]^T$. If $L = [1 \quad 1; 1 \quad -1]$, a 2×2 matrix, then the image of the line l is

$$L\{l\} = \begin{bmatrix} 1 & 1 \\ 1 & -1 \end{bmatrix} \begin{bmatrix} 0.5 \\ 0.5 \end{bmatrix} + \begin{bmatrix} 1 & 1 \\ 1 & -1 \end{bmatrix} \begin{bmatrix} -0.5 \\ 1 \end{bmatrix} = \begin{bmatrix} 1 \\ 0 \end{bmatrix} + \begin{bmatrix} 0.5 \\ -1.5 \end{bmatrix}$$

The original line and its image, after the transformation, are shown in Figure 1.6. Observe that $0.5 + (1-0.5) = 1$.

Axis

Let the operator be of the form:

$$L = \begin{bmatrix} 2 & 1 \\ 1 & 2 \end{bmatrix}$$

The basis for a rectangular coordinate system is $\{\boldsymbol{e}_1, \boldsymbol{e}_2\} = \{[1 \quad 0]^T, [0 \quad 1]^T\}$. After applying the operator, we have

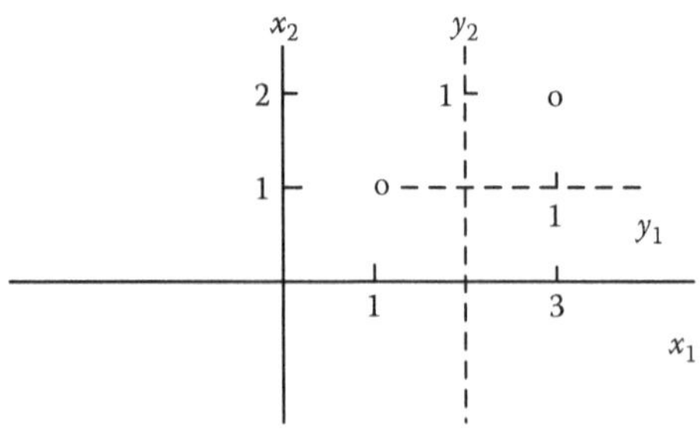

FIGURE 1.5

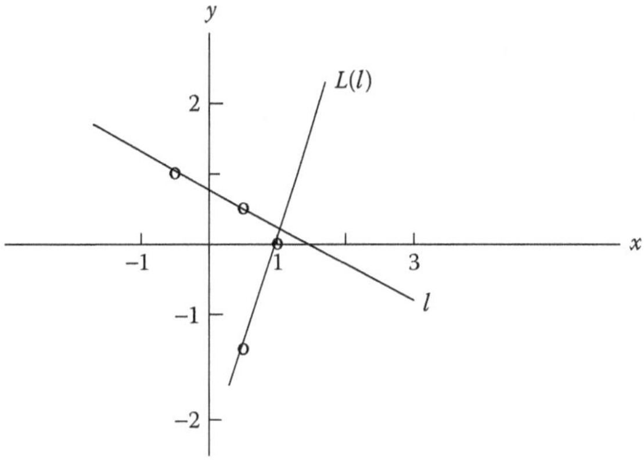

FIGURE 1.6

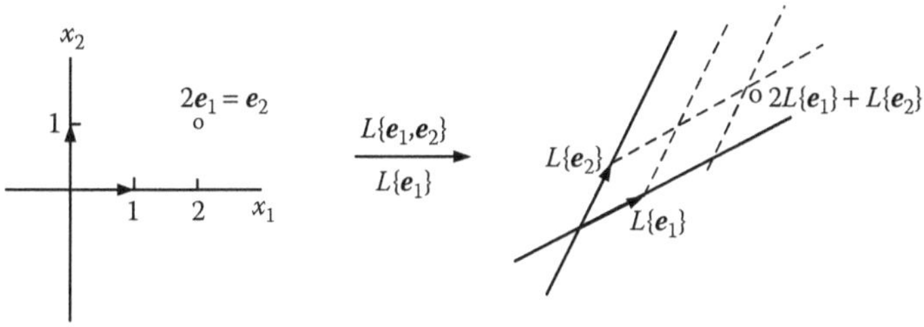

FIGURE 1.7

$$L\{e_1\} = \begin{bmatrix} 2 & 1 \\ 1 & 2 \end{bmatrix}\begin{bmatrix} 1 \\ 0 \end{bmatrix} = \begin{bmatrix} 2 \\ 1 \end{bmatrix}$$

and

$$L\{e_2\} = \begin{bmatrix} 2 & 1 \\ 1 & 2 \end{bmatrix}\begin{bmatrix} 0 \\ 1 \end{bmatrix} = \begin{bmatrix} 1 \\ 2 \end{bmatrix}$$

Therefore,

$$L\{ae_1 + be_2\} = aL\{e_1\} + bL\{e_1\} = a\begin{bmatrix} 2 \\ 1 \end{bmatrix} + b\begin{bmatrix} 1 \\ 2 \end{bmatrix}$$

Figure 1.7 shows the transformation for $a = 2$ and $b = 1$.

1.2 LINEAR INDEPENDENCE, VECTOR SPACES, AND BASIS VECTORS

A set of n vectors $v_1, v_2, \dots, v_n$ is said to be **linearly independent** if

$$\alpha_1 v_1 + \alpha_2 v_2 + \cdots + \alpha_n v_n = 0 \tag{1.13}$$

implies that all the α_i's are 0. If a set of nonzero α_i's can be found so that (1.13) holds, then the set of the vectors is said to be **linearly dependent**. If $\boldsymbol{v}_1, \boldsymbol{v}_2, \ldots, \boldsymbol{v}_n$ is a set of linearly dependent vectors, then one of the vectors can be expressed as a linear combination of the remaining vectors. Hence, we have

$$\boldsymbol{v}_1 = \beta_2 \boldsymbol{v}_2 + \cdots + \beta_n \boldsymbol{v}_n$$

The β_i's are some set of scalars. If a set of vectors is n-dimensional, there can be only n independent vectors. In other words, any set that contains more than n vectors will always be linearly dependent.

Example 1.2.1

Find out if the vectors $\boldsymbol{a}^T = [1 \quad 4 \quad 1]$ and $\boldsymbol{b}^T = [1 \quad 0 \quad 1]$ are linearly dependent or independent.

Solution: Applying (1.13), we obtain

$$\alpha_1 \begin{bmatrix} 1 \\ 4 \\ 1 \end{bmatrix} + \alpha_2 \begin{bmatrix} 1 \\ 0 \\ 1 \end{bmatrix} = \mathbf{0} = \begin{bmatrix} 0 \\ 0 \\ 0 \end{bmatrix} \quad \text{or} \quad \begin{bmatrix} 1\alpha_1 + 1\alpha_2 \\ 4\alpha_1 + 0\alpha_2 \\ 1\alpha_1 + 1\alpha_2 \end{bmatrix} = \begin{bmatrix} 0 \\ 0 \\ 0 \end{bmatrix}$$

Two vectors are equal if their corresponding elements are equal. Hence, $\alpha_1 + \alpha_2 = 0$ and $4\alpha_1 = 0$, which results in $\alpha_1 = \alpha_2 = 0$. This indicates that the vectors are independent. If we add the vector $\boldsymbol{v}_3^T = [0 \quad 1 \quad 0]$, we find that the vectors are dependent (see Problem 1.2.1).

■

Given a set of N vectors $\boldsymbol{V} = \{\boldsymbol{v}_1, \boldsymbol{v}_2, \boldsymbol{v}_3, \ldots, \boldsymbol{v}_N\}$, consider the set of all vectors V that may be formed by the linear combination of the vectors $\boldsymbol{v} = \sum_{i=1}^{N} \alpha_i \boldsymbol{v}_i$ for different sets of α_i's. This set forms a vector space, and the vectors $\boldsymbol{v}_i$ are said to **span** the **vector space** $\mathcal{V}$. In addition, if the vectors $\boldsymbol{v}_i$ are linearly independent, then it is said that they form a **basis** for the vector space. The number N is referred to as the **dimension** of the space. For example, the set of all real vectors of the form $\boldsymbol{v} = [v_1, v_2, v_3, \ldots, v_N]^T$ forms an N-dimensional vector space, denoted by R^N, which is spanned by the basis vectors:

$$\begin{aligned} \boldsymbol{e}_1 &= [1 \quad 0 \quad 0 \quad \cdots \quad 0]^T \\ \boldsymbol{e}_2 &= [0 \quad 1 \quad 0 \quad \cdots \quad 0]^T \\ &\vdots \\ \boldsymbol{e}_N &= [0 \quad 0 \quad 0 \quad \cdots \quad 1]^T \end{aligned}$$

On this basis (basis is not unique), any vector $\boldsymbol{v}$ in R^N may be uniquely decomposed as follows:

$$\boldsymbol{v} = \sum_{i=1}^{N} v_i \boldsymbol{e}_i$$

Definition 1.2.1: Let $\mathfrak{S}$ and $\mathfrak{T}$ denote any two vector spaces. Then the following definitions apply:

1. The vector space consisting of all vectors common to both vector spaces is called the **intersection** of the two spaces and is written as $\mathfrak{S} \cap \mathfrak{T}$ (see Figure P1.2.1).
2. The vector space consisting of all vectors $\boldsymbol{a}+\boldsymbol{b}$, where $\boldsymbol{a}$ lies in $\mathfrak{S}$ and $\boldsymbol{b}$ in $\mathfrak{T}$, is called the sum or the **union** of the two spaces and is written as $\mathfrak{S} \cup \mathfrak{T}$.

▲

An important special case occurs when the two spaces have in common only the $\mathbf{0}$ vector, which is represented by $\mathfrak{S} \cap \mathfrak{T} = \mathbf{0}$. This type of union is known as the **direct sum** of the two spaces and is written as $\mathfrak{S} \oplus \mathfrak{T}$. From Figure P1.2.1, we observe that

$$\dim(\mathfrak{S} \oplus \mathfrak{T}) = \dim(\mathfrak{S}) + \dim(\mathfrak{T})$$

1.2.1 Orthogonal Basis Vectors

Definition 1.2.2: The vector spaces $\mathfrak{S}$ and $\mathfrak{T}$ are said to be **mutually orthogonal** if every vector in $\mathfrak{S}$ is orthogonal to every vector in $\mathfrak{T}$.

▲

Definition 1.2.3: Let $\mathfrak{S}$ be a subspace of $\mathfrak{E}^{(n)}$ $[\mathfrak{S} \in \mathfrak{E}^{(n)}]$. Then the set of all vectors in $\mathfrak{E}^{(n)}$ that are perpendicular to $\mathfrak{S}$ is known as the **orthogonal complement** of $\mathfrak{S}$, denoted by $\mathfrak{S}^{\perp}$.

▲

PROBLEMS

1.1.1 Show that the inner product is distributive over addition (subtraction).
1.1.2 Show that if a vector is multiplied by a scalar, its magnitude changes and not its direction.
1.1.3 Prove (1.8).
1.1.4 Prove (1.10).
1.2.1 Show that the three vectors in Example 1.2.1 are dependent.
1.2.2 Find the dimensions of the general union $\mathfrak{S} \cup \mathfrak{T}$.

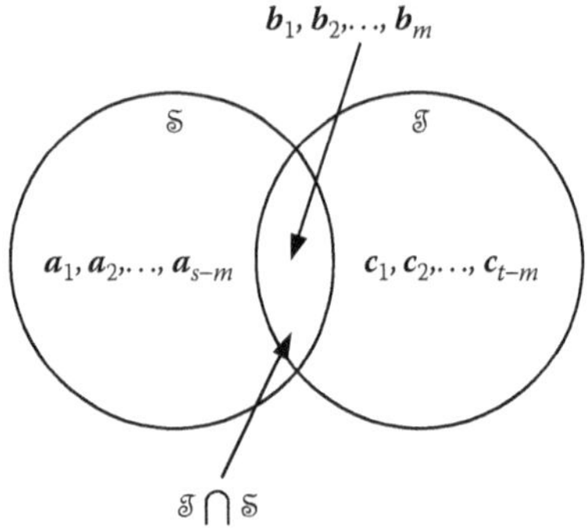

FIGURE P1.2.1

HINTS–SUGGESTIONS–SOLUTIONS

1.1.1

$$\boldsymbol{a}^T(\boldsymbol{b} \pm \boldsymbol{c}) = a_1(b_1 \pm c_1) + a_2(b_2 \pm c_2) + \cdots + a_n(b_n \pm c_n)$$

$$= (a_1 b_1 \pm a_2 b_2 \pm \cdots \pm a_n b_n) + (a_1 c_1 \pm a_2 c_2 \pm \cdots \pm a_n c_n) = \boldsymbol{a}^T\boldsymbol{b} \pm \boldsymbol{a}^T\boldsymbol{c}$$

1.1.2 The magnitude of the vector $\alpha\boldsymbol{b}$ is

$$\|\alpha\boldsymbol{b}\| = \left[(\alpha b_1)^2 + (\alpha b_2)^2 + \cdots + (\alpha b_n)^2\right]^{1/2}$$

$$= \alpha\left[(b_1)^2 + (b_2)^2 + \cdots + (b_n)^2\right]^{1/2} = \alpha\|\boldsymbol{b}\|$$

If $\alpha > 0$, we have extension, and if $\alpha < 1$, we have contraction.

1.1.3 From the law of trigonometry of a triangle with two sides $\boldsymbol{a}$ and $\boldsymbol{b}$ and their angle between them θ, we have $\|\boldsymbol{b}-\boldsymbol{a}\|^2 = \|\boldsymbol{b}\|^2 + \|\boldsymbol{a}\|^2 - 2\|\boldsymbol{b}\|\|\boldsymbol{a}\|\cos\theta$. If the angle is 90°, we have the Pythagoras theorem. We can also write the above expression in the following form:

$$\|\boldsymbol{b}-\boldsymbol{a}\|^2 = (\boldsymbol{b}-\boldsymbol{a})^T(\boldsymbol{b}-\boldsymbol{a}) = \boldsymbol{b}^T\boldsymbol{b} - 2\boldsymbol{a}^T\boldsymbol{b} + \boldsymbol{a}^T\boldsymbol{a} = \boldsymbol{b}^T\boldsymbol{b} + \boldsymbol{a}^T\boldsymbol{a} - 2\|\boldsymbol{b}\|\|\boldsymbol{a}\|\cos\theta$$

Canceling the same quantities from both sides of the above equation, we recognize (1.8).

1.1.4 If $\boldsymbol{a} = \pm\boldsymbol{b}$, the equality holds. For any two vectors, the magnitude of their sum or difference is a positive scalar. Hence, $\|\boldsymbol{b} \pm \boldsymbol{a}\|^2 \geq 0$. Expanding the norm (see Problem 1.1.3), we obtain

$$\|\boldsymbol{b} \pm \boldsymbol{a}\|^2 = \|\boldsymbol{b}\|^2 + \|\boldsymbol{a}\|^2 \pm 2\|\boldsymbol{b}\|\|\boldsymbol{a}\|\cos\theta = \|\boldsymbol{b}\|^2 + \|\boldsymbol{a}\|^2 \pm 2\boldsymbol{a}^T\boldsymbol{b} \geq 0$$

From the above expression, the inequality easily follows.

1.2.1 By equating the vectors, we obtain the set of equations $\alpha_1 + \alpha_2 + 0\alpha_3 = 0$; $4\alpha_1 + 0\alpha_2 + \alpha_3 = 0$ and $\alpha_1 + \alpha_2 + 0\alpha_3 = 0$. Solving the set, we obtain $\alpha_2 = -\alpha_1, \alpha_3 = -4\alpha_1$. Setting arbitrarily $\alpha_1 = 1$, we obtain $\boldsymbol{a} = \boldsymbol{b} + 4\boldsymbol{c}$.

1.2.2

$$\dim(\mathbb{S} \cup \mathfrak{J}) = (s - m) + m + (t - m) = s + t - m$$

or

$$\dim(\mathbb{S} \cup \mathfrak{J}) = \dim(\mathbb{S}) + \dim(\mathfrak{J}) - \dim(\mathbb{S} \cap \mathfrak{J})$$

2 Matrices

2.1 INTRODUCTION

A generalization of a vector, that is, an ordered array of numbers, determines the magnitude and direction of the vector. A further generalization is that of an $n \times m$ matrix $\boldsymbol{A} = \{a_{ij}\} = (a_{ij})$, which is shown below:

$$\boldsymbol{A} = \begin{bmatrix} a_{11} & a_{12} & \cdots & a_{1m} \\ a_{21} & a_{22} & \cdots & a_{2m} \\ \vdots & \vdots & & \vdots \\ a_{n1} & a_{n2} & \cdots & a_{nm} \end{bmatrix} \tag{2.1}$$

with a typical element a_{ij}, where $i = 1, 2, \ldots, n$ and $j = 1, 2, \ldots, m$.

2.2 GENERAL TYPES OF MATRICES

We shall present two- and three-element matrices for convenience. However, the properties are identical for any size.

2.2.1 Diagonal, Identity, and Scalar Matrices

$$\begin{gathered} \text{Diagonal: } \boldsymbol{A} = \begin{bmatrix} a_{11} & 0 \\ 0 & a_{22} \end{bmatrix}; \text{ Identity: } \boldsymbol{I} = \begin{bmatrix} 1 & 0 \\ 0 & 1 \end{bmatrix}; \\ \text{Scalar: } \boldsymbol{A} = \begin{bmatrix} a & 0 \\ 0 & a \end{bmatrix} = a \begin{bmatrix} 1 & 0 \\ 0 & 1 \end{bmatrix} = a\boldsymbol{I} \end{gathered} \tag{2.2}$$

We shall also write diag $\{a_{11}, a_{22}, \ldots, a_{nn}\}$ in short.

2.2.2 Upper and Lower Triangular Matrices

$$\text{Lower triangular: } \boldsymbol{A} = \begin{bmatrix} a_{11} & 0 \\ a_{21} & a_{22} \end{bmatrix}; \text{ Upper triangular: } \boldsymbol{A} = \begin{bmatrix} a_{11} & a_{12} \\ 0 & a_{22} \end{bmatrix} \tag{2.3}$$

2.2.3 Symmetric and Exchange Matrices

$$\text{Symmetric: } \boldsymbol{A} = \begin{bmatrix} a_{11} & a_{12} & a_{13} \\ a_{12} & a_{22} & a_{23} \\ a_{13} & a_{23} & a_{33} \end{bmatrix}, \quad \boldsymbol{A} = \boldsymbol{A}^T$$

$$\text{Exchange: } \boldsymbol{J} = \begin{bmatrix} 0 & 0 & 1 \\ 0 & 1 & 0 \\ 1 & 0 & 0 \end{bmatrix} \tag{2.4}$$

2.2.4 Toeplitz Matrix

An $n \times n$ matrix is said to be Toeplitz if all the elements along each of the diagonals have the same value, that is, $a_{ij} = a_{i+j,j+1}$ for all $i < n$ and $j < n$.

$$\text{Toeplitz: } \boldsymbol{A} = \begin{bmatrix} a_{11} & a_{12} & a_{13} \\ a_{21} & a_{11} & a_{12} \\ a_{23} & a_{21} & a_{11} \end{bmatrix} \tag{2.5}$$

2.2.5 Hankel and Hermitian

A matrix is said to be Hankel if its elements along the diagonals that are perpendicular to the main diagonal are equal. A complex matrix is said to be Hermitian if it is conjugate symmetric.

$$\text{Hankel: } \boldsymbol{A} = \begin{bmatrix} a_{11} & a_{12} & a_{13} \\ a_{12} & a_{13} & a_{23} \\ a_{13} & a_{23} & a_{33} \end{bmatrix}$$

$$\text{Hermitian: } \boldsymbol{A} = \begin{bmatrix} 1 & 1+2j & -j \\ 1-2j & 2 & 4j \\ j & -4j & 6 \end{bmatrix} = (\boldsymbol{A}^*)^T = \boldsymbol{A}^H \tag{2.6}$$

```
>>a = [1 1+2*j -j;1-2*j 2 4*j;j -4*j 6];
>>conj(transpose(a))
1.0000                1.0000 + 2.0000i     0 - 1.0000i
1.0000 - 2.0000i           2.0000          0 + 4.0000i
     0 + 1.0000i           0 - 4.0000i         6.0000
```

2.3 MATRIX OPERATIONS

Table 2.1 presents some of the algebraic operation of matrices together with the MATLAB operations.

TABLE 2.1
Matrix Simple Operations

Operation	MATLAB Symbol	Example
Sum or difference[a]	+ or −	$\begin{bmatrix}1 & 2\\3 & 4\end{bmatrix} \pm \begin{bmatrix}1 & 0\\2 & 1\end{bmatrix} = \begin{bmatrix}1\pm1 & 2\pm0\\3\pm2 & 4\pm1\end{bmatrix}$; $[1 \;\; 2 \;\; 3]-3=[-2 \;\; -1 \;\; 0]$
Product[b]	*	$\underbrace{\begin{bmatrix}1\;4\;0\\1\;2\;3\end{bmatrix}}_{2\times3} * \underbrace{\begin{bmatrix}1 & -2\\3 & 4\\5 & 0\end{bmatrix}}_{3\times2} = \begin{bmatrix}1*1+4*3+0*5 & 1*(-2)+4*4+0*0\\1*1+2*3+3*5 & 1*(-2)+2*4+3*0\end{bmatrix} = \underbrace{\begin{bmatrix}13 & 14\\22 & 6\end{bmatrix}}_{2\times2}$ $3*[2 \;\; -4 \;\; 1+j*4]=[6 \;\; -12 \;\; 3+j12]$
Array product	.*	$\begin{bmatrix}1 & 2\\3 & 4\end{bmatrix} .* \begin{bmatrix}-1 & 0\\2 & -1\end{bmatrix} = \begin{bmatrix}1*(-1) & 2*0\\3*2 & 4*(-1)\end{bmatrix} = \begin{bmatrix}-1 & 0\\6 & -4\end{bmatrix}$ $2.*[2 \;\; -1 \;\; 2+j*2]=[4 \;\; -2 \;\; 4+j4]$ $2*[2 \;\; -1 \;\; 2+j*2]=[4 \;\; -2 \;\; 4+j4]$
Transpose	′	$\begin{bmatrix}1\;2\;3\\4\;5\;6\end{bmatrix}' = \begin{bmatrix}1\;4\\2\;5\\3\;6\end{bmatrix}$; $[1 \;\; 2 \;\; 3+j*4]' = \begin{bmatrix}1.0000\\2.0000\\3.0000-4.0000i\end{bmatrix}$
Exponentiation	.^	$2.\wedge[1 \;\; 2 \;\; 3]=[2^1 \;\; 2^2 \;\; 2^3]=[1 \;\; 4 \;\; 8]$; $2.\wedge[1 \;\; -2 \;\; 3]=[2 \;\; 0.25 \;\; 8]$ $[1 \;\; 4 \;\; 3].\wedge2=[1^2 \;\; 4^2 \;\; 3^2]=[1 \;\; 16 \;\; 9]$ $\begin{bmatrix}0 & 1 & 2\\3 & 4 & 5\end{bmatrix} .\wedge \begin{bmatrix}0 & 1 & 3\\2 & 3 & 0\end{bmatrix} = \begin{bmatrix}0^0 & 1^1 & 2^3\\3^2 & 4^3 & 5^0\end{bmatrix} = \begin{bmatrix}1 & 1 & 8\\9 & 64 & 1\end{bmatrix}$
Backslash	\	$\boldsymbol{A}*\boldsymbol{b}=\boldsymbol{c}, \quad \boldsymbol{b}=\boldsymbol{A}\backslash\boldsymbol{c} \quad (\boldsymbol{A}=\text{matrix}, \boldsymbol{b}=\text{vector}, \boldsymbol{c}=\text{vector})$

[a] The matrices must have the same rows and columns.

[b] The columns of the first and rows of the second matrix must be the same.

Example 2.3.1

Find the solution of the system using the backslash approach in MATLAB and regular inversion approach.

$$5x+2y=1$$

$$2x+2y=4$$

Solution: If we subtract the second equation from the first, we find $3x=-3$ or $x=-1$. Substituting $x=-1$ in any of the two equations, we find $y=3$, and thus, the

TABLE 2.2
Fundamental Matrix Properties

Multiplication	$C = A_{n\times m}B_{m\times k} = \{c_{ij}\}; \quad c_{ij} = \sum_{r=1}^{m} a_{ir}b_{rj}; \quad AA = A^2$
	$A^r = AA\cdots A; \quad (A^r)^q = A^{rq} \quad r>0,\ q>0; \quad A^0 = I; \quad A^rA^q = A^{r+q} = A^qA^r$
Associative law	$A(BC) = (AB)C$
Distributive law	$A(B+C) = AB+AC$
Commutative law	$AB \neq BA$
Distributive law for scalar multiplication	$(a+b)A = aA + bA = A(a+b)$
Scalar multiplication	$a\begin{bmatrix} b_{11} & b_{12} \\ b_{21} & b_{22}\end{bmatrix} = \begin{bmatrix} ab_{11} & ab_{12} \\ ab_{21} & ab_{22}\end{bmatrix} = \begin{bmatrix} a & 0 \\ 0 & a\end{bmatrix}\begin{bmatrix} b_{11} & b_{12} \\ b_{21} & b_{22}\end{bmatrix} = a\begin{bmatrix} 1 & 0 \\ 0 & 1\end{bmatrix}\begin{bmatrix} b_{11} & b_{12} \\ b_{21} & b_{22}\end{bmatrix} = aIA = Aa$
Matrix transposition	$(A^T)^T = A; \ (A+B)^T = A^T + B^T; \ (AB)^T = B^TA^T$
	$A = \begin{bmatrix} A_1 & A_2 \\ A_3 & A_4\end{bmatrix}; \ B = \begin{bmatrix} B_1 & B_2 \\ B_3 & B_4\end{bmatrix}; \ A^TB = \begin{bmatrix} A_1^T & A_3^T \\ A_2^T & A_4^T\end{bmatrix}\begin{bmatrix} B_1 & B_2 \\ B_3 & B_4\end{bmatrix}$
	$= \begin{bmatrix} A_1^TB_1 + A_3^TB_3 & A_1^TB_2 + A_3^TB_4 \\ A_2^TB_1 + A_4^TB_3 & A_2^TB_2 + A_4^TB_4\end{bmatrix}$

Note: Each multiplication of the matrices must obey the number of rows and columns as given at the top of this table.

unknown $\boldsymbol{b} = [x \quad y]^T = [-1 \quad 3]^T$. We obtain the same value of $\boldsymbol{b}$ using MATLAB backslash approach; if we set

$$A = \begin{bmatrix} 5 & 2 \\ 2 & 2 \end{bmatrix}, \ c = \begin{bmatrix} 1 \\ 4 \end{bmatrix}$$

then in MATLAB we write $\boldsymbol{b} = \boldsymbol{A}/\boldsymbol{c}'$. Since the solution of the above system is equal to $\boldsymbol{b} = \boldsymbol{A}^{-1}\boldsymbol{c}$, in MATLAB Command window we write $\boldsymbol{b} = \text{inv}(\boldsymbol{A}) * \boldsymbol{c}$ to find $\boldsymbol{b}$. ■

Example 2.3.2

Let $A = \begin{bmatrix} 1 & -1 & 1 \\ -3 & 2 & 1 \\ -2 & 1 & 0 \end{bmatrix}$ and $B = \begin{bmatrix} 1 & 2 \\ 2 & 4 \\ 1 & 2 \end{bmatrix}$. Find AB, $(AB)^T$, and B^TA^T.

Solution:

$$AB = \begin{bmatrix} 1\cdot 1+(-1\cdot 2)+1\cdot 1 & 1\cdot 2+(-1\cdot 4)+1\cdot 2 \\ -3\cdot 1+2\cdot 2+1\cdot 1 & -3\cdot 2+2\cdot 4+1\cdot 2 \\ -2\cdot 1+1\cdot 2+0\cdot 1 & -2\cdot 2+1\cdot 4+0\cdot 2 \end{bmatrix} = \begin{bmatrix} 0 & 0 \\ 2 & 4 \\ 0 & 0 \end{bmatrix}$$

$$(AB)^T = \begin{bmatrix} 0 & 2 & 0 \\ 0 & 4 & 0 \end{bmatrix}$$

$$B^T A^T = \begin{bmatrix} 1 & 2 & 1 \\ 2 & 4 & 2 \end{bmatrix} \begin{bmatrix} 1 & -1 & 1 \\ -3 & 2 & 1 \\ -2 & 1 & 0 \end{bmatrix} = \begin{bmatrix} 0 & 2 & 0 \\ 0 & 4 & 0 \end{bmatrix} = (AB)^T$$

```
>>A=[1  -1  1;-3  2  1;-2  1  0]; B=[1  2;2  4;1  2];
>>A*B
0  0
2  4
0  0
>>(A*B)'
0  2  0
0  4  0
>>B'*A'
0  2  0
0  4  0
```

■

We can invert a matrix by using the MATLAB function, inv(**A**).

2.4 DETERMINANT OF A MATRIX

2.4.1 Definition and Expansion of a Matrix

The determinant of a square matrix is written as follows:

$$\det(A) = |A| = \begin{vmatrix} a_{11} & a_{12} & \cdots & a_{1n} \\ a_{21} & a_{22} & \cdots & a_{2n} \\ \vdots & \vdots & & \vdots \\ a_{n1} & a_{n2} & \cdots & a_{nn} \end{vmatrix}; \ |A| = \begin{vmatrix} a_{11} & a_{12} \\ a_{21} & a_{22} \end{vmatrix} = a_{11}a_{22} - a_{21}a_{12} \quad (2.7)$$

The determinant of a 3 × 3 matrix, which is true for any dimensions, is given by

$$\begin{vmatrix} a_{11} & a_{12} & a_{13} \\ a_{21} & a_{22} & a_{23} \\ a_{31} & a_{32} & a_{33} \end{vmatrix} = (-1)^{1+1} a_{11} \begin{vmatrix} a_{22} & a_{23} \\ a_{32} & a_{33} \end{vmatrix} + (-1)^{1+2} a_{12} \begin{vmatrix} a_{21} & a_{23} \\ a_{31} & a_{33} \end{vmatrix}$$

$$+ (-1)^{1+3} a_{13} \begin{vmatrix} a_{21} & a_{22} \\ a_{31} & a_{32} \end{vmatrix} \quad (2.8)$$

$$= a_{11}(a_{22}a_{33} - a_{32}a_{23}) - a_{12}(a_{21}a_{33} - a_{31}a_{23})$$

$$+ a_{13}(a_{21}a_{32} - a_{31}a_{22})$$

Note: In the above 3×3 determinant, we selected to expand the first row, but we could have taken any row or column. The exponents correspond to the row and column numbers that correspond to the element of the row we select. The remaining rows and columns produce the minor matrices whose determinant is taken.
Note: The MATLAB function for finding the determinant is det($\boldsymbol{A}$).
Note: An $n \times n$ matrix $\boldsymbol{A}$ can be inverted to find $\boldsymbol{A}^{-1}$, if and only if $\det(\boldsymbol{A}) \neq 0$. In MATLAB we write inv($\boldsymbol{A}$).

Additional properties of the determinant are given below, assuming that both, $\boldsymbol{A}$ and $\boldsymbol{B}$, are $n \times n$ matrices.

$$\begin{aligned} &\det(\boldsymbol{AB}) = \det(\boldsymbol{A})\det(\boldsymbol{B}) = \det(\boldsymbol{B})\det(\boldsymbol{A}) \\ &\det(\boldsymbol{A}^T) = \det(\boldsymbol{A}) \\ &\det(a\boldsymbol{A}) = a^n \det(\boldsymbol{A}) \\ &\det(\boldsymbol{A}^{-1}) = \frac{1}{\det(\boldsymbol{A})} \quad (\boldsymbol{A} \text{ must be invertible}) \end{aligned} \tag{2.9}$$

2.4.2 Trace of a Matrix

A trace of an $n \times n$ matrix $\boldsymbol{A}$ is given by

$$\begin{aligned} \operatorname{tr}(\boldsymbol{A}) = \sum_{i=1}^{n} a_{ii}; \quad \operatorname{tr}(\boldsymbol{A}^{-1}\boldsymbol{BA}) &= \operatorname{tr}[\boldsymbol{A}^{-1}(\boldsymbol{BA})] = \operatorname{tr}[(\boldsymbol{BA})\boldsymbol{A}^{-1}] \\ &= \operatorname{tr}[(\boldsymbol{B}(\boldsymbol{AA}^{-1})] = \operatorname{tr}(\boldsymbol{B}) \end{aligned} \tag{2.10}$$

In the Command window of MATLAB, we write `trace(A)` to obtain the trace of the matrix $\boldsymbol{A}$.

2.4.3 Inverse of a Matrix

Definition 2.4.1: An $n \times n$ matrix $\boldsymbol{A}$ is **nonsingular**, if it possesses a inverse or an equivalent its determinant is not 0, otherwise it is known as **singular**. Its inverse $\boldsymbol{A}^{-1}$ has the property

$$\boldsymbol{A}^{-1}\boldsymbol{A} = \boldsymbol{AA}^{-1} = \boldsymbol{I} \tag{2.11}$$

where:
$\boldsymbol{I}$ is the identity matrix

Additional properties of invertible matrices are as follows:

$$(\boldsymbol{AB})^{-1} = \boldsymbol{B}^{-1}\boldsymbol{A}^{-1}; \; (\boldsymbol{A}^{-1})^T = (\boldsymbol{A}^T)^{-1}; \; (\boldsymbol{A}^{-1})^{-1} = \boldsymbol{A}; \; \left|\boldsymbol{A}^{-1}\right| = \frac{1}{|\boldsymbol{A}|} \tag{2.12}$$

▲

The inverse of a nonsingular matrix can be computed by the formula:

$$A^{-1} = \frac{\text{adj}(A)}{|A|}, \ \text{adj}(A) = C^T, \ C_{ij} = \text{cofactor} = (-1)^{i+j} M_{ij} \tag{2.13}$$

where:

C_{ij} is known as the **cofactor** of the element a_{ij}

M_{ij} is the determinant of the **minor** matrix resulting by eliminating the i row and the j column of the matrix A

Example 2.4.1

Find the inverse of the matrix $A = \begin{bmatrix} 1 & 2 & 3 \\ 2 & 0 & 1 \\ 3 & 4 & 5 \end{bmatrix}$.

Solution: The determinant of the matrix is 6. The first column cofactors are as follows:

$$c_{11} = (-1)^{1+1}|M_{11}| = +\begin{vmatrix} 0 & 1 \\ 4 & 5 \end{vmatrix} = -4$$

$$c_{21} = (-1)^{2+1}|M_{21}| = -\begin{vmatrix} 2 & 3 \\ 4 & 5 \end{vmatrix} = 2$$

$$c_{31} = (-1)^{3+1}|M_{31}| = +\begin{vmatrix} 2 & 3 \\ 0 & 1 \end{vmatrix} = 2$$

Following the same procedure for the rest of the cofactors, we obtain

$$C = \begin{bmatrix} -4 & -7 & 8 \\ 2 & -4 & 2 \\ 2 & 5 & -4 \end{bmatrix}; \ A^{-1} = \frac{\text{adj}(A)}{|A|} = \frac{C^T}{6} = \begin{bmatrix} -0.6667 & 0.3333 & 0.3333 \\ -1.1667 & -0.6667 & 0.8333 \\ 1.3333 & 0.3333 & -0.6667 \end{bmatrix}$$

```
>>a = [1  2  3;2  0  1;3  4  5];
>>inv(a)
ans =
-0.6667   0.3333   0.3333
-1.1667  -0.6667   0.8333
 1.3333   0.3333  -0.6667
```

■

2.5 LINEAR EQUATIONS

A system of equations of the form

$$\begin{aligned} a_{11}x_1 + a_{12}x_2 + \cdots + a_{1m}x_m &= b_1 \\ a_{21}x_1 + a_{22}x_2 + \cdots + a_{2m}x_m &= b_2 \\ &\vdots \\ a_{n1}x_1 + a_{n2}x_2 + \cdots + a_{nm}x_m &= b_n \end{aligned}$$

which can be written in the matrix form:

$$\underbrace{\boldsymbol{A}}_{n\times m}\ \underbrace{\boldsymbol{x}}_{m\times 1} = \underbrace{\boldsymbol{b}}_{n\times 1} \tag{2.14}$$

2.5.1 Square Matrices ($n \times n$)

The unknown vector $\boldsymbol{x}$ can be found using inverse matrix $\boldsymbol{A}$, assuming that $\boldsymbol{A}$ has nonzero determinant (nonsingular matrix). Hence,

$$\boldsymbol{x} = \boldsymbol{A}^{-1}\boldsymbol{b} \tag{2.15}$$

If $\boldsymbol{A}$ is singular, then there may be no solution (the equation is **inconsistent**) or many solutions.

There is a convenient way to represent the vector $\boldsymbol{b}$ as a linear combination of the columns of the matrix $\boldsymbol{A}$ as follows:

$$\boldsymbol{b} = \sum_{i=1}^{m} x_i \boldsymbol{a}_i \tag{2.16}$$

Example 2.5.1 (Regression Line)

Let us consider the data on students' weight and height given in the following table:

Student number:	1	2	3	4	5
Weight (y):	152	164	167	156	150
Height (x):	157	149	160	151	144

We are requested to plot a straight line that is the most representative of the random variations of these two measurements (Figure 2.1).

Solution: The requested line is known as linear **regression line**, and for each pair of values, the line is defined by the equation:

$$y_i = b_0 + b_1 x_i + v_i$$

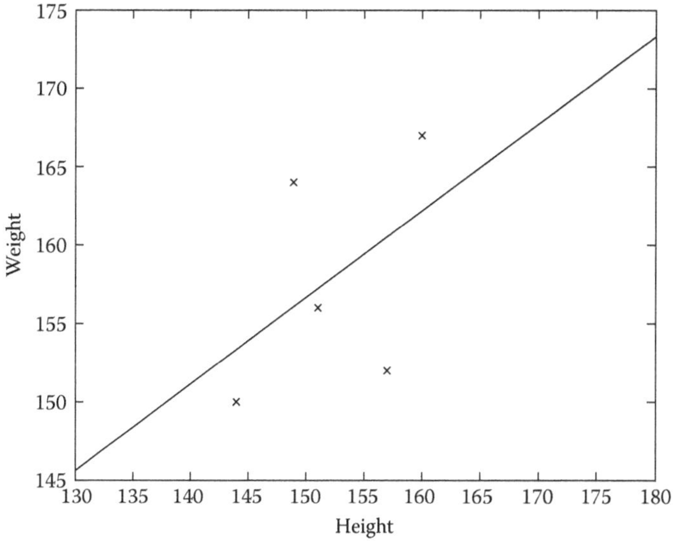

FIGURE 2.1

We, therefore, must find the unknowns b_0 and b_1 so that we find the appropriate line. The constants v_i's are the errors introduced in the data.

To determine the b's, we must use the least squares method such that the quantity

$$\sum_{i=1}^{n}(y_i - b_0 - b_1 x_i)^2$$

becomes the smallest possible. Therefore, we write

$$\frac{\partial}{\partial b_0}\left[\sum_{i=1}^{n}(y_i - b_0 - b_1 x_i)^2\right] = 0 \quad \text{or} \quad nb_0 + \left(\sum_{i=1}^{n} x_i\right) b_1 = \sum_{i=1}^{n} y_i$$

$$\frac{\partial}{\partial b_1}\left[\sum_{i=1}^{n}(y_i - b_0 - b_1 x_i)^2\right] = 0 \quad \text{or} \quad \left(\sum_{i=1}^{n} x_i\right) b_0 + \left(\sum_{i=1}^{n} x_i^2\right) b_1 = \sum_{i=1}^{n} x_i y_i$$

The above two linear equations in b_0 and b_1 are called the **normal equations**. The above equations take the matrix form:

$$\begin{bmatrix} n & \sum_{i=1}^{n} x_i \\ \sum_{i=1}^{n} x_i & \sum_{i=1}^{n} x_i^2 \end{bmatrix} \begin{bmatrix} b_0 \\ b_1 \end{bmatrix} = \begin{bmatrix} \sum_{i=1}^{n} y_i \\ \sum_{i=1}^{n} x_i y_i \end{bmatrix}$$

Using the inverse matrix approach and MATLAB, we obtain the values of the b's, which are shown in Figure 2.1.

```
>>x = [157 149 160 151 144];
>>y = [152 164 167 156 150];
>>b = inv([5 sum(x);sum(x) sum(x.^2)])*[sum(y) ; sum(x.*y)];
>>    %it was found that b0 = 73.4730, b1 = 0.5541;
>>x1 = 130:5:180;
>>y1 = 73.4730+0.5541*x1;
>>plot(x,y,'.k');hold on;plot(x1,y1,'k');
>>xlabel('Height');ylabel('Weight');
```

■

2.5.2 Rectangular Matrices ($n < m$)

If the rows are fewer than the columns, $n < m$, indicates that there are fewer equations than unknowns. If the equations are consistent, there are many vectors that satisfy the equations. Specifically, to find a unique solution, the rank of $\boldsymbol{A}$ must be equal to n. All the $n \times n$ matrices of the matrix $\boldsymbol{A}$ must have nonzero matrix. Then the solution (known as the **minimum norm** solution) is as follows:

$$\boldsymbol{x}_0 = \boldsymbol{A}^H(\boldsymbol{A}\boldsymbol{A}^H)^{-1}\boldsymbol{b} \tag{2.17}$$

where:

H stands for the conjugate transpose of the matrix (**Hermitian**)

For real matrices, we substitute H with T (transpose). The matrix

$$\boldsymbol{A}^+ = \boldsymbol{A}^H(\boldsymbol{A}\boldsymbol{A}^H)^{-1} \tag{2.18}$$

is known as the **pseudoinverse** of the matrix $\boldsymbol{A}$ for the undetermined problem.

Example 2.5.2

Find the solution for the system given below:

$$x_1 + x_2 + x_3 + x_4 = 1$$

$$x_1 - x_2 + x_3 - x_4 = 1$$

Solution: Any 2×2 matrix of

$$\boldsymbol{A} = \begin{bmatrix} 1 & 1 & 1 & 1 \\ 1 & -1 & 1 & -1 \end{bmatrix}$$

has nonzero determinant. This indicates that the rank of $\boldsymbol{A}$ is 2. Hence, we can use (2.17). Therefore, we obtain

$$x = A^T(AA^T)^{-1}\begin{bmatrix}1\\1\end{bmatrix} = \begin{bmatrix}1 & 1\\1 & -1\\1 & 1\\1 & -1\end{bmatrix}\begin{bmatrix}4 & 0\\0 & 4\end{bmatrix}^{-1}\begin{bmatrix}1\\1\end{bmatrix} = \begin{bmatrix}0.5\\0.0\\0.5\\0.0\end{bmatrix} \equiv \begin{bmatrix}x_1\\x_2\\x_3\\x_4\end{bmatrix}$$

$$A = \begin{bmatrix}4 & 0\\0 & 4\end{bmatrix},\ \text{adj}(A) = \begin{bmatrix}4 & 0\\0 & 4\end{bmatrix}^T = \begin{bmatrix}4 & 0\\0 & 4\end{bmatrix},\ |A| = 16$$

$$A^{-1} = \frac{\text{adj}(A)}{|A|} = \frac{1}{16}\begin{bmatrix}4 & 0\\0 & 4\end{bmatrix} = \frac{1}{4}\begin{bmatrix}1 & 0\\0 & 1\end{bmatrix}$$

$$\Rightarrow \begin{bmatrix}1 & 1\\1 & -1\\1 & 1\\1 & -1\end{bmatrix}\frac{1}{4}\begin{bmatrix}1 & 0\\0 & 1\end{bmatrix}\begin{bmatrix}1\\1\end{bmatrix} = \frac{1}{4}\begin{bmatrix}1 & 1\\1 & -1\\1 & 1\\1 & -1\end{bmatrix}\begin{bmatrix}1\\1\end{bmatrix} = \frac{1}{4}\begin{bmatrix}2\\0\\2\\0\end{bmatrix} = \begin{bmatrix}0.5\\0\\0.5\\0\end{bmatrix}$$

```
>>A = [1  1  1  1;1  -1  1  -1];
>>x = A'*inv(A*A')*[1  1]';
x =
0.5000
     0
0.5000
     0
```

■

2.5.3 Rectangular Matrices ($m < n$)

For this case, in general, no solution exists. The equations in this case are **inconsistent** and the solution is said to be **overdetermined**. In this case, we cannot expand an arbitrary vector $\boldsymbol{b}$ as a function of the column vectors of $\boldsymbol{A}$ as in (2.16). Our next action is to find the coefficients x_i that produce the best approximation to $\boldsymbol{b}$:

$$\hat{\boldsymbol{b}} = \sum_{i=1}^{m} x_i \boldsymbol{a}_i \tag{2.19}$$

The most common approach is to use the **least squares** solution. This is equivalent to finding a vector $\boldsymbol{x}$ that minimizes the norm of the error vector:

$$\|\boldsymbol{e}\|^2 = \|\boldsymbol{b} - \boldsymbol{A}\boldsymbol{x}\|^2 \tag{2.20}$$

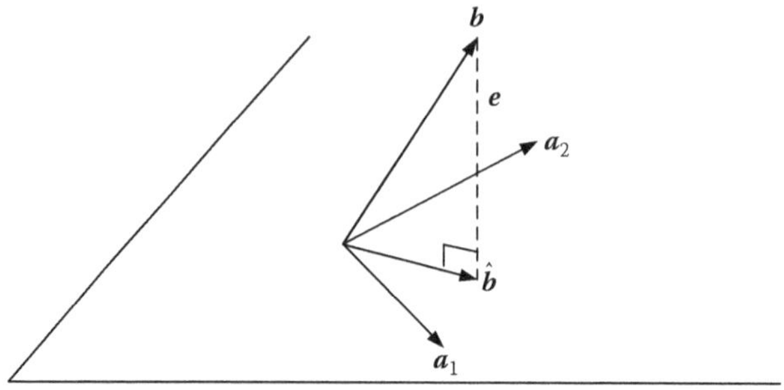

FIGURE 2.2

As shown in Figure 2.2, the least squares solution has the property that the error

$$\boldsymbol{e} = \boldsymbol{b} - \boldsymbol{A}\boldsymbol{x}$$

is **orthogonal** to every vector $\boldsymbol{a}$ on the plane, which is used in the approximation for $\boldsymbol{b}$, that is, the columns of $\boldsymbol{A}$. This orthogonality implies that

$$\boldsymbol{A}^H \boldsymbol{e} = \boldsymbol{0} \tag{2.21}$$

or

$$\boldsymbol{A}^H \boldsymbol{e} = \boldsymbol{A}^H \boldsymbol{b} - \boldsymbol{A}^H \boldsymbol{A}\boldsymbol{x} = \boldsymbol{0} \text{ or } \boldsymbol{A}^H \boldsymbol{A}\boldsymbol{x} = \boldsymbol{A}^H \boldsymbol{b} \tag{2.22}$$

The above two expressions are known as the **normal equations**. For real matrices, H (Hermitian) is substituted by the transposition T. If the columns of $\boldsymbol{A}$ are linearly independent ($\boldsymbol{A}$ has full rank), then the matrix $\boldsymbol{A}^H\boldsymbol{A}$ is invertible and the **least squares solution** is

$$\boldsymbol{x}_0 = (\boldsymbol{A}^H \boldsymbol{A})^{-1} \boldsymbol{A}^H \boldsymbol{b} = \boldsymbol{A}^+ \boldsymbol{b}; \quad \boldsymbol{A}^+ = (\boldsymbol{A}^H \boldsymbol{A})^{-1} \boldsymbol{A}^H \tag{2.23}$$

Matrix $\boldsymbol{A}^+$ is known as **pseudoinverse** for the **overdetermined** system. In addition, the best approximation $\hat{\boldsymbol{b}}$ to $\boldsymbol{b}$ is given by the **projection** of the vector $\boldsymbol{b}$ onto the subspace spanned by the vectors $\boldsymbol{a}_i$ (Figure 2.2):

$$\hat{\boldsymbol{b}} = \boldsymbol{A}\boldsymbol{x}_0 = \boldsymbol{A}(\boldsymbol{A}^H \boldsymbol{A})^{-1} \boldsymbol{A}^H \boldsymbol{b}; \quad \hat{\boldsymbol{b}} = \boldsymbol{P}_A \boldsymbol{b}; \quad \boldsymbol{P}_A = \boldsymbol{A}(\boldsymbol{A}^H \boldsymbol{A})^{-1} \boldsymbol{A}^H \tag{2.24}$$

Matrix $\boldsymbol{P}_A$ is known as the projection matrix. The minimum error square is given by (see Problem 2.5.2)

$$\min\|\boldsymbol{e}\|^2 = \boldsymbol{b}^H\boldsymbol{e} = \boldsymbol{b}^H\boldsymbol{b} - \boldsymbol{b}^H\boldsymbol{A}\boldsymbol{x}_0 \tag{2.25}$$

Example 2.5.3 (Overdetermined Case)

Find the vector $\boldsymbol{x}_0$ for the system as well as the minimum error. The system is

$$2x_1 - x_2 = 1$$

$$2x_1 + x_2 = 1$$

$$x_1 + x_2 = 1$$

Solution: Using MATLAB, we obtain

```
>>a = [2  -1;2  1;1  1];
>>b = [1;1;1];
>>xo = inv(a'*a)*a'*b;
>>bap = a*inv(a'*a)*a'*b;
>>mes = b'*b-b'*bap;
```

The values are $\hat{\boldsymbol{b}}$ = bap = [1.3333; 0.6667; 0.6667]; $\boldsymbol{A}^T\boldsymbol{A} = [6\ 3;3\ 6]$; $(\boldsymbol{A}^T\boldsymbol{A})^{-1} = [0.2222\ -0.1111;-0.1111\ 0.2222]$; $\boldsymbol{x}_o = [0.6667\ 0]$; $\min(\boldsymbol{e}^2) = 0.3333$.

■

2.5.4 Quadratic and Hermitian Forms

A **quadratic form** of a real symmetric $n \times n$ matrix $\boldsymbol{A}$ is the scalar defined by (see Problem 2.5.3)

$$\begin{aligned} Q_A(\boldsymbol{x}) &= \boldsymbol{x}^T\boldsymbol{A}\boldsymbol{x} = \sum_{i=1}^{n}\sum_{j=1}^{n} a_{ij}x_ix_j \\ Q_A(\boldsymbol{x}) &= \boldsymbol{x}^H\boldsymbol{A}\boldsymbol{x} = \sum_{i=1}^{n}\sum_{j=1}^{n} a_{ij}x_i^*x_j,\ x = \text{complex} \end{aligned} \tag{2.26}$$

Example 2.5.4

Find the projection of the vector $\boldsymbol{a} = [1\ 4\ -2]^T$ onto the plane spanned by $\boldsymbol{x}_1 = [1\ 2\ 1]^T$ and $\boldsymbol{x}_2 = [2\ 1\ 0]^T$.

Solution: Using MATLAB, we obtain

TABLE 2.3
Some Additional Properties of Matrices

$A^T = A$	A is a symmetric matrix.
$A^T A = I; A^H A = I$	The columns of $n \times n$ matrix A are orthogonal if the rows (columns) are orthogonal, that is, $a_i^T a_i = 1, a_i^T a_j = 0$.
AB	AB is symmetric if $AB = BA$.
A^{-1}	A^{-1} is symmetric if A is symmetric.
$A = A^H$	Matrix A is Hermitian.
$A^H = A^{-1}$	Matrix A is unitary.
$A = -A^T$	The $n \times n$ matrix A is skew symmetric.
$Q_A(x) = x^T A x = \sum_{i=1}^{n}\sum_{j=1}^{n} a_{ij}x_i x_j$	Q_A (scalar) is the **quadratic form** of a real symmetric $n \times n$ matrix A.
$Q_A(x) > 0$	A is **positive definite**.
$Q_A(x) \geq 0$	A is **positive semidefinite**. $Q_A(x) \leq 0$ is negative semidefinite. $Q_A(x) < 0$ is negative definite.
Grammian	An $n \times n$ matrix A is Grammian if it is symmetric and positive semidefinite.
$A = P^T P$	If P is nonsingular, then A ($n \times n$) is positive definite.
$P = P^2$	P is an **idempotent** matrix and also a **projection** matrix.
P	P projects vectors orthogonally onto a subspace if and only if it is an idempotent symmetric matrix.
$P_A = A(A^T A)^{-1} A^T$	A is an $n \times m$ matrix with $\text{rank}(A) = m < n$. Then, P_A is idempotent and symmetric and consequently an orthogonal projection matrix.

Note: T can be substituted with H.

$$a = [1 \;\; 4 \;\; -2]^T,\; x_1 = [1 \;\; 2 \;\; 1]^T,\; x_2 = [2 \;\; 1 \;\; 0]^T$$

$$X = \begin{bmatrix} 1 & 2 \\ 2 & 1 \\ 1 & 0 \end{bmatrix},\; X^T X = \begin{bmatrix} 6 & 4 \\ 4 & 5 \end{bmatrix},\; \left[X^T X\right]^{-1} = \begin{bmatrix} 0.3571 & -0.2857 \\ -0.2857 & 0.4286 \end{bmatrix}$$

$$P = X(X^T X)^{-1} X^T = \begin{bmatrix} 0.9286 & 0.1429 & -0.2143 \\ 0.1429 & 0.7143 & 0.4286 \\ -0.2143 & 0.4286 & 0.3571 \end{bmatrix}$$

$$P = PP,\; \hat{a} = \text{projected} = Pa = \begin{bmatrix} 1.9286 \\ 2.1429 \\ 0.7857 \end{bmatrix}$$

The coefficients of the linear combination of the column vectors of $\boldsymbol{X}$ that represent the projected vector of $\boldsymbol{a}$ is given by

$$\boldsymbol{b} = (\boldsymbol{X}^T\boldsymbol{X})^{-1}\boldsymbol{X}^T\boldsymbol{a} = \begin{bmatrix} 0.7857 \\ 0.5714 \end{bmatrix}, \quad \boldsymbol{a} = 0.7857\boldsymbol{x}_1 + 0.5714\boldsymbol{x}_2$$

The error is given by

$$\boldsymbol{e} = \boldsymbol{a} - \hat{\boldsymbol{a}} = (\boldsymbol{I} - \boldsymbol{P})\boldsymbol{a} = \begin{bmatrix} -0.9286 \\ 1.8571 \\ -2.7857 \end{bmatrix}, \hat{\boldsymbol{a}}^T\boldsymbol{e} = 2.2204\text{e}-015 \cong 0$$

It is apparent that the error vector is perpendicular to the projected one. The error vector is also perpendicular to $\boldsymbol{x}_1$ and $\boldsymbol{x}_2$, which span the plane on which the projected vector lies. For example, $\boldsymbol{x}_1^T\boldsymbol{e} = 8.8818\text{e}-016$ as it was expected.

■

2.6 EIGENVALUES AND EIGENVECTORS

Let $\boldsymbol{A}$ be an $n \times n$ matrix and consider the following set of linear equations:

$$\boldsymbol{Aq} = \lambda\boldsymbol{q} \quad \text{or} \quad (\boldsymbol{A} - \lambda\boldsymbol{I})\boldsymbol{q} = \boldsymbol{0} \tag{2.27}$$

For the above equation to have nonzero vector $\boldsymbol{q}$, the determinant of the matrix $\boldsymbol{A} - \lambda\boldsymbol{I}$ must be 0, $|\boldsymbol{A} - \lambda\boldsymbol{I}| = 0$. The determinant gives an nth-order polynomial with respect to λ, and thus, its solution gives n roots of the unknown λ, λ_i for $i=1, 2, 3,\ldots, n$. These values are known as **eigenvalues** and the corresponding vectors are the **eigenvectors**.

For example, for a 2×2 matrix, we write

$$\begin{bmatrix} a_{11} & a_{12} \\ a_{21} & a_{22} \end{bmatrix}\begin{bmatrix} q_1 \\ q_2 \end{bmatrix} = \begin{bmatrix} \lambda q_1 \\ \lambda q_2 \end{bmatrix} \quad \text{or} \quad \left(\begin{bmatrix} a_{11} & a_{12} \\ a_{21} & a_{22} \end{bmatrix} - \begin{bmatrix} \lambda & 0 \\ 0 & \lambda \end{bmatrix}\right)\begin{bmatrix} q_1 \\ q_2 \end{bmatrix} = \begin{bmatrix} 0 \\ 0 \end{bmatrix} \quad \text{or}$$

$$\begin{bmatrix} a_{11} - \lambda & a_{12} \\ a_{21} & a_{22} - \lambda \end{bmatrix}\begin{bmatrix} q_1 \\ q_2 \end{bmatrix} = \begin{bmatrix} 0 \\ 0 \end{bmatrix} \quad \text{or} \quad \det\left(\begin{bmatrix} a_{11} - \lambda & a_{12} \\ a_{21} & a_{22} - \lambda \end{bmatrix}\right) = 0 \quad \text{or}$$

$$\lambda^2 - (a_{11} + a_{22})\lambda + (a_{11}a_{22} - a_{12}a_{21}) = 0$$

which gives us two values for the unknown λ, λ_i for $i=1,2$, which are known as the **eigenvalues**.

Let us assume that the a's in the above expression have the following values: $a_{11} = a_{22} = 2, a_{12} = a_{21} = 1$. Therefore, the above polynomial becomes: $\lambda - 4\lambda + 3 = 0$. Hence, the roots of the polynomial (eigenvalues) are as follows:

$$\lambda = \frac{4 \pm \sqrt{4^2 - 4\times 3}}{2\times 1} \Rightarrow \lambda_1 = 1, \lambda_2 = 3$$

```
>>roots([1 -4 3]);%roots() is a MATLAB function that
       %gives the roots of the polynomials;
       %observe the arrangement of the coefficients of
       %the polynomial;
ans =
3
1
```

2.6.1 Eigenvectors

It is desired to find the eigenvectors of the matrix $\boldsymbol{A} = \begin{bmatrix} 2 & 1 \\ 1 & 2 \end{bmatrix}$ that has the eigenvalues 1 and 3. From (2.27), we obtain

$$\begin{bmatrix} 2 & 1 \\ 1 & 2 \end{bmatrix}\begin{bmatrix} q_1 \\ q_2 \end{bmatrix} = 1 \times \begin{bmatrix} q_1 \\ q_2 \end{bmatrix} \Rightarrow \begin{bmatrix} 2q_1 + q_2 \\ q_1 + 2q_2 \end{bmatrix} = \begin{bmatrix} q_1 \\ q_2 \end{bmatrix}$$

$$\Rightarrow 2q_1 + q_2 = q_1 \Rightarrow q_1 = -q_2 \Rightarrow \boldsymbol{q}_1 = \begin{bmatrix} -1 \\ 1 \end{bmatrix}$$

The normalized eigenvector $\boldsymbol{q}_1$ corresponding to the eigenvalue $\lambda = 1$ is

$$\boldsymbol{q}_1 = \frac{1}{\sqrt{1^2 + (-1)^2}}\begin{bmatrix} -1 \\ 1 \end{bmatrix} = \begin{bmatrix} -0.7071 \\ 0.7071 \end{bmatrix}$$

Following the same procedure, the second eigenvector corresponding to the second eigenvalue $\lambda = 3$ is

$$\boldsymbol{q}_1 = \frac{1}{\sqrt{1^2 + 1^2}}\begin{bmatrix} 1 \\ 1 \end{bmatrix} = \begin{bmatrix} 0.7071 \\ 0.7071 \end{bmatrix}$$

The above results can be found using MATLAB as follows:

```
>> A = [2    1;1   2];
>>[V,D] = e.g(A)
>>V
V =
-0.7071       0.7071
 0.7071       0.7071
>>D
D =
1  0
0  3
```

Based on the above results, we can analyze the matrix $\boldsymbol{A}$ as follows:

$$\boldsymbol{A} = \boldsymbol{V}\boldsymbol{D}\boldsymbol{V}^{-1} \tag{2.28}$$

2.6.2 Properties of Eigenvalues and Eigenvectors

We will use the following data produced with the help of MATLAB to check some of the properties given below:

```
>>a = [2  1  1;1  2  1;1  1  2];
>>[v,d] = eig(a);
>>v
v =

0.4082      0.7071    0.5774
0.4082     -0.7071    0.5774
-0.8165     0         0.5774

>>d
d =

1.0000     0          0
0          1.0000     0
0          0          4.0000
```

Property 1. The nonzero eigenvectors $\boldsymbol{q}_1, \boldsymbol{q}_2,\ldots,\boldsymbol{q}_n$ corresponding to distinct eigenvalues $\lambda_1, \lambda_2,\ldots, \lambda_n$ are linearly independent.

Property 2. The eigenvalues of a Hermitian ($\boldsymbol{A}^H = \boldsymbol{A}; \boldsymbol{A}^H = \boldsymbol{A}^{*T}$) matrix are real (see Problem 2.6.2).

Property 3. A Hermitian matrix is positive definite, if its eigenvalues are positive. If its eigenvalues are greater than or equal to 0, the matrix is semidefinite.

Property 4. The **sum** of the n eigenvalues of an $n \times n$ matrix $\boldsymbol{A}$ is equal to the sum of its diagonal terms (or trace of $\boldsymbol{A}$). The product of the n eigenvalues is equal to the determinant of $\boldsymbol{A}$ (see Problem 2.6.3).

```
>>trace(a)
6
>>trace(d)
6.0000
```

Property 5. The eigenvectors corresponding to distinct eigenvalues of a Hermitian matrix are orthogonal (see Problem 2.6.4).

Property 6 (spectral theorem). Any Hermitian $n \times n$ matrix $\boldsymbol{A}$ may be decomposed as

$$\boldsymbol{A} = \boldsymbol{Q}\Lambda\boldsymbol{Q}^H = \sum_{i=1}^{n} \lambda_i \boldsymbol{q}_i \boldsymbol{q}_i^H \tag{2.29}$$

where there is a set of orthogonal eigenvectors with their corresponding eigenvalues.

$$A = \begin{bmatrix} 2 & 1 & 1 \\ 1 & 2 & 1 \\ 1 & 1 & 2 \end{bmatrix}$$

$$[Q,L] = \text{eig}(A)$$

$$Q = \begin{bmatrix} 0.4082 & 0.7071 & 0.5774 \\ 0.4082 & -0.7071 & 0.5774 \\ -0.8165 & 0 & 0.5774 \end{bmatrix}; L = \begin{bmatrix} 1 & 0 & 0 \\ 0 & 1 & 0 \\ 0 & 0 & 4 \end{bmatrix}$$

```
>>1*Q(:,1)*Q(:,1)'+1*Q(:,2)*Q(:,2)'+4*Q(:,3)*Q(:,3)'
ans =
2.0000 1.0000 1.0000
1.0000 2.0000 1.0000
1.0000 1.0000 2.0000
```

Property 7. A real $n \times n$ matrix $\boldsymbol{A}$ having linearly independent eigenvectors with its corresponding eigenvalues can be expressed in the form (see Problem 2.6.5):

$$\boldsymbol{A} = \boldsymbol{Q}\Lambda\boldsymbol{Q}^{-1} \tag{2.30}$$

```
>> Q*L*inv(Q)
ans =
2.0000 1.0000 1.0000
1.0000 2.0000 1.0000
1.0000 1.0000 2.0000
```

Property 8. The inverse of Hermitian matrix $\boldsymbol{A}$ is of the form (see Problem 2.6.6):

$$\boldsymbol{A}^{-1} = \sum_{i=1}^{n} \frac{1}{\lambda_i} \boldsymbol{q}_i \boldsymbol{q}_i^H \tag{2.31}$$

```
>>1*Q(:,1)*Q(:,1)'+1*Q(:,2)*Q(:,2)'+0.25*Q(:,3)*Q(:,3)'
ans =
0.7500    -0.2500   -0.2500
-0.2500    0.7500   -0.2500
-0.2500   -0.2500    0.7500
The above is exactly the inv(A) as it should be.
```

Property 9. $\boldsymbol{B}$ is an $n \times n$ matrix with eigenvalues λ_i and eigenvectors $\boldsymbol{q}_i$. $\boldsymbol{A}$ is a matrix related to $\boldsymbol{B}$ as follows:

$$\boldsymbol{A} = \boldsymbol{B} + a\boldsymbol{I}$$

Then, $\boldsymbol{A}$ and $\boldsymbol{B}$ have the same eigenvectors and the eigenvalues of $\boldsymbol{A}$ are $\boldsymbol{a}+\lambda_i$ (see Problem 2.6.7).

Property 10. For a symmetric (eigenvectors form an orthonormal set) positive definite ($\boldsymbol{A} > 0$, eigenvalues are positive) matrix $\boldsymbol{A}$, the equation

$$\boldsymbol{x}^T \boldsymbol{A}\boldsymbol{x} = 1$$

defines an ellipse in n dimensions whose axes are in the direction of the eigenvectors $\boldsymbol{q}_i$ of $\boldsymbol{A}$ with the half-length of these axes being equal to $1/\sqrt{\lambda_i}$.

Property 11. The maximum value of the eigenvalues of an $n \times n$ matrix $\boldsymbol{A}$ has the following relations:

$$\lambda_{\max} \le \sum_{i=1}^{n} \lambda_i = \mathrm{tr}(\boldsymbol{A})$$

$$\lambda_{\max} \le \max_i \sum_{j=1}^{n} a_{ij}$$

Property 12. The eigenvalues of $\boldsymbol{A}^k$ are equal to λ_i^k and its eigenvectors are the same. Hence,

$$\Lambda^k = \left(\boldsymbol{Q}^{-1}\boldsymbol{A}\boldsymbol{Q}\right)\left(\boldsymbol{Q}^{-1}\boldsymbol{A}\boldsymbol{Q}\right)\cdots\left(\boldsymbol{Q}^{-1}\boldsymbol{A}\boldsymbol{Q}\right) = \boldsymbol{Q}^{-1}\boldsymbol{A}^k\boldsymbol{Q}$$

```
>>L^3
ans =
1.0000        0          0
0        1.0000          0
0             0    64.0000
>>inv(Q)*A^3*Q
ans =
1.0000        0     0.0000
0.0000   1.0000     0.0000
0.0000        0    64.0000
```

Example 2.6.1

Identify the ellipse for the matrix $\boldsymbol{A} = \begin{bmatrix} 2 & 1 \\ 1 & 2 \end{bmatrix}$.

Solution: The eigenvalues are $\lambda_1 = 1$ and $\lambda_2 = 3$ and the eigenvectors are $\boldsymbol{q}_1 = [1 \quad -1]^T$ and $\boldsymbol{q}_2 = [1 \quad 1]^T$. Figure 2.3 shows the ellipse with the eigenvectors and the minor and major axes.

■

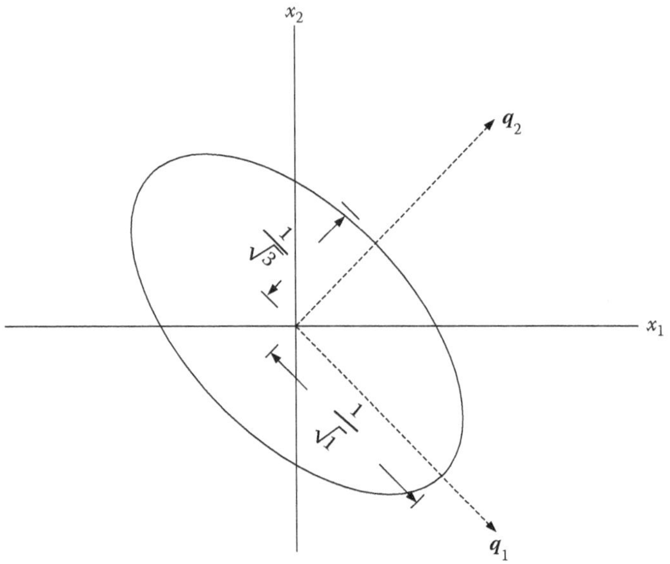

FIGURE 2.3

PROBLEMS

2.1.1 Find the matrix $\boldsymbol{A} = \boldsymbol{J}\boldsymbol{J}$.

$$\text{If } \boldsymbol{C} = \begin{bmatrix} 1 & 1 \\ 3 & 4 \end{bmatrix} \text{and } \boldsymbol{C}^{-1} = \text{inverse of } \boldsymbol{C} = \begin{bmatrix} 4 & -1 \\ -3 & 1 \end{bmatrix}, \text{ find } \boldsymbol{C}\boldsymbol{C}^{-1}.$$

2.4.1 Compute the determinant of the matrix $\boldsymbol{A}$ along the first column and also find the trace.

$$\boldsymbol{A} = \begin{bmatrix} 1 & 2 & 3 \\ 2 & 0 & 1 \\ 3 & 4 & 5 \end{bmatrix}$$

2.4.2 Compute the inverse of the matrices:

$$\boldsymbol{A} = \begin{bmatrix} 1 & -2 \\ 2 & 4 \end{bmatrix}; \boldsymbol{B} = \begin{bmatrix} 2 & 4 \\ 1 & 2 \end{bmatrix}$$

2.5.1 Find the solution of the system

$$\begin{bmatrix} -1 & 0 & 2 & 1 \\ -1 & 1 & 0 & -1 \end{bmatrix} [x_1 \ x_2 \ x_3 \ x_4]^T = [1 \quad 1]^T$$

2.5.2 Prove (2.25).

2.5.3 Show that $A = \begin{bmatrix} 1 & 0 \\ 0 & 2 \end{bmatrix}$ is positive definite and $B = \begin{bmatrix} 4 & 0 \\ 0 & 0 \end{bmatrix}$ is semidefinitive.

2.5.4 Prove that the coefficients of expansion of a vector projected on a subspace is equal to $b = (X^T X)^{-1} X^T Y$.

2.6.1 Find the eigenvalues and the corresponding eigenvectors for the matrix

$$A = \begin{bmatrix} 2 & 1 \\ 1 & 2 \end{bmatrix}$$

2.6.2 Verify Property 2 of the eigenvalues.

2.6.3 Verify Property 4 of the eigenvalues using a 2×2 matrix.

2.6.4 Verify Property 5 of the eigenvalues.

2.6.5 Verify Property 7 of the eigenvalues.

2.6.6 Verify Property 8 of the eigenvalues.

2.6.7 Verify Property 9 of the eigenvalues.

HINTS–SUGGESTIONS–SOLUTIONS

2.1.1

$$A = JJ = I \cdot C^{-1}C = I \Rightarrow CC^{-1}C = IC = C$$

Hence, $J^{-1}JJ = J^{-1}I = J^{-1}$ or $J = J^{-1}$.

2.4.1

$$|A| = (-1)^{1+1}(1)\begin{vmatrix} 0 & 1 \\ 4 & 5 \end{vmatrix} + (-1)^{2+1}(2)\begin{vmatrix} 2 & 3 \\ 4 & 5 \end{vmatrix} + (-1)^{3+1}3\begin{vmatrix} 2 & 3 \\ 0 & 1 \end{vmatrix}$$

$$= (0 \cdot 5 - 4 \cdot 1) - 2(2 \cdot 5 - 3 \cdot 4) + 3(2 \cdot 1 - 0 \cdot 3) = 6;\ \text{tr}(A) = 1 + 0 + 5 = 6$$

2.4.2

$$A^{-1} = \frac{1}{8}\begin{bmatrix} 4 & 2 \\ -2 & 1 \end{bmatrix};\ \det(B) = 0 \Rightarrow \text{there is no inverse}$$

2.5.1

$$x = [-0.5000\ \ 0.3333\ \ 0.3333\ \ -0.1667]^T$$

2.5.2

$$\min\|e\|^2 = e^H e = (b^H - x^H A^H)e = b^H e - x^H 0 = b^H b - b^H A x_0$$

2.5.3 For any vector x different than 0,

$$Q_A(x) = [x_1\ \ x_2]\begin{bmatrix} 1 & 0 \\ 0 & 2 \end{bmatrix}\begin{bmatrix} x_1 \\ x_2 \end{bmatrix} = x_1^2 + 2x_2^2 > 0$$

Therefore, matrix A is nonnegative.

$$Q_B(x) = [x_1\ \ x_2]\begin{bmatrix} 4 & 0 \\ 0 & 2 \end{bmatrix}\begin{bmatrix} x_1 \\ x_2 \end{bmatrix} = 4x_2^2 > 0.$$

However, for $\boldsymbol{x} = [0\ x_2], Q_B = 0$, and thus, $\boldsymbol{A}$ is positive semidefinite.

2.5.4

$$\boldsymbol{e}^T\boldsymbol{X} = \left[(\boldsymbol{I}-\boldsymbol{P})\boldsymbol{Y}\right]^T\boldsymbol{X} = \boldsymbol{Y}^T(\boldsymbol{I}-\boldsymbol{P})^T\boldsymbol{X} = \boldsymbol{Y}^T(\boldsymbol{I}^T-\boldsymbol{P}^T)\boldsymbol{X} = \boldsymbol{Y}^T\boldsymbol{X}-\boldsymbol{Y}^T\boldsymbol{P}\boldsymbol{X}$$

$$= \boldsymbol{Y}^T\boldsymbol{X}-\boldsymbol{Y}^T\boldsymbol{X}(\boldsymbol{X}^T\boldsymbol{X})^{-1}\boldsymbol{X}^T\boldsymbol{X} = \boldsymbol{Y}^T\boldsymbol{X}-\boldsymbol{Y}^T\boldsymbol{X} = 0;\ \boldsymbol{P}\boldsymbol{X} = \boldsymbol{X}(\boldsymbol{X}^T\boldsymbol{X})^{-1}\boldsymbol{X}^T\boldsymbol{X}$$

$$= \boldsymbol{X}\boldsymbol{X}^{-1}\boldsymbol{X}^{-T}\boldsymbol{X}^T\boldsymbol{X} = \boldsymbol{X};\ \hat{\boldsymbol{Y}} = \boldsymbol{P}\boldsymbol{Y} = \boldsymbol{X}(\boldsymbol{X}^T\boldsymbol{X})^{-1}\boldsymbol{X}^T\boldsymbol{Y} = \boldsymbol{X}\boldsymbol{b} \Rightarrow \boldsymbol{b} = (\boldsymbol{X}^T\boldsymbol{X})^{-1}\boldsymbol{X}^T\boldsymbol{Y}$$

The columns of $\boldsymbol{X}$ span the subspace in which the projection $\hat{\boldsymbol{Y}}$ of $\boldsymbol{Y}$ lies.

2.6.1

$$\begin{vmatrix} 2-\lambda & 1 \\ 1 & 2-\lambda \end{vmatrix} = (2-\lambda)^2 - 1 = 0,\ \lambda_1 = 3, \lambda_2 = 1$$

$$\begin{bmatrix} 2 & 1 \\ 1 & 2 \end{bmatrix}\begin{bmatrix} q_{11} \\ q_{12} \end{bmatrix} = \begin{bmatrix} 3q_{11} \\ 3q_{12} \end{bmatrix} \Rightarrow 2q_{11} + q_{12} = 3q_{11}$$

$\Rightarrow q_{11} = q_{12}$ which satisfies the second equation $q_{11} + 2q_{12} = 3q_{12}$

$$\Rightarrow \boldsymbol{q}_1 = \frac{1}{\sqrt{8}}\begin{bmatrix} 2 \\ 2 \end{bmatrix} = \text{normalized eigenvector arbitrarily setting } q_{11} = 2$$

Similarly, $\boldsymbol{q}_2 = \dfrac{1}{\sqrt{8}}\begin{bmatrix} 2 \\ -2 \end{bmatrix}$

2.6.2

$$\boldsymbol{A}\boldsymbol{q}_i = \lambda_i \boldsymbol{q} \Rightarrow \boldsymbol{q}_i^H\boldsymbol{A}\boldsymbol{q}_i = \lambda_i \boldsymbol{q}_i^H\boldsymbol{q}_i \quad (1)$$

$$\Rightarrow \left(\boldsymbol{q}_i^H\boldsymbol{A}\boldsymbol{q}_i\right)^H = \left(\lambda_i\boldsymbol{q}_i^H\boldsymbol{q}_i\right)^H \Rightarrow \boldsymbol{q}_i^H\boldsymbol{A}^H\boldsymbol{q}_i = \lambda_i^*\boldsymbol{q}_i^H\boldsymbol{q}_i$$

$$\Rightarrow \boldsymbol{q}_i^H\boldsymbol{A}\boldsymbol{q}_i = \lambda_i^*\boldsymbol{q}_i^H\boldsymbol{q}_i$$

Comparing with (1) ($\boldsymbol{A}$ is Hermitian) $\lambda_i = \lambda_i^*$

2.6.3

$$\begin{bmatrix} a & b \\ c & d \end{bmatrix};\ \det\left(\begin{bmatrix} a-\lambda & b \\ c & d-\lambda \end{bmatrix}\right) = \lambda^2 - (\text{tr})\lambda + \det;\ \lambda_{1,2} = \frac{\text{tr} \pm [(\text{tr})^2 - 4\det]^{1/2}}{2}$$

$$\lambda_1 + \lambda_2 = \frac{\text{tr} + [(\text{tr})^2 - 4\det]^{1/2}}{2} + \frac{\text{tr} - [(\text{tr})^2 - 4\det]^{1/2}}{2} = \frac{2\text{tr}}{2} = \text{tr};\ \lambda_1\lambda_2 = \det$$

2.6.4

$Aq_i = \lambda_i q_i,\ Aq_j = \lambda_j q_j \quad (1)$

$\Rightarrow q_j^H A q_i = \lambda_i q_j^H q_i \quad (2)$

$\Rightarrow q_i^H A q_j = \lambda_j q_i^H q_j \quad (3)$

$\Rightarrow$ conjugate transpose (3)

$\Rightarrow q_j^H A q_i = \lambda_j q_j^H q_i \quad$ (4) (A is Hermitian and λ_i's are real),

subtract (4) from (2) $\Rightarrow (\lambda_i - \lambda_j) q_j^H q_i = 0$, *if* $\lambda_i - \lambda_j \neq 0$

$\Rightarrow q_j^H q_i = 0 =$ orthogonal

2.6.5

$Aq_k = \lambda_k q_k,\ k = 1, 2, \ldots, n \Rightarrow$ the n equations are $A[q_1\ q_2 \cdots q_n]$

$= [\lambda_1 q_1\ \lambda_2 q_2 \cdots \lambda_n q_n] \quad (1)$

$\Rightarrow Q = [q_1\ q_2 \cdots q_n],\ \Lambda = \text{diag}\{\lambda_1, \lambda_2, \ldots, \lambda_n\}$

Then, (1) becomes $AQ = Q\Lambda \Rightarrow AQQ^{-1} = Q\Lambda Q^{-1} \Rightarrow A = Q\Lambda Q^{-1}$

2.6.6

$$A^{-1} = \left(Q\Lambda Q^H\right)^{-1} = \left(Q^H\right)^{-1} \Lambda^{-1} Q^{-1} = Q\Lambda^{-1} Q^H = \sum_{i=1}^{n} \frac{1}{\lambda_i} q_i q_i^H$$

2.6.7

If q_i is the eigenvector of B, then $Aq_i = Bq_i + aq_i = \lambda_i q_i + aq_i = (\lambda_i + a) q_i$

3 Processing of Discrete Deterministic Signals
Discrete Systems

3.1 DISCRETE-TIME SIGNALS

Discrete-time signals are seldom found in nature. Numerous discrete signals are used to verify results or theories, for example, the daily change of the stock market in economics, the daily maximum temperature at a particular place for the climate change, and so on. However, most of the signals used by engineers and scientists are the outputs of transducers that, by nature, are continuous signals. Therefore, we are forced to digitize these signals and proceed to manipulate them at their digitized form. Hence, a continuous-time signal $f(t)$ is digitized (sampled by a constant amount of time T) and becomes a set of numbers $\{f(nT)\}$. The sampling theorem tells us that for those signals with finite spectrum (**band-limited** signals) and maximum frequency ω_N(known as the **Nyquist** frequency), we must use a sampling frequency $\omega_s = 2\pi/T$ or $T = 2\pi/\omega_s$, which should be twice as large as the Nyquist frequency, or, equivalently, the sampling time T must be less than one-half of the Nyquist time, $T_N = 2\pi/\omega_N$. In this chapter, we shall assume that all the detected signals are band limited. This is a reasonable assumption, because almost all of the signals have been transferred through a finite-frequency filter (low-pass filter, prefiltering), transducer.

3.1.1 Time-Domain Representation of Basic Continuous and Discrete Signals

A set of basic continuous and discrete signals is included in Table 3.1.

Example 3.1.1

Draw the signal $f(nT) = 2\delta(n+2T) + 2u(n-T) - u(n-5T)$.

Solution: Figure 3.1 shows the above signal.

■

TABLE 3.1
Continuous and Discrete-Time Signals

Delta function	$\delta(t)=0 \quad t\neq 0;\quad \delta(t-t_0)=0 \quad t\neq t_0$
	$\int_{-t_0}^{t_0} f(t)\delta(t)dt = f(0);\ \int_{t_1}^{t_2} f(t)\delta(t-t_0)dt = f(t_0) \quad t_1<t<t_2$
	$\int_{t_1}^{t_2}\delta(t-t_0)dt=1 \quad f(t)=1 \quad t_1<t<t_2$
	$\delta(nT)=\begin{cases}1 & n=0\\ 0 & n\neq 0\end{cases}$
Comb function (sampling function)	$\text{comb}_T=\sum_{n=-\infty}^{\infty}\delta(t-nT)$
Unit step function	$u(t)=\begin{cases}1 & t\geq 0\\ 0 & t<0\end{cases}$
	$u(nT)=\begin{cases}1 & n\geq 0\\ 0 & n<0\end{cases}$
Exponential function	$x(t)=e^{at}u(t)$
	$x(nT)=e^{-anT}u(nT)=(e^{-aT})^n u(nT)=b^n u(nT)$

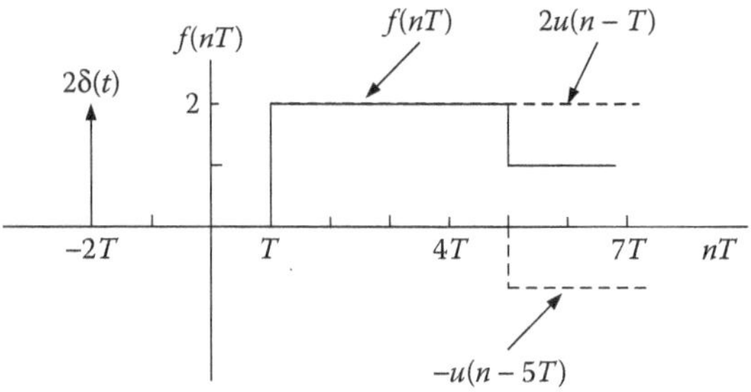

FIGURE 3.1

3.2 TRANSFORM-DOMAIN REPRESENTATION OF DISCRETE SIGNALS

3.2.1 Discrete-Time Fourier Transform

Any nonperiodic discrete-time signal $x(t)$ with finite energy has a discrete-time Fourier transform (DTFT), which is found by an approximation of the Fourier transform (FT). Any discrete signal that comes from discretizing a continuous signal has a **periodic** FT.

The transform is found as follows:

$$X(e^{j\omega})=\mathfrak{F}\{x_s(t)\}=\mathfrak{F}\{x(nT)\}=\int_{-\infty}^{\infty}x(t)e^{-j\omega t}dt=\sum_{n=-\infty}^{\infty}\int_{nT-T}^{nT}x(t)dt$$
$$\cong T\sum_{n=-\infty}^{\infty}x(nT)e^{-j\omega nT} \tag{3.1}$$

where the exact integral from $nT-T$ to nT has been replaced by the approximate area $T \times x(nT)$. When $T=1$, the above equation takes the form:

$$X(e^{j\omega}) = \sum_{n=-\infty}^{\infty} x(n)e^{-j\omega n} \tag{3.2}$$

The relation $X[e^{j(\omega+2\pi)}] = X(e^{j\omega}e^{j2\pi}) = X(e^{j\omega})$ indicates that the **spectrum of a discrete signal is periodic**.

Example 3.2.1

Plot the amplitude and phase spectrum of the DTFT for the discrete signal $x(n)=0.95^n u(n)$, an exponential one.

Solution: The DTFT of the signal is as follows:

$$X(e^{j\omega}) = \sum_{n=0}^{\infty}(0.95)^n e^{-j\omega n} = \sum_{n=0}^{\infty}(0.95e^{-j\omega})^n = \frac{1}{1-0.95e^{-j\omega}}$$

$$X(e^{j\omega}) = A(\omega)e^{j\phi(\omega)} = \frac{1}{\sqrt{(1-0.95\cos\omega)^2+(0.95\sin\omega)^2}}e^{-j\tan^{-1}\left(\frac{0.95\sin\omega}{1-0.95\cos\omega}\right)}$$

The second summation above constitutes a geometric series of the form $1+a^1+a^2+a^3+\cdots=1/(1-a)$, where, in this case, $a=0.95\exp(-j\omega)$. The absolute value of a must be less than 1. The amplitude and phase are plotted in Figure 3.2a and b, respectively. If we had expended the frequency further, we would have seen the periodic format of the spectrum.

■

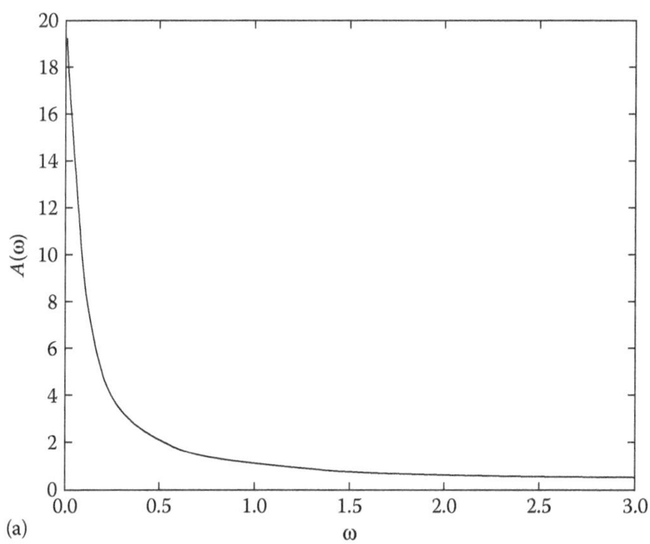

FIGURE 3.2

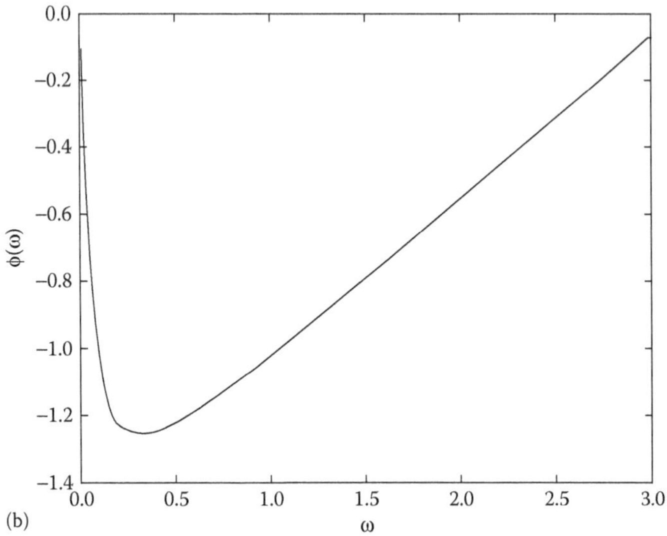

(b)

FIGURE 3.2 (Continued)

3.2.2 The Discrete FT

As a practical matter, we are always restricted to manipulate only a finite time-length signal. Let us assume that the number of the data in the time sequence is N at T time-length apart. The transform sequence is characterized by the same number N of frequency data that are $2\pi/NT$ apart. Hence, we define the **discrete FT** (DFT) of a sequence of N samples $\{x(nT)\}$ by the relation:

$$X\left(k\frac{2\pi}{NT}\right) \equiv X(k\Omega_b) \equiv X(e^{j\Omega_b}) = \mathfrak{F}_D\{x(nT)\} = T\sum_{n=0}^{N-1} x(nT)e^{-j\Omega_b T\,kn} \tag{3.3}$$

$$\Omega_b = \frac{2\pi}{NT}, \quad 0 \le n \le N-1$$

where:

N is the number of samples (even number)
T is the sampling time interval
$(N-1)T$ is the signal length in the time domain
$\Omega_b = (\omega_s/N) = (2\pi/NT) =$ the frequency sampling interval (frequency bin)
$e^{-j\Omega_b T} = N$th principal root of unity

The inverse DFT (IDFT) is given by

$$\begin{aligned} x(nT) &= \mathfrak{F}_D^{-1}\left\{X\left(k\frac{2\pi}{NT}\right)\right\} \equiv \mathfrak{F}_D^{-1}\{X(k\Omega_b)\} = \frac{1}{NT}\sum_{k=0}^{N-1} X\left(k\frac{2\pi}{NT}\right)e^{jk\frac{2\pi}{N}n} \\ &= \frac{1}{NT}\sum_{k=0}^{N-1} X(k\Omega_b)e^{jk\Omega_b Tn} \qquad 0 \le k \le N-1 \qquad \Omega_b = \frac{2\pi}{NT} \end{aligned} \tag{3.4}$$

For $N = 1$, the above two equations become

$$\begin{aligned} X\left(k\frac{2\pi}{N}\right) &= \sum_{n=0}^{N-1} x(n)e^{-jn\frac{2\pi}{N}k} \qquad 0 \le k \le N-1 \\ x(n) &= \frac{1}{N}\sum_{k=0}^{N-1} X\left(k\frac{2\pi}{N}\right)e^{jk\frac{2\pi}{N}n} \qquad 0 \le n \le N-1 \end{aligned} \tag{3.5}$$

Note: *Both sequences, time and frequency, are discrete.*

Both the sequences are periodic (see Problem 3.2.1). This indicates that the DFT is associated with periodic sequences in both the time and frequency domains. To find the DFT of a time sequence and the inverse, we use the following two MATLAB functions:

```
>>X = fft(x,N);
>>x = ifft(X,N);
```

$\{X\}$ is an N-point sequence (vector) and $\{x(n)\}$ is the input time sequence (vector). The middle point $N/2$ of $\{X\}$ is known as the **fold-over** frequency. Plotting the magnitude of $\{X\}$ versus frequency ($k2\pi/N$), where $2\pi/N$ is the bin width, we find that π is the middle of the frequency spectrum, and π to 2π is the reflection from 0 to π. Similarly, we find that π/T is the fold-over frequency when T is positive different than 0. This indicates that to approximate the spectrum of a continuous signal, we must make the sampling time T as small as possible.

Example 3.2.2

Find the DFT for the function $x(n) = 0.9^n u(n)$ for $N = 4$.

Solution: Since the sample time is unity, we use (3.5). Hence, we find

$$X\left[0\left(\frac{2\pi}{4}\right)\right] = \sum_{n=0}^{4-1} 0.9^n e^{-j\frac{2\pi}{4}0n} = 1 + 0.9 + 0.9^2 + 0.9^3 = 3.4390$$

$$X\left[1\left(\frac{2\pi}{4}\right)\right] = \sum_{n=0}^{4-1} 0.9^n e^{-j\frac{2\pi}{4}1n} = 1 + 0.9e^{-j\frac{2\pi}{4}1\times1} + 0.9^2 e^{-j\frac{2\pi}{4}1\times2} + 0.9^3 e^{-j\frac{2\pi}{4}1\times3}$$

$$= 0.1900 - 0.1710j$$

$$X\left[2\left(\frac{2\pi}{4}\right)\right] = \sum_{n=0}^{4-1} 0.9^n e^{-j\frac{2\pi}{4}2n} = 1 + 0.9e^{-j\frac{2\pi}{4}2\times1} + 0.9^2 e^{-j\frac{2\pi}{4}2\times2} + 0.9^3 e^{-j\frac{2\pi}{4}2\times3}$$

$$= 0.1810 - 0.0000j$$

$$X\left[3\left(\frac{2\pi}{4}\right)\right] = \sum_{n=0}^{4-1} 0.9^n e^{-j\frac{2\pi}{4}3n} = 1 + 0.9e^{-j\frac{2\pi}{4}3\times1} + 0.9^2 e^{-j\frac{2\pi}{4}3\times2} + 0.9^3 e^{-j\frac{2\pi}{4}3\times3}$$

$$= 0.1900 + 0.1710j$$

The amplitude spectrum is 3.4390, 0.2556, 0.1810, and 0.2556 for $k = 0, 1, 2,$ and 3. The phase spectrum is 0.0000, −0.7328, 0, and 0.7328 for $k = 0, 1, 2,$ and 3. The phase spectrum is given in radians. Since the frequency bin is $\Omega_b = 2\pi/NT = 2\pi/4\times 1 = \pi/2$, the amplitude and frequency spectrums are located at 0, $\pi/2$, π (fold-over frequency), and $3\pi/2$ radians per unit length.

■

3.2.3 Properties of DFT

To abbreviate the notation, we write

$$X(k\Omega_b) = X(k);\ \ x(nT) = x(n);\ \ e^{-j2\pi/N} = W$$

Using the above notation, the DFT properties are as follows:

Linearity	$af(n)+bh(n) \overset{\mathfrak{F}_D}{\leftrightarrow} aF(k)+bH(k)$
Symmetry	$(1/N)F(n) \overset{\mathfrak{F}_D}{\leftrightarrow} f(-k)$
Time shifting	$f(n-i) \overset{\mathfrak{F}_D}{\leftrightarrow} F(k)e^{-j2\pi ki/N} = F(k)W^{ki}$
Frequency shifting	$f(n)e^{jni} \overset{\mathfrak{F}_D}{\leftrightarrow} F(k-i)$
Time convolution	$y(n) \triangleq f(n) * h(n) \overset{\mathfrak{F}_D}{\leftrightarrow} F(k)H(k)$
Frequency convolution	$f(n)h(n) \overset{\mathfrak{F}_D}{\leftrightarrow} \frac{1}{N}\sum_{x=0}^{N-1} F(x)H(n-x)$
Parseval's theorem	$\sum_{n=0}^{N-1} \lvert f(n)\rvert^2 = \frac{1}{N}\sum_{k=0}^{N-1} \lvert F(k)\rvert^2$
Time reversal	$f(-n) \overset{\mathfrak{F}_D}{\leftrightarrow} F(-k)$
Delta function	$\delta(n) \overset{\mathfrak{F}_D}{\leftrightarrow} 1$
Central ordinate	$f(0) = \frac{1}{N}\sum_{k=0}^{N-1} F(k);\ \ F(0) = \sum_{n=0}^{N-1} f(n)$

Example 3.2.3

Find the DFT of the signal $x(t) = \exp(-t)u(t)$.

Solution: Because there is no practical way to digitize the infinite signal, we select for this example the range $0 \le t \le 4$ and two sampling times $T = 1$ and $T = 0.5$. The results from the DFT will be compared with the exact FT spectrum of the signal. The FT of the signal is

$$X(\omega) = \int_{-\infty}^{\infty} e^{-t}u(t)e^{-j\omega t}dt = \int_{0}^{\infty} e^{-(1+j\omega)t}dt = \frac{1}{1+j\omega} = \frac{1}{(1+\omega^2)^{1/2}}e^{-j\tan^{-1}(\omega)} = A(\omega)e^{j\phi(\omega)}$$

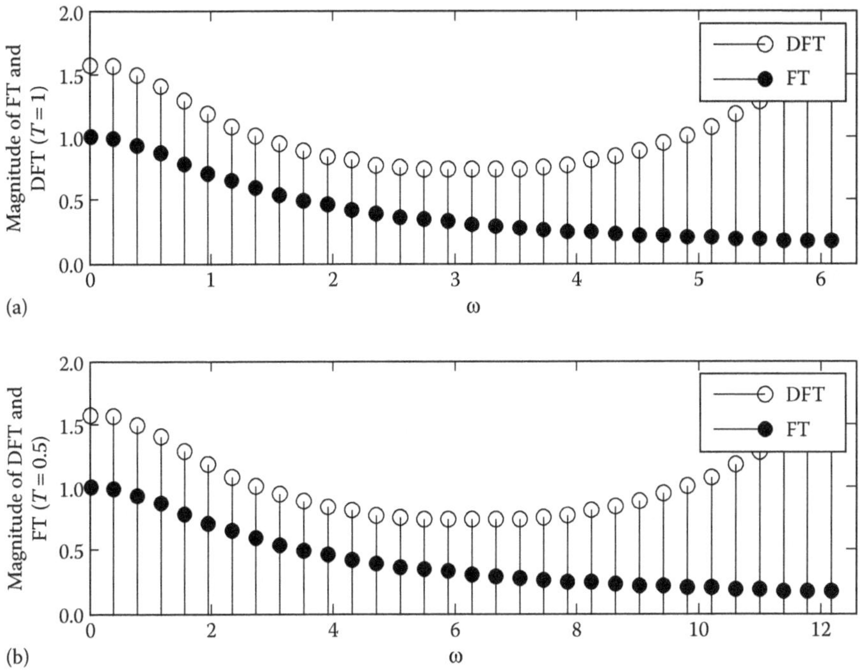

FIGURE 3.3

To find the DFT with $T = 1$ and compare the results with the exact FT, we use the following m-file example3_2_3 Book MATLAB program. At the Command window, having the Book MATLAB files in the MATLAB path (see Appendix 1), you write example3_2_3 and on return create Figure 3.3.

```
%example3_2_3
t = 0:1:4;
x = exp(-t);%x = vector with 5 elements;
dftx1 = fft(x,32);%asks for 32 frequency bins, MATLAB
     %will pad x with 32-5 zeros;
w = 0:2*pi/32:2*pi-(2*pi/32);
FT1 = 1./sqrt(1+w.^2);
subplot(2,1,1);%subplot(2,1,1) produces a picture with
     %two rows 1 column and the first row;
stem(w,abs(dftx1),'k');
hold on;%will produce a second plot on the same figure;
stem(w,FT1,'k','filled');%produces black ('k')lines
     %which are filled;
xlabel('\omega');ylabel('Magnitude of FT1 and DFT, T = 1');
title('(a)');axis([0 2*pi 0 2]); legend('DFT','FT');
nt = 0:0.5:4;
x1 = exp(-t);
dftx2 = fft(x1,32);
```

```
w1 = 0:4*pi/32:4*pi-(4*pi/32);
FT2 = 1./sqrt(1+w.^2);
subplot(2,1,2);stem(w1,abs(dftx2),'k');
hold on;
stem(w1,FT2,'k','filled');
xlabel('\omega');ylabel('Magnitude of DFT and FT,T = 0.5');
title('(b)');axis([0 4*pi 0 2]);legend('DFT','FT');
```

■

Note:

1. From Figure 3.3a, we observe the following: (a) The amplitude of the FT of the continuous signal decreases for ever as it should. (b) The DFT of $x(t)$ has folding frequency at $\pi/1$ ($T=1$ in this case). (c) The approximation of the DFT to the exact DFT is up to fold-over frequency. (d) For the present assumed values, the two spectra are considerable different. (e) If the DFT command was *fft*(x,64), the lines would be twice as many but the accuracy would remain the same. If, however, we had padded with zeros the vector $\boldsymbol{x}$, the accuracy would stay the same. (f) The accuracy would have increased if we extended the time from $0 \le t \le 4$ to $0 \le t \le 10$, for example.
2. From Figure 3.3b, we observe that by decreasing the sampling time, in this case to 0.5, the accuracy increases. If, in addition, we had increased the sampling part of the time function, then the accuracy would have increased even further. We also observe that the time function has a spectrum with an infinite range. Therefore, it is practically impossible to produce an infinitely small sampling time, thus producing an extremely large number of vector samples. However, if the signal was band limited with a maximum frequency ω_N (Nyquist frequency), then we would have been able to sample the signal at twice the Nyquist frequency and recapture the exact spectrum. Observe also that by reducing the sampling time by 1/2, the fold-over frequency doubles to 2π.
3. Note also that we have plotted the values of the continuous spectrum, FT, at the corresponding frequencies of those found using the DFT. This gives us a better visual understanding of the accuracy of the sampled signal to the exact one.

3.3 THE *z*-TRANSFORM

The z-transform of a sequence is a transformation of a discrete-time sequence from the time domain to the frequency domain (z-domain). The z-transform is defined by

$$X(z) = \mathcal{Z}\{x(n)\} = \sum_{n=-\infty}^{\infty} x(n) z^{-n} \tag{3.6}$$

For the inverse transform, we shall use Table 3.3 that is given below.

Example 3.3.1

Obtain the z-transform of the function $x(n) = 0.8^n u(n)$.

Solution: Applying (3.6), we find

$$X(z) = \sum_{n=0}^{\infty} 0.8^n z^{-n} = \sum_{n=0}^{\infty} (0.8z^{-1})^n = \frac{1}{1-0.8z^{-1}}$$

For the last step, we used the geometric series property $(1 + x + x^2 + \cdots) = 1/(1-x)$ only if $|x| < 1$. This guarantees the convergence of the series in the z-domain. To obtain the region of convergence (ROC) in the z-domain, we must have $|0.8z^{-1}| = (0.8/|z|) < 1$ or $|z| > 0.8$. Hence, the ROC for this problem is all the complex space outside the circle with radius 0.8. The ROC is important if we are to find the inverse transform using integration in the complex plane. If we set $z = \exp(j\omega t)$ in (3.6), we obtain the DTFT for the time function $x(n)$.

■

Tables 3.2 and 3.3 show the properties and the transform pairs of the z-transform.

Example 3.3.2

Find the inverse z-transform of the function:

$$F(z) = \frac{1}{(1-0.2z^{-1})(1+0.2z^{-1})} = \frac{z^2}{z^2 - 0.04}$$

Solution: The equation can be written in the form:

$$\frac{F(z)}{z} = \frac{z}{(z-0.2)(z+0.2)} = \frac{A}{z-0.2} + \frac{B}{z+0.2}$$

TABLE 3.2
Properties of the z-Transform

1. $\mathcal{Z}\{ax(n) + by(n)\}$	$aX(z) + bY(z)$
2. $\mathcal{Z}\{y(n-m)\}$	$z^{-m}Y(z) + \sum_{i=1}^{m} y(-i)z^{-(m-i)}$
3. $\mathcal{Z}\{y(n+m)\}$	$z^{m}Y(z) - \sum_{n=0}^{m-1} y(n)z^{m-n}$
4. $\mathcal{Z}\{a^n y(n)\}$	$Y(z/a)$
5. $\mathcal{Z}\{y(n)\}$	$(z^N/z^N - 1)Y_1(z)$; $y_{(1)}$ is the first period of a periodic sequence $y(n) = y(n+N)$
6. $\mathcal{Z}\{y(n) * h(n)\}$	$Y(z)H(z)$
7. $\mathcal{Z}\{ny(n)\}$	$-z(d/dz)Y(z)$
8. $\mathcal{Z}\{n^m y(n)\}$	$[-z(d/dz)]^m Y(z)$
9. $y(n_0)$	$z^{n_0} Y(z)\big\|_{z\to\infty}$; n_0 is the initial value of the sequence $Y(z) = \sum_{n=0}^{\infty} y(n_0+n)z^{-n}$
10. $\lim_{z\to\infty} y(n)$	$\lim_{z\to 1}(1-z^{-1})Y(z)$

TABLE 3.3
Common z-Transform Pairs

Entry Number	$f(n)$, $f(nT)$ for $n \geq 0$	$F(z) = \sum_{n=0}^{\infty} f(n) z^{-n}$	Radius of Convergence $\lvert z \rvert > R$
1	$\delta(n)$	1	0
2	$\delta(n-m)$	z^{-m}	0
3	1	$z/z-1$	1
4	n	$z/(z-1)^2$	1
5	n^2	$z(z+1)/(z-1)^3$	1
6	n^3	$z(z^2+4z+1)/(z-1)^4$	1
7	a^n	$z/(z-a)$	$\lvert a \rvert$
8	na^n	$az/(z-a)^2$	$\lvert a \rvert$
9	$n^2 a^n$	$az(z+a)/(z-a)^3$	$\lvert a \rvert$
10	$a^n/n!$	$e^{a/z}$	0
11	$(n+1)a^n$	$z^2/(z-a)^2$	$\lvert a \rvert$
12	$[(n+1)(n+2)a^n]/2!$	$z^3/(z-a)^3$	$\lvert a \rvert$
13	$[(n+1)(n+2)\cdots(n+m)a^n]/m!$	$z^{m+1}/(z-a)^{m+1}$	$\lvert a \rvert$
14	$\sin(n\omega T)$	$z\sin\omega T/[z^2 - 2z\cos(\omega T)+1]$	1

15	$\cos(n\omega T)$	$z[z-\cos(\omega T)]/z^2-[2z\cos(\omega T)]+1$	1
16	$a^n\sin(n\omega T)$	$az\sin(\omega T)/z^2-[2az\cos(\omega T)]+a^2$	$\lvert a\rvert^{-1}$
17	$a^{nT}\sin(n\omega T)$	$a^T z\sin(\omega T)/[z^2-2a^T z\cos(\omega T)+a^{2T}]$	$\lvert a\rvert^{-T}$
18	$a^n\cos(n\omega T)$	$z[z-a\cos(\omega T)]/z^2-[2az\cos(\omega T)]+a^2$	$\lvert a\rvert^{-1}$
19	$e^{-anT}\sin(n\omega T)$	$ze^{-aT}\sin(\omega T)/z^2-2e^{-aT}z\cos(\omega T)+e^{-2aT}$	$\lvert z\rvert>\lvert e^{-aT}\rvert$
20	$e^{-anT}\cos(n\omega T)$	$z[z-e^{-aT}\cos(\omega T)]/\{z^2-[2e^{-aT}z\cos(\omega T)]+e^{-2aT}\}$	$\lvert z\rvert>\lvert e^{-aT}\rvert$
21	$n(n-1)/2!$	$z/(z-1)^3$	1
22	$n(n-1)(n-2)/3!$	$z/(z-1)^4$	1
23	$[n(n-1)(n-2)\cdots(n-m+1)/m!]a^{n-m}$	$z/(z-a)^{m+1}$	1
24	e^{-anT}	$z/(z-e^{-aT})$	$\lvert e^{-aT}\rvert$
25	ne^{-anT}	$ze^{-aT}/(z-e^{-aT})^2$	$\lvert e^{-aT}\rvert$

The unknown constants are found as follows:

$$A = \left.\frac{z}{z+0.2}\right|_{z=0.2} = 0.5,\quad B = \left.\frac{z}{z-0.2}\right|_{z=-0.2} = 0.5 \Rightarrow F(z) = \frac{1}{2}\left[\frac{z}{z+0.2} + \frac{z}{z-0.2}\right]$$

Using Table 3.3, we obtain

$$f(n) = \begin{cases} \frac{1}{2}\left[0.2^n + (-0.2)^n\right] & n \geq 0 \\ 0 & n < 0 \end{cases}$$

For a small number of values for the function $f(n)$ in the above example, we can use the following Book MATLAB function example3_3_2.

```
%example3_3_2
num = [1 0 0];
den = [1 0 -0.04];
f = dimpulse(num,den,5);%the number 5 requests 5 values
    %of the time function f(n)
```

The 5 values that the above Book MATLAB function gives are $f = [1\quad 0\quad 0.04\quad 0\quad 0.0016]$.

■

3.4 DISCRETE-TIME SYSTEMS

3.4.1 Linearity and Shift Invariant

Two of the most important properties of the systems are **linearity** and **time invariance or shift invariance** (LTI). The system is linear if two inputs are applied to the system and its output is the sum of the outputs corresponding to the inputs. The system is time independent (or shift invariant) if the input is shifted in time and the output is shifted by the same amount.

3.4.2 Causality

A system is causal if its output at time t_0 is due only to the inputs for $t \leq t_0$. In other words, for a causal system, it is not possible for changes in the output to precede the excitations to the input.

3.4.3 Stability

If the output of a system is bounded for any bounded input is called a stable system. A system with such a property is said to be **stable** bounded-input–bounded-output (BIBO) sense.

A discrete time-invariant system is a physical device or an algorithm that transforms an input (or excitation) signal, $\{x(n)\}$, into another is called the output signal, $\{y(n)\}$.

Every system, including the discrete system, is defined by their output, known as the **impulse response** $h(n)$ whose z-transform $H(z)$ is known as the **transfer function**, when the input signal is the delta function. For the discrete system, the delta function is defined as follows:

$$\delta(n) = \begin{cases} 1 & n = 0 \\ 0 & n \neq 0 \end{cases} \tag{3.7}$$

The following example presents the way we can transform a continuous system to a discrete and find its impulse response.

Example 3.4.1

Deduce an expression for the impulse response for the circuit shown in Figure 3.4a with the voltage source as input and the current as output. Use the z-transform technique.

Solution: The controlling equation is

$$L\frac{di(t)}{dt} + Ri(t) = v(t)$$

Figure 3.4b shows the approximation of a first-order derivative at point *A*. In the figure, *AD* is the tangent at point *A*. The approximate tangent is *AB*. It is apparent that as *T* approaches 0, the line *AB* approaches *AD* as it should be. Therefore, the above equation is written in discrete form as follows:

$$L\frac{i(nT) - i(nT - T)}{T} + Ri(nT) = v(nT) \quad \text{or}$$

$$i(nT) - a_1 i(nT - T) = b_0 v(nT) \quad a_1 = \frac{L}{L + RT}, \; b_0 = \frac{T}{L + RT}$$

$$i(n) - a_1 i(n - 1) = b_0 v(n) \qquad T = 1$$

Taking the z-transform (Tables 3.2 and 3.3) of the third equation, we obtain

$$I(z) = b_0 V(z) + a_1 z^{-1} I(z) \;\; \text{or} \;\; I(z) = \frac{b_0}{1 - a_1 z^{-1}} V(z) = H(z)V(z) \;\; \text{or}$$

$$\text{Transfer function} = \frac{\text{Output}}{\text{Input}} = \frac{I(z)}{V(z)} = H(z) = \frac{b_0}{1 - a_1 z^{-1}}$$

From Table 3.3, we find that the inverse transform of the last equation, that is, the impulse response of the system, is

$$h(n) = b_0 (a_1)^n \qquad n \geq 0$$

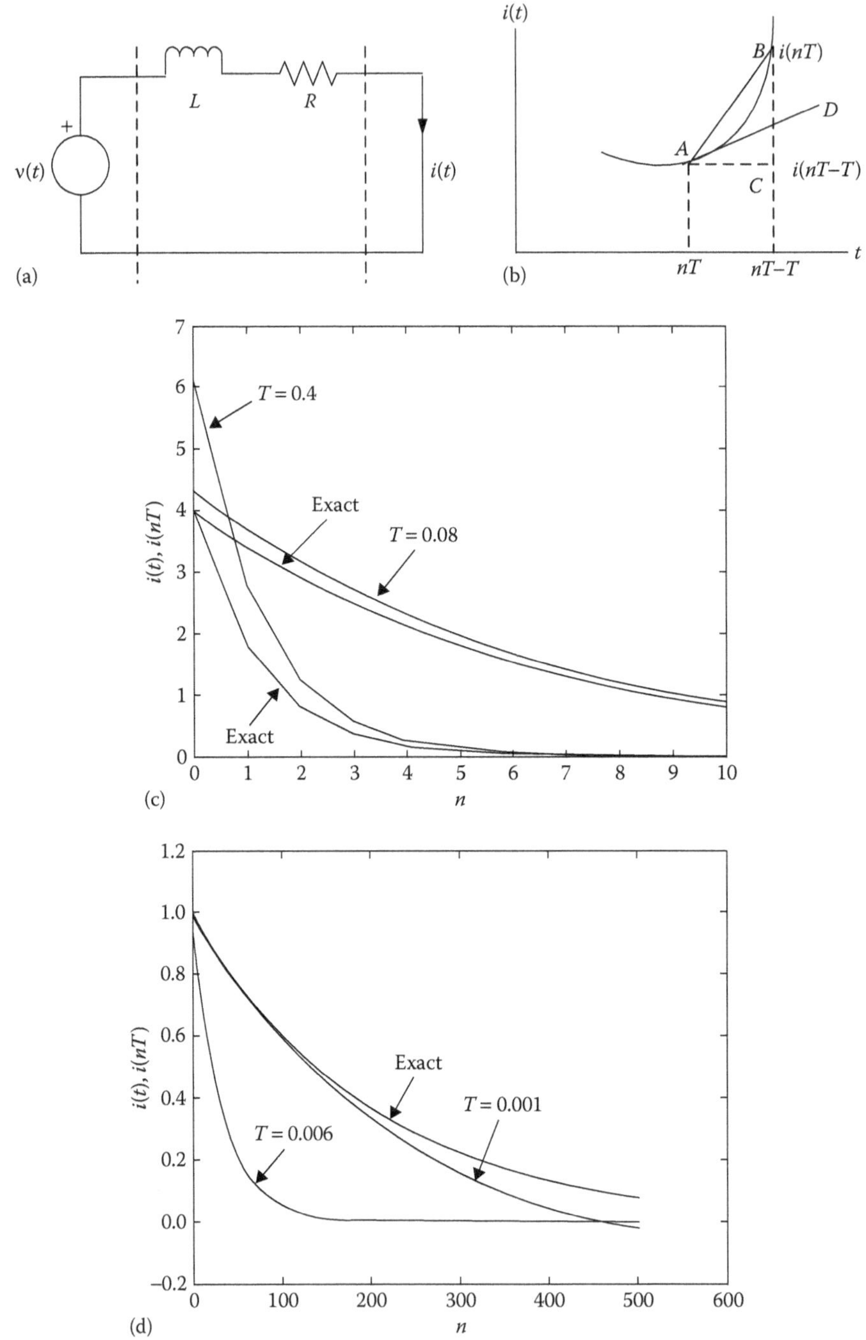
+
v(t)
L
R
i(t)
(a)
i(t)
B
i(nT)
D
A
i(nT−T)
C
t
nT
nT−T
(b)
T = 0.4
Exact
T = 0.08
Exact
i(t), i(nT)
n
(c)
Exact
T = 0.001
T = 0.006
i(t), i(nT)
n
(d)

FIGURE 3.4

Let us assume the following values for the system components: $L = 2$, $R = 1$, and $T = 1$. Hence, $a_1 = 2/3$ and $b_0 = 1/3$. Therefore, the first 10 values of the impulse response are as follows:

0.3333 0.2222 0.1481 0.0988 0.0658 0.0439 0.0293 0.0195 0.0130 0.0087

Next, we solve by iteration the difference equation $i(n) = (2/3)i(n-1) + (1/3)v(n)$, with $i(-1) = 0$.

$$i(0) = \frac{2}{3}i(-1) + \frac{1}{3}1 = \frac{1}{3} = 0.3333$$

$$i(1) = \frac{2}{3}i(0) = \frac{2}{3}\frac{1}{3} = 0.2222 \qquad \delta(\neq 0) = 0$$

$$i(2) = \frac{2}{3}i(1) = \frac{2}{3}\frac{2}{3}\frac{1}{3} = 0.1481$$

$$\vdots$$

■

Let us try to find the current through and the voltage across the capacitor of an RC in a series circuit. Applying Kirchhoff's law, we obtain

$$Ri(t) + \frac{1}{C}\int_0^t i(t')dt' = v(t) \Rightarrow \frac{dq(t)}{dt} + \frac{1}{RC}q(t) = \frac{1}{R}v(t) \quad \left[i(t) = \frac{dq(t)}{dt}\right]$$

With $R = 0.5$, $C = 1$, and $v(t) = \exp(-t)$, we obtain

$$\frac{dq(t)}{dt} + 2q(t) = 2e^{-t}$$

The homogeneous solution, $v(t)=0$, is $q_h(t)=Ae^{-2t}$. The particular solution is assumed to be equal to $q_p(t) = Be^{-t}$. Inserting this solution into a complete equation, we obtain $B = 2$. Hence, the total solution is $q(t) = Ae^{-2t} + 2 \Rightarrow q(0) = 0 = A + 2 \Rightarrow A = 2$. Therefore, $i(t) = [dq(t)/dt] = 4e^{-2t}$. The discretized charge of the capacitor is $q(nT) = 2 - 2e^{-2nT}$. Hence, the discrete form of the current is

$$i(nT) = \frac{q(nT) - q(nT - T)}{T} = \frac{2 - 2e^{-2nT} - 2 + 2e^{-2(n-1)T}}{T} = \frac{2}{T}\left[e^{-2(n-1)} - e^{-2nT}\right]$$

Figure 3.4c shows the exact and discretized values of the current for two different Ts. We also observe that the voltage across the capacitor $[q(t)/C]$ approaches the value of two volts as the time goes to infinity.

In Example 3.4.1, we found that the output in the z-domain is the product of the input and the transfer function. Based on Table 3.3, we observe that the inverse z-transform of a product of functions is equal to their convolution in the time domain. Hence, the inverse z-transform of the product $Y(z) = H(z)X(z)$ is

$$y(nT) = T\sum_{m=-\infty}^{\infty} x(mT)h(nT - mT) \qquad n = 0, \pm 1, \pm 2, \ldots \qquad m = 0, \pm 1, \pm 2, \ldots$$
$$y(n) \triangleq x(n) * h(n) = \sum_{m=-\infty}^{\infty} x(m)h(n - m) = \sum_{m=-\infty}^{\infty} x(n - m)h(m) \tag{3.8}$$

The above equation indicates what procedure must be executed to find $y(n)$. (1) We select another domain, in this case m-domain, for both the signals $x(n)$ and $h(n)$. (2) For one of the signals, we substitute m in place of n. This means that the signal appears identical to both domains. (3) We shift the other function by n and flip it (see the minus sign in front of m) in the m-domain. (4) Next, we multiply the two sequences term by term and then add the result. (5) The result is the output at time n. (6) We repeat the same procedure for all n's and thus obtain the output for $-\infty$ to ∞.

Example 3.4.2

Obtain the convolution of the two functions $x(n) = u(n)$ and $h(n) = 0.9^n u(n)$.

Solution: We first flip and shift the unit function. If the shift is done toward to the left, by setting $n < 0$, the two functions do not overlap and the product is 0 for all the negative values of n. Therefore, we obtain the output if n is greater or equal to 0. Then, for this case (3.8) takes the form:

$$y(n) = \sum_{m=0}^{\infty} u(n-m)0.9^m = \sum_{m=0}^{n} 0.9^m = 1 + 0.9 + 0.9^2 + \cdots + 0.9^n = \frac{1-0.9^{n+1}}{1-0.9}$$

where we used the formula for finite geometric series. The step function $u(n - m)$ is 0 for $m > n$ and it is equal to 1 for $m < n$.

■

Table 3.4 gives the most important properties of the convolution operation of discrete signals.

TABLE 3.4
Useful Convolution Properties

Commutative	$x(n) * y(n) = y(n) * x(n)$
Associative	$x(n) * [y(n) * z(n)] = [x(n) * y(n)] * z(n) = x(n) * y(n) * z(n)$
Distributive	$x(n) * [y(n) + z(n)] = x(n) * y(n) + x(n) * z(n)$
Identity sequence	$x(n) * \delta(n) = \delta(n) * x(n) = x(n)$
Shifted delta function	$x(n) * \delta(n-k) = x(n-k)$

3.4.3 Transform-Domain Representation

The output, without initial conditions, in the z-domain of a discrete system is equal to the product of its transfer function and its input. Hence,

$$Y(z) = X(z)H(z) H(z) = \text{System function or transfer function} \tag{3.9}$$

If the poles of the transfer function are inside the unit circle, the system is stable and $H(e^{j\omega})$ provides its frequency response.

A general realizable discrete LTI and causal system (the impulse response of the system is 0 for $t < 0$) can be represented by the following difference equation:

$$y(n) + \sum_{m=1}^{p} a_m\, y(n-m) = \sum_{m=0}^{q} b_m\, x(n-m) \tag{3.10}$$

where:

$y(n)$ is the output
$x(n)$ is the input
a_m and b_m are the constants specifying each particular system

To obtain the output of a discrete system using MATLAB, we write

```
>>y = filter(b,a,x);%b = row vector of the b's; a = row vector
>>     %of the a's (see (3.10)); x = row input vector {x(n)};
>>     %y = output vector {y(n)};
```

Taking the z-transform of (3.10), and remembering that the transform is with respect to the time n and referring to Table 3.2 for shifting functions, we obtain the transfer function for the general system:

$$Y(z) + \sum_{m=1}^{p} a_m Y(z) z^{-m} = \sum_{m=0}^{q} b_m X(z) z^{-m}$$

or

$$H(z) = \sum_{m=0}^{\infty} h(m) z^{-m} = \frac{Y(z)}{X(z)} = \frac{\sum_{m=0}^{q} b_m z^{-m}}{1 + \sum_{m=1}^{p} a_m z^{-m}} = \frac{B(z)}{A(z)} \tag{3.11}$$

The above system is known as the **autoregressive moving average** (ARMA) system. It is made up of two systems (filters): a finite impulse response (FIR) and an infinite response system (IIR). The example below presents a first-order ARMA system.

Example 3.4.3

Using (3.10), define a first-order ARMA system and plot its block-format diagram.

Solution: In Equation 3.11, we set $m = 0,1$ and, thus, we find the equation:

$$y(n) + a_1 y(n-1) = b_0 x(n) + b_1 x(n-1)$$

Taking the z-transform of the above equation, we obtain the transfer function:

$$H(z) \triangleq \frac{Y(z)}{X(z)} = \frac{b_0 + b_1 z^{-1}}{1 + a_1 z^{-1}}$$

Set $z = \exp(-j\omega)$ in the above equation and use the following Book MATLAB program to obtain the frequency spectrum of the transfer function (Figure 3.5). Using $a_1 = 2$, $b_0 = 3$, and $b_1 = 2$, we obtain

```
>>w = 0:0.1:2*pi;
>>H = (3+2*exp(j*w))./(1+2*exp(j*w));
>>subplot(2,1,1);
>>plot(w,abs(H),'k');%abs() is a MATLAB function that gives
>>                   %the absolute value of the complex function;
>>ylabel('abs(H)');
>>subplot(2,1,2);
>>plot(w,angle(H),'k');%angle() is a MATLAB function;
>>xlabel('\omega');ylabel('Angle in rad.');
```

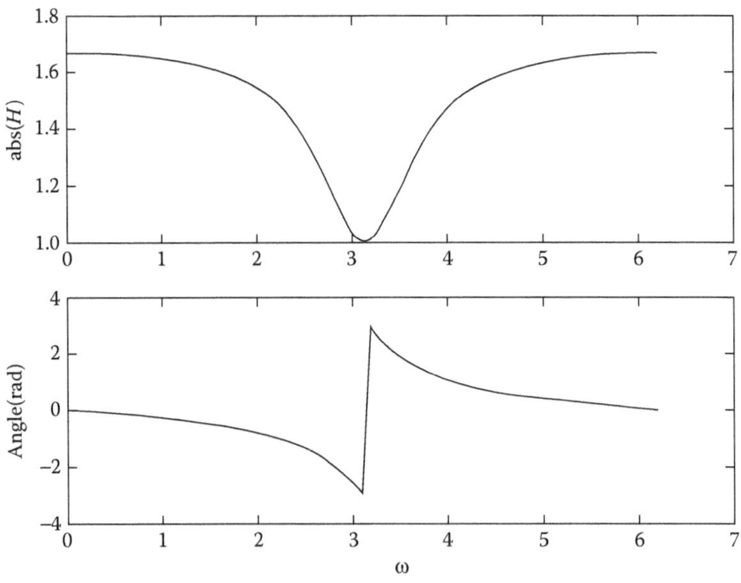

FIGURE 3.5

If we had continued the frequency beyond 2π, we would reproduce the diagram every 2π; in other words, the spectrum is periodic every 2π. We also observe that the curve falls off from 0 to π. This indicates that the given system is a low-pass filter.

To be able to create block diagrams, we must use the fundamental building blocks that are shown in Figure 3.6. The block diagram representation is shown in Figure 3.7.

■

If in Equation 3.11 we set the a's equal to zero, then the system in time and z-domain is given, respectively, by

$$y(n) = \sum_{m=0}^{q} b_m x(n-m)$$
$$H(z) = B(z) \tag{3.12}$$

The above system is known as the **FIR** system or **nonrecursive**.

If in the same equation we set b's equal to 0 besides b_0, we obtain a system called the **infinite impulse response** (IIR) system or **recursive**. The system in time and z-domain is given by

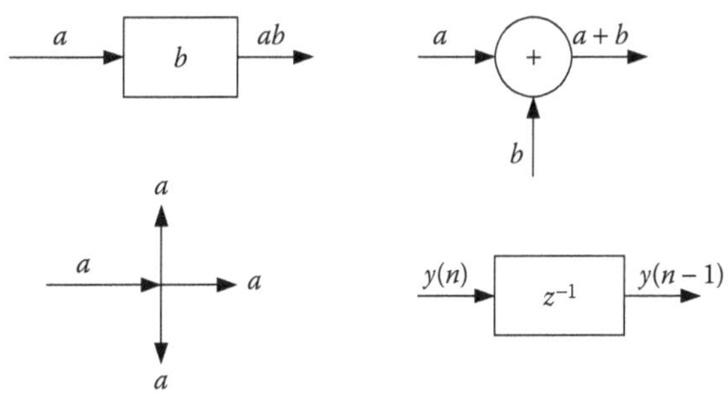

FIGURE 3.6

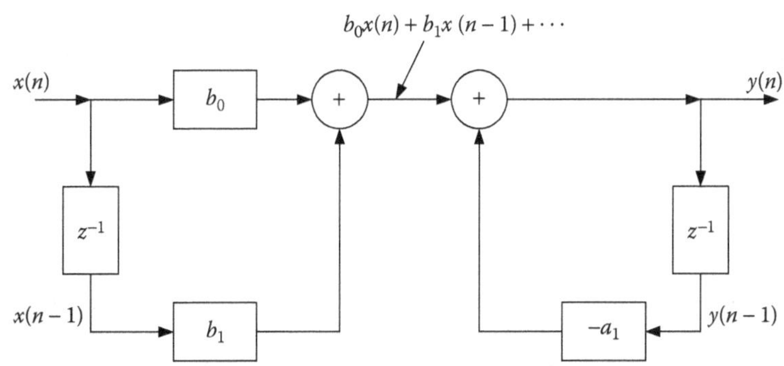

FIGURE 3.7

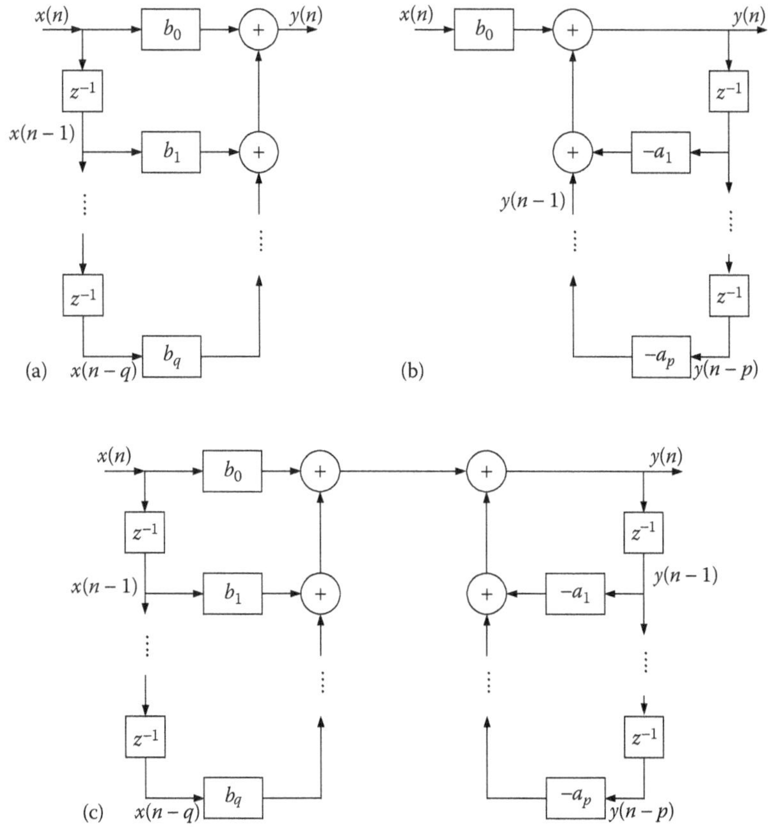

FIGURE 3.8

$$y(n)+\sum_{m=1}^{p}a_m y(n-m)=b_0 x(n)$$

$$H(z)=\frac{b_0}{1+\sum_{m=1}^{p}a_m z^{-m}}=\frac{b_0}{A(z)} \tag{3.13}$$

Figure 3.8 shows the above three systems in their block-format realization. In this chapter, we are mostly concerned with only FIR filters because they are inherently stable. Figure 3.8a shows a FIR system, Figure 3.8b shows an IIR system, and Figure 3.8c shows an ARMA filter.

PROBLEMS

3.1.1 Use MATLAB to plot $\exp(-0.1n)$, $\exp(-0.25n)$, and $f(n) = \exp(-0.1n) - \exp(-0.25n)$.

3.2.1 Show that sequences $\{x(n)\}$ and $\{X(k)\}$ of the DFT are periodic with a period N.

3.2.2 Use the Book MATLAB m-file, example3_2_3, and verify all the observations given in the Note section following Example 3.2.3. The reader should change the appropriate constants to verify the results.

3.2.3 Find the fold-over frequency for the function $f(t) = \exp(-0.2t)u(t)$ for $0 \le n \le 255$, $T = 0.1$, and $T = 0.4$.

3.3.1 Find the z-transform for the function $f(t) = A\exp(-at)u(t)$ sampled every T seconds.

3.4.1 Find the convolution of the two discrete functions: $x(n) = 0.9^n u(n)$ and the pulse having 7 elements $h(n) = u(n) - u(n-6)$.

HINTS–SUGGESTIONS–SOLUTIONS

3.2.1 If we replace k in (3.5) by $k+N$, we find that $X(k)$ is periodic with period N. Similarly, if we introduce $n+N$ instead of n in the second equation, we also find that n is periodic with period N.

3.3.1

$$\{f(nT)\} = A, Ae^{-aT}, Ae^{-2aT}, \ldots; \; F(z) = A\left[1 + \frac{e^{-aT}}{z} + \left(\frac{e^{-aT}}{z}\right)^2 + \cdots\right]$$

$$= \frac{A}{1-\left(e^{-aT}/z\right)} = \frac{Az}{z-e^{-aT}}$$

The ROC is defined as

$$\left|\frac{e^{-aT}}{z}\right| = e^{-aT}\left|z^{-1}\right| < 1 \text{ or } |z| > e^{-aT}$$

3.4.1 Use the following Book MATLAB program to produce Figure P3.4.1

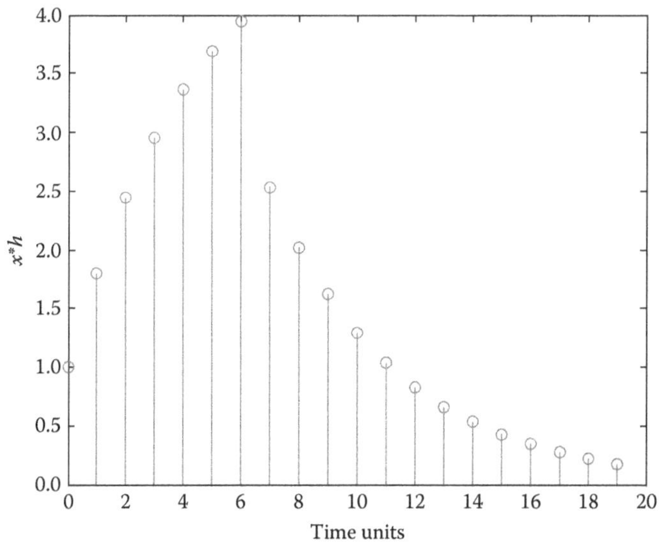

FIGURE P3.4.1

```
%prob3_4_1
x = [1 1 1 1 1 1];
n = 0:20;
h = 0.8.^n;
y = conv([x zeros(1,15)],h);%conv(x,h) is a MATLAB function
      %that gives the convolution between two vectors
      % of equal number of elements, if they are not
      %equal we pad the smaller with zeros;
m = 0:40;
stem(m,y,'k');xlabel('Time units');ylabel('Output y');
```

We can also use the following Book MATLAB program:

```
%prob3_4_1a
for n = 0:6
      for m = 0:n
          y(m+1) = 0.8^(n-m);
      end;
      o(n+1) = sum(y);
end;
for n = 8:20
      for m = 0:6
            y1(m+1) = 0.8^(n+m-6);
      end;
      o1(n) = sum(y1);
end;
o11 = o1(8:20);k = 0:19;stem(k,[o o11],'k');
xlabel('Time units');ylabel('Conv. x*h');
```

4 Discrete-Time Random Processes

4.1 DISCRETE RANDOM SIGNALS, PROBABILITY DISTRIBUTIONS, AND AVERAGES OF RANDOM VARIABLES

Most signals in practice are not deterministic but random. However, they can be described by precise mathematical analysis whose tools are contained in the theory statistical analysis. In this chapter, we are dealing only with discrete random signals. This can always be accomplished by sampling the continuous signals at sampling rates at least twice their highest frequency, thus avoiding aliasing. Remember that the signals we receive are band limited because all signals must be detected by a physical (not ideal) transducer, such as a voltmeter or a receiver, which measures the electrocardiogram, the movement during an earthquake, and so on. All these physical transducers cannot respond to a delta function excitation, and hence, they pass only frequencies up to a specific value.

It is customary to use uppercase letters for random variables. However, in this chapter, we use for both cases, deterministic and random, lowercase letters, and their identity will be understood by the context. A **discrete random signal** $\{x(n)\}$ is a sequence of **indexed random variables** (rv's) assuming the values:

$$\{x(0),\ x(1),\ x(2),\ldots\} \tag{4.1}$$

The random sequence with values $\{x(n)\}$ is discrete with respect to sampling index n. In our case, we assume that the rv at any time n takes continuous values, and hence, it is a continuous rv at any time n. What we really say is that we can associate at each n an infinite set of values (continuous) of the rv $x(n)$. This type of sequence is also known as **time series**. In case we study a continuous random signal, we will assume that we have sampled it at high enough rate so that we construct a time series that is free of aliasing in the frequency domain.

A particular rv $x(n)$ is characterized by its **probability density function** (pdf), $f(x(n))$, at the specific time n by

$$f(x(n)) = \frac{\partial F(x(n))}{\partial x(n)} \tag{4.2}$$

and its **cumulative density function** (cdf), $F(x(n))$, is given by

$$F(x(n)) = \mathrm{pr}\{x(n) \le x(n)\} = \int_{-\infty}^{x(n)} f(y(n))dy(n) \tag{4.3}$$

In the above equation, the first $x(n)$ from left is an rv and the second is the value it takes. $f(x(n))$ is the pdf such as the normal and the exponential. $\mathrm{pr}\{x(n) \le x(n)\}$ specifies the probability of the rv $x(n)$ at time n to be less than or equal to the value $x(n)$ at time n. As the value $x(n)$ at n approaches infinity, the cdf approaches 1.

Similarly, for **multivariate** distributions, we have

$$F(x(n_1),\ldots,x(n_k)) = \mathrm{pr}\{x(n_1) \le x(n_1),\ldots,x(n_k) \le x(n_k)\}$$
$$f(x(n_1),\ldots,x(n_k)) = \frac{\partial^k F(x(n_1),\ldots,x(n_k))}{\partial x(n_1)\cdots\partial x(n_k)} \tag{4.4}$$

If, for example, we want to check the accuracy of reading a dial by a person, we will have two readings: one due to the person and another due to the instruments. A simultaneous plot of these two readings, each one associated with a different orthogonal axis, will produce a scattering diagram but with a linear dependence. The closer the points fall on a straight line, the more reliable the person's readings are. This example presents a case of a bivariate distribution.

To obtain a formal definition of a discrete-time stochastic process, we consider an experiment with a finite or infinite number of unpredictable outcomes from a sample space, $S(z_1, z_2,\ldots)$, each one occurring with a probability $\mathrm{pr}\{z_i\}$. Next, by some rule we assign a deterministic sequence $\{x(n, z_i)\}, -\infty < n < \infty$, to each element z_i of the sample space. The sample space, the probabilities of each outcome, and the sequences constitute a **discrete-time stochastic process** or a **random sequence**. From this definition, we obtain the following four interpretations:

- $x(n,z)$ is an rv if n is fixed and z is variable.
- $x(n,z)$ is a sample sequence called realization if z is fixed and n is variable.
- $x(n,z)$ is a number if both n and z are fixed.
- $x(n,z)$ is a stochastic process if both n and z are variable.

Each time we run an experiment under identical conditions, we create a sequence of rv's $\{x(n)\}$, which is known as a **realization** and constitutes an **event**. A realization is one member of a set called the ensemble of all possible results from the repetition of an experiment. It is obvious that the more realizations we create, the better we can find some of the characteristics of the rv. The following Book MATLAB m-file produces 5 realizations shown in Figure 4.1.

Book m-File: ch4_realizations

```
%Book MATLAB m-file: ch4_realizations
for n = 1:5
    x(n,:) = randn(1,50);%x = 5x50 matrix with each row
    having
    %zero mean normal (Gaussian) rv signals;
end;
m = 0:49;
for i = 1:5
```

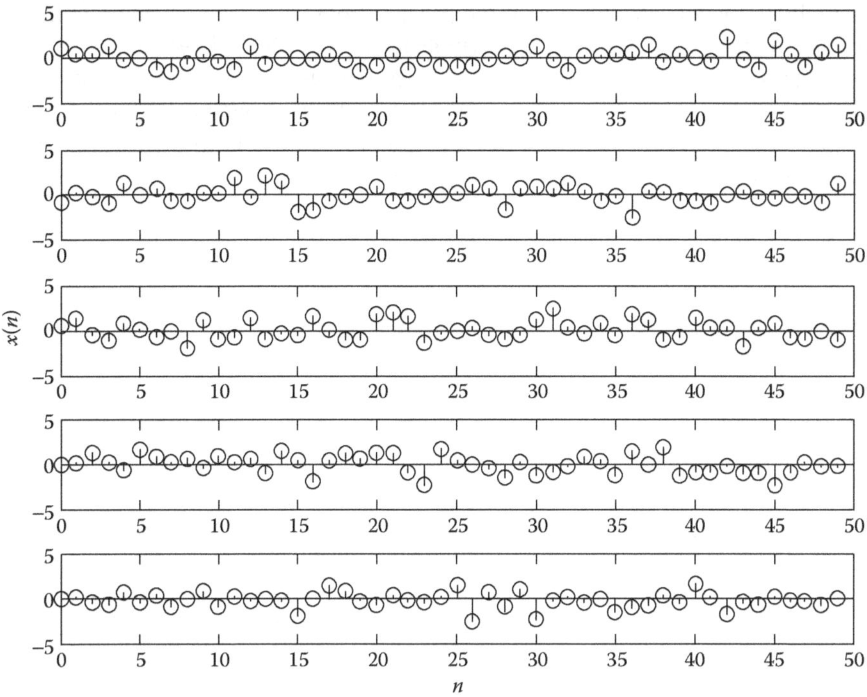

FIGURE 4.1

```
        subplot(5,1,i);stem(m,x(i,:),'k');%plots five rows
        of matrix x;
end;
xlabel('n');ylabel('x(n)');
```

4.1.1 Stationary and Ergodic Processes

It is seldom in practice that we will be able to create an ensemble of a random process with numerous realizations so that we can find some of its statistical characteristics, for example, mean value, variance, and so on. To find these statistical quantities, we need to know the pdf of the process, which, most of the times, is not possible to know. Therefore, we will restrict our studies to processes that are easy to study and handle mathematically.

The process that produces an ensemble of realizations and whose statistical characteristics do not change with time is called **stationary**. For example, the pdf of the rv's $x(n)$ and $x(n+k)$ of the process $\{x(n)\}$ are the same independently of the values of n and k.

Since we will be unable to produce ensemble averages in practice, we are left with only one realization of the stochastic process. To overcome this difficulty, we assume that the process is **ergodic**. This characterization permits us to find the desired statistical characteristics of the process from only one realization at hand. We refer to those statistical values as **sample mean**, **sample variance**, and so on. We must have in mind that these values are the approximate to the ensemble values. One of the main objectives of the statisticians is to find the ways and give to these sample values some confidence limits around the sample values within which, with a high degree of probability, the ensemble value exists.

4.1.2 Averages of RV

4.1.2.1 Mean Value

The ensample **mean value** or **expectation value** $m(n)$ at time n of a rv $x(n)$ having a pdf $f(x(n))$ is given by

$$m(n) = E\{x(n)\} = \int_{-\infty}^{\infty} x(n) f(x(n)) dx(n) \tag{4.5}$$

where:

$E\{.\}$ stands for expectation operator that is defined by the integration above

We can use the frequency formula to approximate the ensemble average by using the realizations. It is obvious that the larger the number of the realizations we use, the better approximation of the mean value we obtain to the ensemble average. We can find the mean value using the approximation formula:

$$m(n) = \lim_{N \to \infty} \left\{ \frac{1}{N} \sum_{i=1}^{N} x_i(n) \right\} \tag{4.6}$$

where:

N is the number of realizations

$x_i(n)$ is the outcome at time n and of the ith realization

The above equation indicates that at a particular time n, we add all the values $x(n)$ belonging to different realizations and divide by the number of the realizations used (in this case, N realizations).

For an ergotic process, the sample mean approximation (estimator) is found using the formula:

$$\hat{m} = \frac{1}{N} \sum_{n=0}^{N-1} x(n) \tag{4.7}$$

It turns out that (see Problem 4.1.1) the sample average (mean) that is found using (4.7) is equal to the ensemble average of the mean. We call the estimate an **unbiased** estimator.

Example 4.1.1

Using the results of the Book m-file: ch4_realizations, find the mean value using the frequency formula and for $n = 12, n = 17$, and $n = 21$.

Solution: Running the program, we obtain $x(12) = [-1.3335\ 1.8482\ -0.6885\ 0.2685\ 0.1391]$, $x(17) = [-0.2494\ -1.6805\ 1.6197\ -1.9171\ -0.1435]$, $x(21) = [-1.0290\ 0.8369\ 1.7357\ 1.3808\ -0.6614]$. The corresponding means are $m(12) = 0.0468$, $m(17) = -0.4742$, and $m(21) = 0.4526$. We must note that if we rerun the program, we will find different vectors for the same times. Since the program gives a matrix $\boldsymbol{x}$ with dimensions 5 × 50 (5 rows by 50 columns), we use the MATLAB expression $\boldsymbol{x}(:,n)$ to get the vectors. This expression says: Produce the vector that is its n column. We can also think of all the rows of the column n.

■

4.1.2.2 Correlation

The **cross-correlation** between two random sequences is defined by

$$r_{xy}(m,n) = E\{x(m),y(n)\} = \iint x(m)y(n)f(x(m),y(n))dx(m)dy(n) \tag{4.8}$$

The limits of the integrals are from $-\infty$ to ∞. If $x(n) = y(n)$, the correlation is called **autocorrelation**.

4.1.2.2.1 Frequency Interpretation of the Correlation

Having an ensemble of realizations, the frequency interpretation of the autocorrelation function is found using the formula:

$$r_{xx}(m,n) = \lim_{N\to\infty}\left\{\frac{1}{N}\sum_{i=1}^{N} x_i(m)x_i(n)\right\} \tag{4.9}$$

The Book m-function that produces the mean and autocorrelation using the frequency interpretation formulas is as follows:

Book m-Function

```
function[mx,rx] = fr_mean_autocor(M,N)
        %to store the function we write only the name: fr_mean_
        %autocor;
        %at the MATLAB command window we write:[a,b] = fr_mean_
        %autocor(20,%50);
        %N = number of time instances;easily modified for
        %other pdf's;M is the rows (realizations)of the matrix;
x = randn(M,N);%randn = MATLAB function producing zero mean
        %Gaussian distributed white noise; x = MxN matrix;
        %sum(x,1) = MATLAB function that sums all the rows;
        %sum(x,2) = MATLAB function that sums all the columns;
mx = sum(x,1)/M;
for i = 1:N
        rx(i) = sum(x(:,1).*x(:,i))/M;
end;
```

The reader is encouraged to run the above program, for example, ($M = 10$, $N = 60$) and ($M = 400$, $N = 60$). The reader will observe that the mean value at each n will decrease and the autocorrelation will tend to a delta function at the origin.

4.1.2.2.2 Sample Autocorrelation Function

The sample autocorrelation function is found from only one realization at hand. The formula is

$$\hat{r}_{xx}(m) = \frac{1}{N-|m|}\sum_{n=0}^{N-|m|-1} x(n)x\big(n+|m|\big) \quad m = 0, 1, \ldots, N-1 \tag{4.10}$$

The absolute value of m ensures the symmetry of the sample autocorrelation sequence at $n = 0$. Although the formula gives an **unbiased** autocorrelation sequence, it sometimes produces autocorrelation matrices (discussed below) that do not have inverses. Therefore, it is customary in practice to use anyone of the biased formulas:

$$\hat{r}_{xx}(m) = \frac{1}{N}\sum_{n=0}^{N-|m|-1} x(n)x\big(n+|m|\big) \quad 0 \le m \le N-1$$

or (4.11)

$$\hat{r}_{xx}(m) = \frac{1}{N}\sum_{n=0}^{N-1} x(n)x(n-m) \quad 0 \le m \le N-1$$

Book m-Function for Biased Autocorrelation Function

```
function[r] = lms_sample_biased_autoc(x,lg)
%this function finds the biased autocorrelation function
%with lag from 0 to lg;it is recommended that lg is 20-30% of
%N;N is the total number of elements of the observed vector;
N = length(x);%x = data;lg stands for lag number;
for m = 1:lg
        for n = 1:N+1-m
                x1(n) = x(n-1+m)*x(n);
        end;
        r(m) = sum(x1)/N;
end;
```

Book m-Function for Unbiased Autocorrelation Function

```
function[r] = lms_sample_unbiased_autoc(x,lg)
%this function finds the biased autocorrelation function
%with lag from 0 to lg; it is recommended that lg is 20-30% of
%N;N is the total number of elements of the observed vector;
N = length(x);%x = data; lg stands for lag number;
for m = 1:lg
```

```
        for n = 1:N+1-m
                x1(n) = x(n-1+m)*x(n);
        end;
        r(m) = sum(x1)/(N-m);
end;
```

Figure 4.2 shows the plots of an unbiased autocorrelation function (shown by circles) and a biased autocorrelation function (shown by dots).

We can also use MATLAB functions to find the biased and unbiased autocorrelation of a signal. The function is

```
>>r = xcorr(x,y,'biased');%for biased case;xcorr()is a MATLAB
>>      %function;
>>r = xcorr(x,y,'unbiased');%for unbiased case;
>>%x,y are N length vectors; r is a 2N-1 symmetric
>>%cross-correlation vector; for unequal vectors we padd
>>%with zeros the shorter;
```

If we do not use any of the words, biased or unbiased, the correlation will be unbiased and the results are not divided by N. The reader will learn more about the `xcorr` function by entering in the MATLAB Command window `help xcorr`.

4.1.2.3 Covariance

A **covariance** (autocovariance) of a random sequence is defined by

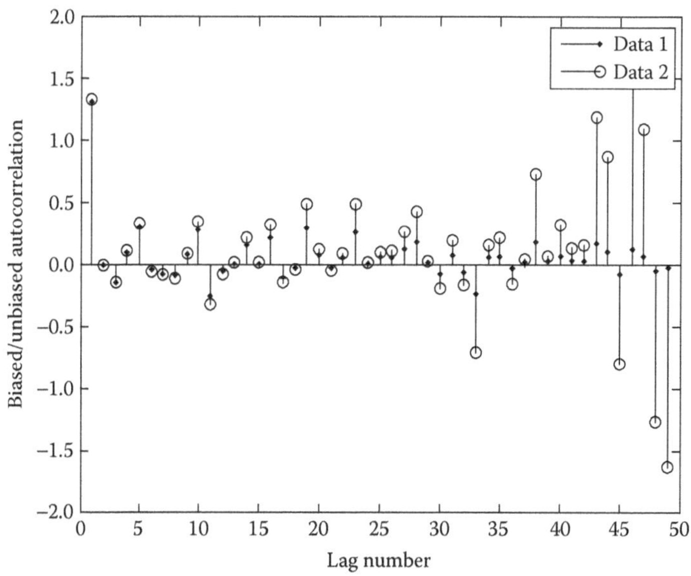

FIGURE 4.2

$$\begin{aligned} c_{xx}(m,n) &= E\{[x(m)-m_m][x(n)-m_n]\} \\ &= E\{x(m)x(n)\} - m_m E\{x(n)\} - m_n E\{x(m)\} + m_m m_n \\ &= r_{xx}(m,n) - m_m m_n \end{aligned} \tag{4.12}$$

If we set the second x with y, we produce the cross-covariance of the two different random sequences. If $c_{xx}(m,n)=0$ for all m's and n's, then

$$r_{xx}(m,n) = m_m m_n \tag{4.13}$$

Two random sequences are orthogonal if their cross-correlation function is 0. Hence,

$$r_{xy}(m,n) = 0$$

Example 4.1.2

The output $y(n)$ of the linear time-invariant (LTI) system is given by the convolution of the input $x(n)$ and its impulse response $h(n)$ (deterministic function), or

$$y(n) = \sum_{m=-\infty}^{\infty} h(m)x(n-m)$$

The cross-correlation between the input and the output is given by

$$\begin{aligned} r_{xy}(k,l) = E\{x(k)y(l)\} &= E\left\{x(k)\sum_{m=-\infty}^{\infty} h(m)x(l-m)\right\} \\ &= \sum_{m=-\infty}^{\infty} h(m)E\{x(k)x(l-m)\} \\ &= \sum_{m=-\infty}^{\infty} h(m)r_{xx}(k,l-m) \end{aligned}$$

■

The variance is found by setting $m=n$ in (4.12). Therefore, we obtain

$$\begin{aligned} &c_{xx}(n,n) \triangleq \sigma_n = E\left\{[x(n)-m_n]^2\right\} = E\left\{x^2(n)\right\} - m_n^2 = r_{xx}(n,n) - m_n^2 \\ &c_{xx}(n,n) \triangleq \sigma_n = E\left\{x^2(n)\right\} = r_{xx}(n,n) \qquad m_n = 0 \end{aligned} \tag{4.14}$$

4.1.2.3.1 Sample Variance

The estimator of the variance, when we use one sample from the ensemble, is given by

$$\hat{\sigma}^2 = \frac{1}{N}\sum_{n=1}^{N}[x(n)-\hat{m}]^2; \quad \hat{\sigma}^2 = \frac{1}{N}\sum_{n=1}^{N} x(n)^2 \quad \hat{m} = 0 \tag{4.15}$$

To obtain the variance, we can also use the MATLAB function **var(data _ vector)**. The standard deviation that is equal to the square root of the variance can be found using the MATLAB function **std(data _ vector)**.

4.1.2.3.2 Independent and Uncorrelated rv's

If the join pdf of two rv's can be separated into two pdf's, $f_{xy}(m,n) = f_x(m)f_y(n)$, then the rv's are statistically **independent**. Therefore,

$$E\{x(m)x(n)\} = E\{x(m)\}E\{x(n)\} = m_m m_n \tag{4.16}$$

The above equation is necessary and sufficient condition for the two rv's $x(m)$, $x(n)$ to be uncorrelated. Note that independent rv's are always uncorrelated. However, the converse is not necessarily true. If the mean value of any two uncorrelated rv's is 0, then the rv's are called **orthogonal**. In general, two rv's are called orthogonal if their correlation is 0.

4.2 STATIONARY PROCESSES

For a wide-sense (or weakly) stationary (WSS) process, the joint pdf satisfies the relation

$$f(x(n),x(m)) = f(x(n+k)x(m+k)) \tag{4.17}$$

for any k, m, and n (Table 4.1).

4.2.1 Autocorrelation Matrix

In this case, we use a 3×3 autocorrelation matrix for elucidating the procedure. This is easily extended to any matrix dimensions. Let $\boldsymbol{x} = [x(0)\ x(1)\ x(2)]^T$ is a vector representing a finite random sequence (in this case three elements); then the autocorrelation matrix is given by

TABLE 4.1
Properties of WSS Processes

1. $m_n = m_{n+k} = m = \text{constant}$.
2. The autocorrection $r_x(m,n)$ depends on the difference $m - n$.
3. The variance is less than infinity.
4. The autocorrection is symmetric: $r(n) = r(-n)$.
5. For lag $l = 0$, $r_x(0) = E\{|x(n)|^2\} \geq 0$.
6. $r_x(0) \geq |r_x(l)|$.

Note: See Problem 4.2.1.

$$
\begin{aligned}
\boldsymbol{R}_x = E\{\boldsymbol{x}\boldsymbol{x}^T\} &= E\left\{\begin{bmatrix} x(0) \\ x(1) \\ x(2) \end{bmatrix}\begin{bmatrix} x(0) & x(1) & x(2) \end{bmatrix}\right\} \\
&= E\begin{Bmatrix} x(0)x(0) & x(0)x(1) & x(0)x(2) \\ x(1)x(0) & x(1)x(1) & x(1)x(2) \\ x(2)x(0) & x(2)x(1) & x(2)x(2) \end{Bmatrix} \\
&= \begin{bmatrix} E\{x(0)x(0)\} & E\{x(0)x(1)\} & E\{x(0)x(2)\} \\ E\{x(1)x(0)\} & E\{x(1)x(1)\} & E\{x(1)x(2)\} \\ E\{x(2)x(0)\} & E\{x(2)x(1)\} & E\{x(2)x(2)\} \end{bmatrix} \\
&= \begin{bmatrix} r_x(0) & r_x(-1) & r_x(-2) \\ r_x(1) & r_x(0) & r_x(-1) \\ r_x(2) & r_x(1) & r_x(0) \end{bmatrix}
\end{aligned}
\tag{4.18}
$$

The covariance matrix is given by (see Problem 4.2.4)

$$
\boldsymbol{C} = E\left\{(\boldsymbol{x}-\boldsymbol{m}_x)(\boldsymbol{x}-\boldsymbol{m}_x)^T\right\} = \boldsymbol{R}_x - \boldsymbol{m}_x\boldsymbol{m}_x^T \tag{4.19}
$$

Example 4.2.1

Find (1) the unbiased autocorrelation with lag 15 of a 50-term sequence, (2) the biased autocorrelation with the same settings, and (3) a 3×3 autocorrelation matrix. Use a sequence of independent rv's having Gaussian (normal) distribution and zero mean value.

Solution: The sequence $\{x(n)\}$ of 50 terms, a vector, is obtained by using the MATLAB function `randn(1,50)`. To find the autocorrelation, we use either the Book m-files or the MATLAB function `xcorr()`. Figure 4.3 shows one realization of an unbiased autocorrelation function. To produce the 3×3 matrix, we first created the vector $\boldsymbol{x} = [0.7444 \;\; -0.0745 \;\; -0.1010]$ from the first three elements of the correlation function and then we used the MATLAB function `toeplitz(x',x)` to produce the matrix:

$$
\hat{\boldsymbol{R}}_x = \begin{bmatrix} 0.7444 & -0.0745 & -0.1010 \\ -0.0745 & 0.7444 & -0.0745 \\ -0.1010 & -0.0745 & 0.7444 \end{bmatrix}
$$

Note the symmetry of the matrix along the diagonals from left to right. This type of a matrix is called **Toeplitz**.

■

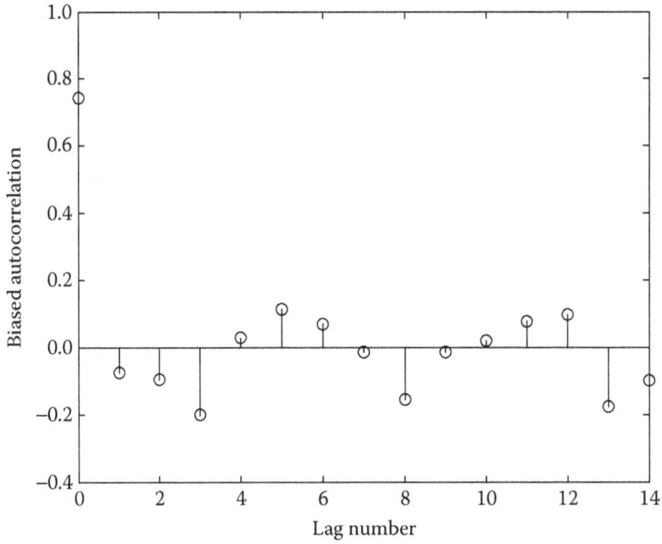

FIGURE 4.3

Note: If we have a row vector $\boldsymbol{x}$ and we need to create another row vector $\boldsymbol{y}$ with the elements of x from m to n only, we write

```
>>y = x(1,m:n);
```

Note: If we have a column vector $\boldsymbol{x}$ and we need to create another column vector $\boldsymbol{y}$ with the elements of x from m to n only, we write

```
>>y = x(m:n,1);
```

Note: If we have produced the autocorrelation vector $\boldsymbol{r}$ and we want to create the autocorrelation matrix $\boldsymbol{R}$ with the first 5 elements of $\boldsymbol{r}$, at the Command window we write:

```
>>R = Toeplitz(r(1,1:5));
```

Note: To create a subscript to `xlabel()` and `ylabel()` MATLAB functions, we write `xlabel('r _ {xx}'}`, which will produce the following expression on the x-axis of the figure: x_{xx}. For a superscript, we write `xlabel('r^{12})  :  x`^{12}.

Example 4.2.2

Let $\{v(n)\}$ be a zero-mean, uncorrelated Gaussian random sequence with variance $\sigma_v^2(n) = \sigma^2 = \text{constant}$. (1) Characterize the random sequence $\{v(n)\}$ and (2) determine the mean and the autocorrelation of the sequence $\{x(n)\}$ if $x(n) = v(n) + av(n-1)$, in the range $-\infty < n < \infty$; a is a constant.

Solution: (1) The variance of $\{v(n)\}$ is constant and, hence, is independent of the time, n. Since $\{v(n)\}$ is an uncorrelated sequence, it is also independent due to the fact that it is Gaussian sequence. Therefore, we obtain $c_v(l,n) = r_v(l,n) - m_l m_n = r_v(l,n)$ or $\sigma^2 = r_v(n,n) = \text{constant}$. Hence, $r_v(l,n) = \sigma^2\delta(l-n)$, which implies that $\{v(n)\}$ is a WSS process. (2) $E\{x(n)\} = 0$ since $E\{v(n)\} = E\{v(n-1)\} = 0$. Hence,

$$
\begin{aligned}
r_x(l,n) &= E\{[v(l) + av(l-1)][v(n) + av(n-1)]\} \\
&= E\{v(l)v(n)\} + aE\{v(l-1)v(n)\} + aE\{v(l)v(n-1)\} + a^2E\{v(l-1)v(n-1)\} \\
&= r_v(l,n) + ar_v(l-1,n) + ar_v(l,n-1) + a^2r_v(l-1,n-1) \\
&= \sigma^2\delta(l-n) + a\sigma^2\delta(l-n+1) + a^2\sigma^2\delta(l-n) \\
&= (1+a^2)\sigma^2\delta(r) + a\sigma^2\delta(r+1) + a\sigma^2\delta(r-1),\ l-n=r
\end{aligned}
$$

Since the mean of $\{x(n)\}$ is 0, a constant, and its autocorrelation is a function of the lag factor $r = l - n$, it is a WSS process.

■

4.2.2 Purely Random Process (White Noise)

A discrete process is a purely random process if the rv's $\{x(n)\}$ of a sequence are mutually independent and identically distributed (iid) variables. Since the mean and $\text{cov}\{x(m),x(m+k)\}$ do not depend on time, the process is said to be WSS. This process is also known as **white noise** (WN) and is given by

$$
c_{xx}(k) = \begin{cases} E\{[x(m)-m_x][x(m+k)-m_x]\} & k=0 \\ 0 & k=\pm1,\pm2,\pm3,\cdots \end{cases} \tag{4.20}
$$

4.2.3 Random Walk

Let $\{x(n)\}$ be a purely random process with mean m_x and variance σ_x^2. A process $\{y(n)\}$ is a random walk if

$$
y(n) = y(n-1) + x(n) \qquad y(0) = 0 \tag{4.21}
$$

For $n = 1$, $y(1) = x(1)$; for $n = 2$, $y(2) = x(2) + x(1)$, and therefore, the solution is

$$
y(n) = \sum_{n=i}^{n} x(i) \tag{4.22}
$$

The mean is easily found to be $E\{y(n)\} = nm_x$, and the covariance is

$$
\text{cov}\{y(n)y(n)\} = E\{[y(n)-m_y][y(n)-m_y]\} = n\left[E\{x^2\} - m_x^2\right] = n\sigma_x^2 \tag{4.23}
$$

It is important to note that the difference $x(n) = y(n) - y(n-1)$ is purely random and hence stationary.

4.3 SPECIAL RANDOM SIGNALS AND PDF'S

In this section, we present the most important types of rv distributions which, it is recommended, are used as additive noise in later chapters dealing with the adaptive filtering.

4.3.1 White Noise

If the pdf of a WSS discrete random sequence satisfies the relation

$$f(x(0),\, x(1),\cdots) = f(x(0))f(x(1)f(x(2))\cdots \tag{4.24}$$

is a purely random sequence whose elements $x(n)$ are **statistically independent and identically distributed**. Therefore, the **zero-mean iid** sequence has the following correlation function (see Problem 4.3.1):

$$r_{xx}(m-n) = E\{x(m)x(n)\} = \sigma_x^2\delta(n-m) \tag{4.25}$$

where:

σ_x^2 is the variance of the rv's at each n
$\delta(m-n)$ is the discrete-delta function

For $m \neq n$, $\delta(m-n) = 0$, and hence, (4.25) becomes

$$r_{xx}(k) = \begin{cases} \sigma_x^2\delta(k) & k = 0 \\ 0 & k \neq 0 \end{cases} \tag{4.26}$$

For example, a random process consisting of a sequence of uncorrelated Gaussian rv's is a WN process referred to as **white Gaussian noise** (WGN). The MATLAB has a special function that will produce WGN. For example, if we write `x = randn(1,500)` will produce a sequence (vector x) that its elements are normally distributed and variance 1. We can then use the MATLAB function `hist(x,20)` to produce the pdf of the time series $\{x(n)\}$. The function divides the range between the maximum and minimum values of the sequence in 20 sections. Then plot the number of x's, whose values fall within each section, versus the range of values of x's.

It is instructive to be able to create different types of rv distributions to be able to check their influence on filtering.

4.3.2 Gaussian Distribution (Normal Distribution)

The pdf of a Gaussian rv $x(n)$ at time n is given by

$$f(x(n)) \triangleq N(m_{xn},\sigma_{xn}) = \frac{1}{\sqrt{2\pi\sigma_{xn}^2}}e^{-\left\{\frac{[x(n)-m_{xn}]^2}{2\sigma_{xn}^2}\right\}} \tag{4.27}$$

The algorithm to produce normalized Gaussian distribution is as follows:

1. Generate two rv's u_1 and u_2 from uniform distribution (0,1).
2. $x_1 = [-2\ln(u_1)]^{1/2}\cos(2\pi u_2)$ $\{\text{or } x_2 = [-2\ln(u_1)]^{1/2}\sin(2\pi u_2)\}$.
3. Keep x_1 or x_2.

Book m-Function for Producing a Normal Distribution

```
function[x] = lmsnormalpdf(m,s,N)
%in the command window we write:[x] = lmsnormalpdf(m,s,N);
%m = mean value;s = standard deviation;N = number of
%realizations;
for i = 1:N
        u1 = rand;%uniform distributed rv 0<u1<1;
        u2 = rand;
        z(i) = sqrt(-2*log(u1))*cos(2*pi*u2);%MATLAB function log()
        %gives the natural logarithm; log10()
        %gives the logarithm with base 10;
end;
x = s*z+m;
```

A discrete-time random process $\{x(n)\}$ is said to be Gaussian if every finite collection of samples of $x(n)$ are jointly Gaussian. A Gaussian random process has the following properties: (1) It is completely defined by its mean vector and covariance matrix, (2) any linear operation on the time variables produces another Gaussian random process, (3) all higher moments can be expressed by the first and second moments of the distribution (mean, covariance), and (4) WN is necessarily generated by iid samples (independence implies uncorrelated rv's and vice versa).

For example, the distribution of the rv's x = randn(1,2000) + 2 * randn (1,2000) − 0.5 * randn(1,2000) + 0.8 * randn(1,2000) + randn(1,2000) is shown in Figure 4.4 using the Book m-file MATLAB program:

Book m-File: normal_hist

```
%Book m-file: normal_hist;
x = randn(1,2000)+2*randn(1,2000)-0.5*randn(1,2000)...
+0.8*randn(1,2000)+randn(1,2000);
[m,z] = hist(x,20);%calculates counts in bins and
        %bin coordinates for 20 bins;
w = max(x)/length(z);%calculates bin length;
pb = m/(2000*w);%probability in each bin;
colormap([1 1 1]);%creates white bars, for other
        %colors see help colormap;
bar(z,pb);%plots the histogram;
```

Observe that the summation of normal rv's, linear operation, is also normally distributed.

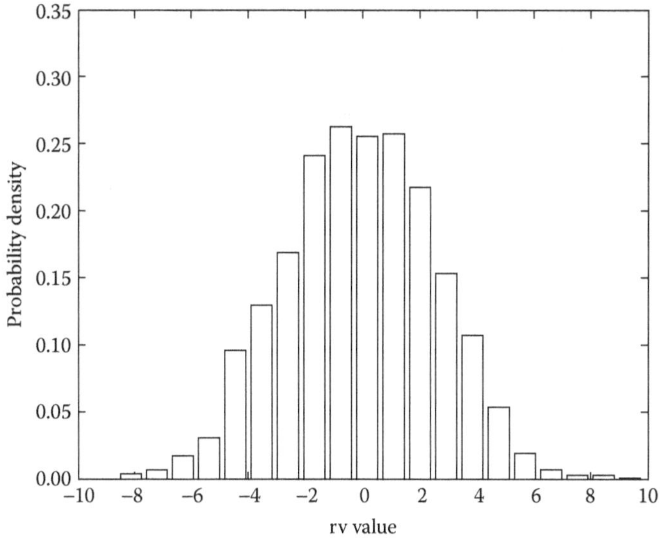

FIGURE 4.4

Example 4.3.1

If the pdf is $N(2.5,4)$ (mean 2.4 and variance 4), find $\mathrm{pr}\{x \le 6.5\}$.

Solution:

$$\mathrm{pr}\{x \le 6.5\} = \frac{1}{\sqrt{2\pi}}\frac{1}{2}\int_{-\infty}^{6.5} e^{-[(x-2.5)^2/8]}dx \quad (1)$$

Set $y=(x-2.5)/2$ in (1) to obtain: for $x=-\infty \Rightarrow y=-\infty$, for $x=6.5 \Rightarrow y=2$ and $dx=2dy$. Hence, (1) becomes

$$\mathrm{pr}\{x \le 6.5\} = \mathrm{pr}\{y \le 2\} = \frac{1}{\sqrt{2\pi}}\int_{-\infty}^{0} e^{-(y^2/2)}dy + \frac{1}{\sqrt{2\pi}}\int_{0}^{2} e^{-(y^2/2)}dy = 0.5 + \mathrm{erf}(2)$$

The error function is tabulated, but MATLAB can give very accurate values by introducing sampling values of <0.0001. For this sampling value, MATLAB gives the result: `erf(2)=(1/sqrt(2*pi)*0.0001*sum(-y.^2/2)=0.47727`, where the vector $\boldsymbol{y}$ is given by $\boldsymbol{y}=0:0.0001:2$. Tables give $\mathrm{erf}(2)=0.47724$. Therefore, $\mathrm{pr}\{\mathrm{x} \le 6.5\}=0.97727$.

■

Book m-Function for Normalized Normal pdf

```
function[x] = lms_normalized_normal_pdf(m,s,N)
z = sqrt(-2*log(rand(1,N))).*cos(2*pi*rand(1,N));
x = s*z+m;
```

We can also use Monte Carlo approach to obtain the normal distribution of rv's. The following Book m-function produces the desired results.

Book m-Function for Monte Carlo Normal pdf

```
function[x,y] = lms_monte_carlo_normal_pdf(m,s,N,M)
      %m = mean value;s = standard deviations;
      %N = number of variables; M = umber of summable
      %normal variables;
for n = 1:N
      x(n) = sum(randn(1,M))/sqrt(M);
end;
y = m+s*x;
```

The above Monte Carlo processing was based on the **central limit theorem**, which states as follows:

For N independent rv's $X_1, X_2, \ldots, X_N$ with mean m_i and variance σ_i^2, respectively,

$$y_N = \frac{\sum_{i=1}^{N}(x_i - m_i)}{\sqrt{\sum_{i=1}^{N}\sigma_i^2}}, \qquad \lim N \to \infty \Rightarrow y \equiv N(0,1) \tag{4.28}$$

The central limit theorem has the following interpretation:

> The properly normalized sum of many uniformly small and negligibly independent rv's tends to be a standard normal (Gaussian) rv. If a random phenomenon is the cumulative effect of many uniformly small sources of uncertainty, it can be reasonably modeled as normal rv.

4.3.3 Exponential Distribution

The pdf of the exponential distribution is given by

$$f(x(n)) = \begin{cases} \frac{1}{a}e^{-x(n)/a} & 0 \le x < \infty,\ a > 0 \\ 0 & \text{otherwise} \end{cases} \tag{4.29}$$

The exponential distribution is generally denoted by exp(a) and its standard form by exp(1). The mean value of the distribution is a and its variance is a^2.

The algorithm to produce exponential distribution is as follows:

1. Generate u from the uniform distribution (0,1).
2. $x = -a\ln(u)$.
3. Keep x.

Book m-Function for Producing an Exponential Distribution

```
function[x,m,sd] = lms_exp_pdf(a,N)
       %N = number of rv's needed to produce
       %the pdf;a = mean value of x;
       %variance of x is a^2;mean of x = a;
       %standard deviation = a;
for i = 1:N
       u = rand;
       x(i) = -a*log(u);
end;
m = mean(x);sd = std(x);
```

4.3.4 Lognormal Distribution

Let $X(n)$ be a normal distributed rv with mean m and variance σ^2, $N(m,\sigma^2)$. Then, $Y(n)=e^{X(n)}$ has the lognormal distribution with pdf:

$$f(y(n))=\begin{cases}\dfrac{1}{\sqrt{2\pi}\,\sigma y(n)}\exp\left\{-\dfrac{[\ln y(n)-m]^2}{2\sigma^2}\right\} & 0\le y(n)<\infty \\ 0 & \text{otherwise}\end{cases} \tag{4.30}$$

The algorithm to produce lognormal pdf is as follows:

1. Generate Z from a normal distribution with zero mean and unit variance, $N(0,1)$.
2. $X=m+\sigma Z$.
3. $Y=e^X$.
4. Keep Y.

Book m-Function to Produce Lognormal PDF

```
function[y,x] = lms_lognormal_pdf(m,s,N)
       %[y] = lognormal pdf; m = mean;s = standard deviation;
       %N = number of samples;
for i = 1:N
       r1 = rand;r2 = rand;
       z(i) = sqrt(-2*log(r1))*cos(2*pi*r2);
end;
x = m+s*z;%x normaly distributed;
y = exp(x);
```

Other statistical characteristics of the lognormal distribution are as follows:

Range: $0\le x<\infty$
Mean: $\exp(m_x)\exp(s^2/2)$
Variance: $\exp(2m_x)\exp(s^2)[\exp(s^2)-1]$

4.3.5 Chi-Square Distribution

If $z_1, z_2, z_3, \ldots, z_n$ are $N(0,1)$, then

$$y = \sum_{i=1}^{n} z_i^2 \tag{4.31}$$

has chi-square distribution with $n-1$ degrees of freedom, and it is denoted by $\chi^2(n)$. If $z_1, z_2, z_3, \ldots, z_n$ are normally distributed with $N(m,\sigma)$, then

$$y = \frac{1}{\sigma^2}\sum_{i=1}^{n}(z_i - m)^2 \tag{4.32}$$

is chi-square distributed with $n-1$ degrees of freedom and it is denoted by $\chi^2(n)$.

Book m-Function for Chi-Squared Distributed rv

```
function[y] = lms_monte_carlo_chi_squared_pdf(r,N)
      %y is chi-squared disributed rv;r = degrees
      %of freedom, MUST BE EVEN NUMBER;number
      %of chi squared rvs y;
for n = 1:N
      y(n) = sum(randn(1,r).^2);
end;
```

Other statistical characteristics of the chi-squared distribution are as follows:

Range: $0 \le y < \infty$
Mean: r
Variance: $2r$

4.4 WIENER–KHINCHIN RELATIONS

For a WSS process $\{x(n)\}$, the correlation function asymptotically goes to 0, and therefore, we can find its spectrum using the discrete-time Fourier transform (DTFT). Hence, the **power spectral density (PSD)** is given by

$$S_x(e^{j\omega}) = \sum_{k=-\infty}^{\infty} r_{xx}(k)e^{-j\omega k} \tag{4.33}$$

This function is periodic with period 2π since $\exp(j\omega) = \exp[j(\omega + 2\pi)]$. Therefore, the autocorrelation sequence is given by (see Problem 4.4.1)

$$r_{xx}(k) = \frac{1}{2\pi}\int_{-\pi}^{\pi} S_x(e^{j\omega})e^{j\omega k}d\omega \tag{4.34}$$

For a real symmetric correlation function, the power spectrum is even function. In addition, the power spectrum of a WSS process is also nonnegative. The mathematical expressions of the above assertions are as follows:

$$\begin{aligned} S_x(e^{j\omega}) &= S_x(e^{-j\omega}) = S_x^*(e^{j\omega}) \\ S_x(e^{j\omega}) &\geq 0 \end{aligned} \tag{4.35}$$

Example 4.4.1

Find the PSD of the sequence $x(n) = \cos(0.2 \times 2\pi n) + \text{randn}(1,64)$ for $n = 0, 1, 2, \ldots, 63$.
Solution: The following Book m-file produces Figure 4.5.

Book m-File: ex4_4_1

```
%Book m-file: ex4_4_1;
n = 0:63;s = cos(0.1*2*pi*n);v = randn(1,64);
x = cos(0.1*2*pi*n)+rand(1,64);
r = xcorr(x,'biased');% the biased autocorrelation
            %is divided by N, in this case by 64;
fs = fft(s); fr = fft(r,64);
w = 0:2*pi/64:2*pi-(2*pi/64);
subplot(3,2,1);stem(n,s,'k');xlabel('n');
ylabel('s(n)');
subplot(3,2,2);stem(n,v,'k');xlabel('n');
ylabel('v(n)');
subplot(3,2,3);stem(n,x,'k');xlabel('n');
ylabel('x(n)');
subplot(3,2,4);stem(n,r(1,64:127),'k');
```

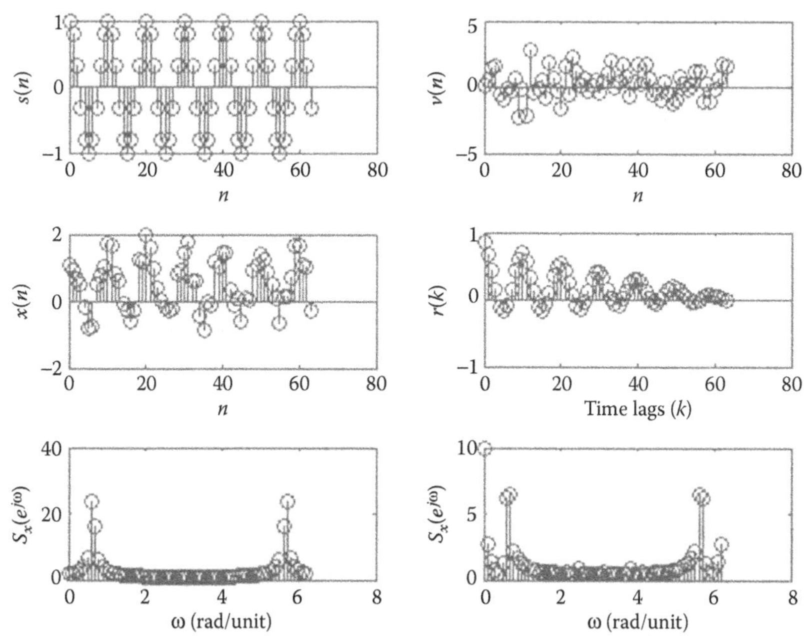

FIGURE 4.5

```
xlabel('k, time lags');ylabel('r(k)');
subplot(3,2,5);stem(w,abs(fs),'k');
xlabel('\omega rad/unit');
ylabel('S_s(e^{j\omega})');
subplot(3,2,6);stem(w,abs(fr),'k');
xlabel('\omega rad/unit');
ylabel('S_x(e^{j\omega})');
```

■

A comparison between the PSD's of the noisy signal x and its autocorrelation is shown in Figure 4.6.

The summary of correlation and spectral properties of stationary random sequences is given in Table 4.2.

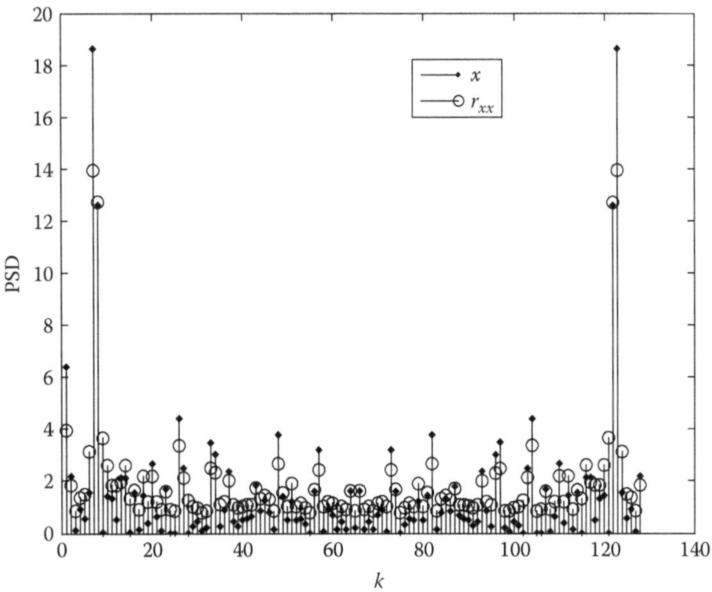

FIGURE 4.6

TABLE 4.2
Summary of Correlation and Spectral Properties of Stationary Sequences

Definitions	
Mean value	$m_x = E\{x(n)\}$
Autocorrelation	$r_{xx}(m) = E\{x(n)x^*(n-m)\}$
Autocovariance	$c_{xx}(m) = E\{[x(n)-m_x][x(n-m)-m_x]^*\}$
Cross-correlation	$r_{xy}(m) = E\{x(n)y^*(n-m)\}$
Cross-covariance	$c_{xy}(m) = E\{[x(n)-m_x][x(n-m)-m_x]^*\}$
Power spectral density	$S_x(e^{j\omega}) = \sum_{m=-\infty}^{\infty} r_{xx}(m)e^{-j\omega m}$
Cross-power spectral density	$S_{xy}(e^{j\omega}) = \sum_{m=-\infty}^{\infty} r_{xy}(m)e^{-j\omega m}$

(Continued)

TABLE 4.2
(Continued) Summary of Correlation and Spectral Properties of Stationary Sequences

Interrelations	
$c_{xx}(m) = r_{xx}(m) - m_x m_x^* = r_{xx}(m) - \lvert m_x \rvert^2$	
$c_{xy}(m) = r_{xy}(m) - m_x m_y^*$	
Properties	
Autocorrelation	***Auto-PSD***
$r_{xx}(m) =$ nonnegative definite	$S_x(e^{j\omega}) \geq 0$ and real
$r_{xx}(m) = r_{xx}^*(-m)$	$S_x(e^{j\omega}) = S_x(e^{-j\omega})$ [real $x(n)$]
$\lvert r_{xx}(m) \rvert \leq r_{xx}(0)$	$S_x(z) = S_x^*(1/z^*)$
Cross-Correlation	***Cross-PSD***
$r_{xy}(m) = r_{xy}^*(-m)$	
$\lvert r_{xy}(m) \rvert \leq [r_{xx}(0) r_{yy}(0)]^{1/2} \leq (1/2)[r_{xx}(0) + r_{yy}(0)]$	$S_{xy}(z) = S_{yx}^*(1/z^*)$

4.5 FILTERING RANDOM PROCESSES

LTI filters are used in many signal processing applications. Since the input signals of these filters are usually random processes, we need to determine how the statistics of these signals are modified as a result of filtering.

Let $x(n)$, $y(n)$, and $h(n)$ be the filter input, the filter output, and the filter impulse response, respectively. It can be shown that if $x(n)$ is a WSS process, then the filter output autocorrelation $r_{yy}(k)$ is related to the filter input autocorrelation $r_{xx}(k)$ as follows (see Problem 4.5.1):

$$r_{yy}(k) = \sum_{l=-\infty}^{\infty} \sum_{m=-\infty}^{\infty} h(l) r_{xx}(m - l + k) h(m)$$

$$= r_{xx}(k) * h(k) * h(-k) = r_{xx}(k) * r_{hh}(k) \tag{4.36}$$

$$h(k) * h(-k) = \text{Convolution between } h(k) \text{ and its reflected form } h(-k)$$

$$= \text{Autocorrelation of } h(k)$$

The right-hand side of the above expression shows convolution of three functions. We can take the convolution of two of the functions and the resulting function is then convolved with the third function. Based on the convolution property discussed in Chapter 3, the results are independent of the order we operate on the functions.

From the z-transform table, we know that the z-transform of the convolution of two functions is equal to the product of their z-transform. Remembering the definition of the z-transform, we find the relationship (the order of summation does not change the results)

$$\mathfrak{Z}\{h(-k)\} = \sum_{k=-\infty}^{\infty} h(-k)z^{-k} = \sum_{m=\infty}^{-\infty} h(m)(z^{-1})^{-m} = H(z^{-1}) \tag{4.37}$$

since the summation is the same regardless of the direction in which we sum the series.

Therefore, the z-transform of (4.36) becomes

$$R_y(z) = \mathfrak{Z}\{r_{xx}(k) * h(k)\}\mathfrak{Z}\{h(-k)\} = R_x(z)H(z)H(z^{-1}) \tag{4.38}$$

If we set $z = e^{j\omega}$ in the definition of the z-transform of a function, we obtain the spectrum of the function. Therefore, the Wiener–Khinchin theorem (4.38) becomes

$$S_y(e^{j\omega}) = S_x(e^{j\omega})\left|H(e^{j\omega})\right|^2 \tag{4.39}$$

Note: The above equation indicates that the power spectrum of the output random sequence is equal to the power spectrum of the input sequence modified by the square of the absolute value of the spectrum of the filter transfer function.

Example 4.5.1

A first-order **finite impulse response** (FIR) filter is defined in the time domain by the difference equation $y(n) = x(n) + 0.8x(n-1)$. If the input signal is $N(0,1)$, find the PSD of the filter output.

Solution: Taking into consideration the z-transform properties of linearity and time shifting, the z-transform of the difference equation is $Y(z)=(1+0.8z^{-1})X(z)$. Therefore, the transfer function is given by $H(z) = \text{Output/Input} = Y(z)/X(z) = 1+0.8z^{-1}$. The absolute value square of the spectrum of the transfer function is

$$\begin{aligned}\left|H(e^{j\omega})\right|^2 &= (1+0.8e^{j\omega})(1+0.8e^{-j\omega}) = 1.64 + 0.8(e^{j\omega} + e^{-j\omega}) \\ &= 1.64 + 0.8(\cos\omega + j\sin\omega + \cos\omega - j\sin\omega) = 1.64 + 1.6\cos\omega\end{aligned}$$

Therefore, the PSD of the output is given by

$$S_y(e^{j\omega}) = S_x(e^{j\omega})\left|H(e^{j\omega})\right|^2 = S_x(e^{j\omega})(1.64 + 1.6\cos\omega)$$

Figure 4.7 shows the requested results. Remember that the spectrum is valid in the range $0 \le \omega \le \pi$. We remind the reader that if the sequence was coming from a continuous function sampled at times T, the spectrum range would have been $0 \le \omega\,(\text{rad/s}) \le \pi / T$. The following Book m-file produces Figure 4.7.

Book m-File: ex4_5_1

```
%Book m-file: ex4_5_1;
x = randn(1,128);%x is N(0,1);
rx = xcorr(x,'biased');
sx = fft(rx(1,128:255),128);
w = 0:2*pi/128:2*pi-(2*pi/128);
```

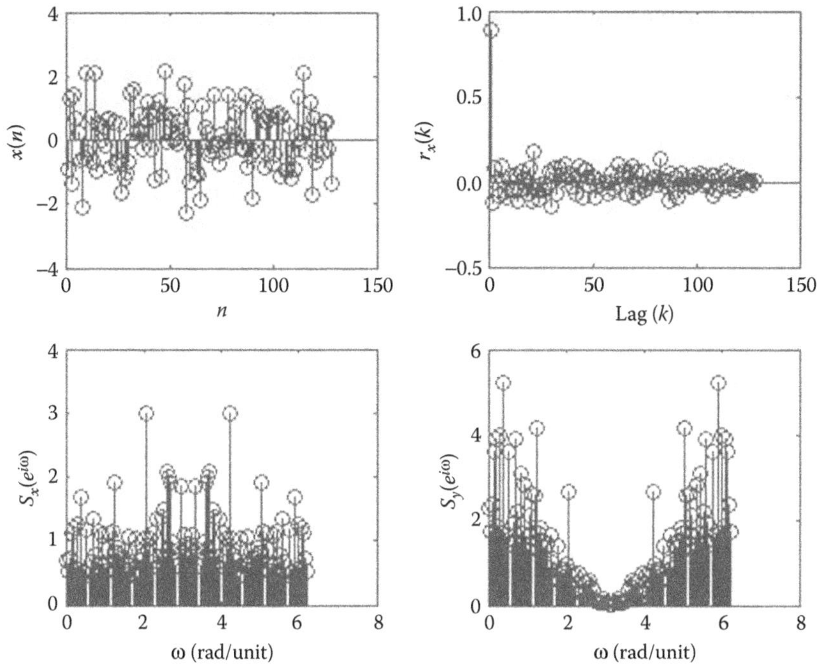

FIGURE 4.7

```
sy = abs(sx).*(1.64+1.6*cos(w));
subplot(2,2,1);stem(x,'k');xlabel('n');
ylabel('x(n)');
subplot(2,2,2);stem(rx(1,128:255),'k');xlabel('k, lag');
ylabel('r_x(k)');
subplot(2,2,3);stem(w,abs(sx),'k');
xlabel('\omega rad/unit');ylabel('S_x(e^{j\omega})');
subplot(2,2,4);stem(w,abs(sy),'k');
xlabel('\omega rad/unit');ylabel('S_y(e^{j\omega})');
```

The reader should observe that the filter is a low-pass one since it attenuates the high frequencies (frequencies close to π).

■

4.6 SPECIAL TYPES OF RANDOM PROCESSES

4.6.1 Autoregressive Process

The autoregressive (AR) filter with p poles is given by

$$H(z) = \frac{b(0)}{1 + \sum_{k=1}^{p} a(k) z^{-k}} = \frac{b(0)}{A(z)} \tag{4.40}$$

Using (4.39), the PSD of the output of this type of a filter, when the input $\{v(n)\}$ is a WN with zero mean and variance σ_v^2, is

$$S_y(e^{j\omega}) = \sigma_v^2\left[H(z)H(z^{-1})\right]_{z=e^{j\omega}} = \sigma_v^2 \frac{b(0)^2}{\left|A_x(e^{j\omega})\right|^2} \tag{4.41}$$

It can be shown that the following correlation relationship exists for the AR process with a white input noise having zero mean value zero and variance σ_v^2:

$$r_{yy}(k) + \sum_{m=1}^{p} a(m) r_{yy}(k-m) = \sigma_v^2 b(0)^2 \delta(k) \tag{4.42}$$

The above equation can be written in the form:

$$\begin{bmatrix} r_{yy}(0) & r_{yy}(-1) & \cdots & r_{yy}(-p) \\ r_{yy}(1) & r_{yy}(0) & \cdots & r_{yy}(-p+1) \\ \vdots & \vdots & & \vdots \\ r_{yy}(p) & r_{yy}(p-1) & \cdots & r_{yy}(0) \end{bmatrix} \begin{bmatrix} 1 \\ a(1) \\ \vdots \\ a(p) \end{bmatrix} = \sigma_v^2 b(0)^2 \begin{bmatrix} 1 \\ 0 \\ \vdots \\ 0 \end{bmatrix} \tag{4.43}$$

For real sequences, the autocorrelation is symmetric: $r_{yy}(k) = r_{yy}(-k)$.

Example 4.6.1

Find the AR coefficients $a(1)$ to $a(5)$ if the autocorrelation of the observed signal $\{y(n)\}$ is given. Let $\{v(n)\}$ be a WN with mean 0 and variance 1, which is the input to the filter $y(n) - 0.8y(n-1) + 0.4y(n-2) = v(n)$.

Solution: The results are shown in Figure 4.8 and the Book m-file is given below. Two of the figures were plotted in a continuous way to show the results better. First we find the a's using the system:

$$\begin{bmatrix} r_{yy}(1) & r_{yy}(0) & \cdots & r_{yy}(-p+1) \\ \vdots & \vdots & & \vdots \\ r_{yy}(p) & r_{yy}(p-1) & \cdots & r_{yy}(0) \end{bmatrix} \begin{bmatrix} 1 \\ a(1) \\ \vdots \\ a(p) \end{bmatrix} = \begin{bmatrix} 0 \\ \vdots \\ 0 \end{bmatrix} \tag{4.44}$$

$$\Rightarrow \begin{bmatrix} r_{yy}(0) & \cdots & r_{yy}(-p+1) \\ \vdots & & \vdots \\ r_{yy}(p-1) & \cdots & r_{yy}(0) \end{bmatrix} \begin{bmatrix} a(1) \\ \vdots \\ a(p) \end{bmatrix} = \begin{bmatrix} r_{yy}(1) \\ \vdots \\ r_{yy}(p) \end{bmatrix}$$

Then, we find $b(0)^2$ from the following equation [see (4.43) first row], assuming that the variance of the input to the system has variance 1:

$$[r_{yy}(0)\ \ r_{yy}(-1) \cdots r_{yy}(-p)][1\ \ a(1) \cdots a(p)]^T = b(0)^2 \tag{4.45}$$

Book m-File: ex4_6_1

```
%Book m-file: ex4_6_1;
x = (rand(1,1000)-0.5)*3.5;%uniformly
```

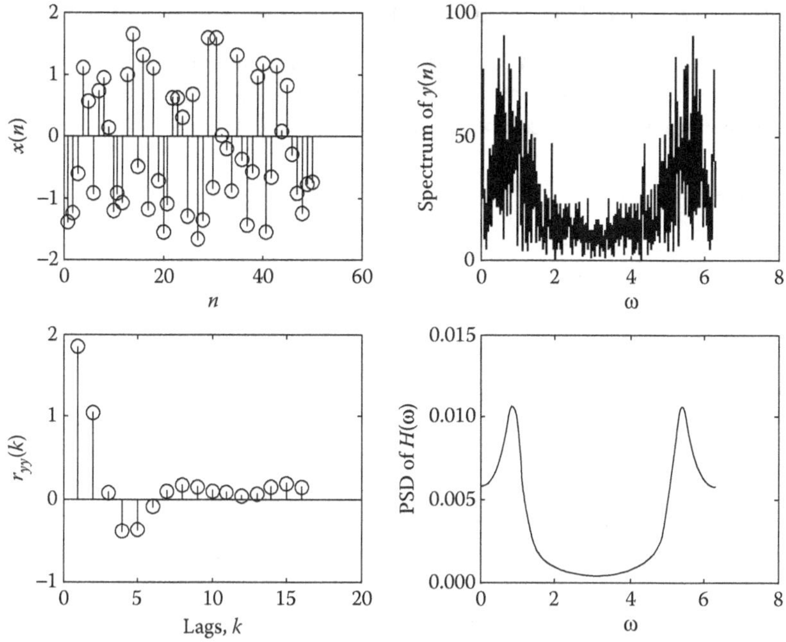

FIGURE 4.8

```
        %distributed WGN of zero mean
        %and unit variance;
y = filter(1,[1 -0.8 0.4],x);% y = observed signal;
[ry1,lags] = xcorr(y,y,15,'biased');
        %ry1 = autocorrelation of y;
ry = ry1(1,16:31);%one sided;
Ry = toeplitz(ry(1,1:6));%Ry = autocorelation
        %matrix of the output;
Rya = Ry(2:6,2:6);%creates a 5x5 matrix from Ry matrix;
a = inv(Rya)*(Ry(2:6,1)*(-1));%solves for a's from
        %the lower part of the
        %system which is given in
        %the example;
b02 = [Ry(1,1:6)]*[1 a']';
H = b02./(fft([1 a'],512));%spectrum of the
        %system function;
SH = H.*conj(H)/512;%PSD of the transfer function;
nn = 0:511;w = 2*pi*nn/512;
subplot(2,2,1);stem(x(1,1:50),'k');xlabel('n');
ylabel('x(n)');
subplot(2,2,2);plot(w,abs(fft(y(1,1:512),512)),'k');
xlabel('\omega');ylabel('Spectrum of y(n)')
subplot(2,2,3);stem(ry,'k');xlabel('Lags');
ylabel('r_{yy}(k)');
subplot(2,2,4);plot(w,SH,'k');xlabel('\omega');
ylabel('PSD of H(\omega)');
```

■

4.7 NONPARAMETRIC SPECTRA ESTIMATION

The spectra estimation problem in practice is based on a finite-length record $\{x(1),\ldots, x(N)\}$ of a second-order stationary random process. However, harmonic processes have line spectra and appear in applications either alone or combined with noise.

4.7.1 PERIODOGRAM

The **periodogram spectral estimator** of the data segment $\{x(0), x(1),\ldots, x(N-1)\}$, finite number of data, is based on the following formula:

$$\hat{S}_x(e^{j\omega}) = \frac{1}{N}\left|\sum_{n=0}^{N-1} x_w(n)e^{-j\omega n}\right|^2 = \frac{1}{N}\left|X(e^{-j\omega})\right|^2 \tag{4.46}$$

where:

$X(e^{-j\omega})$ is the DTFT of the windowed sequence

$$x_w(n) = x(n)w(n) \qquad 0 \le n \le N-1 \tag{4.47}$$

In (4.46), the signal can be multiplied by a unit value square window $\{1, 1, 1,\ldots,1\}$. The PSD of the signal, $\hat{S}_x(e^{j\omega})$, is periodic with period $2\pi, -\pi \le \omega \le \pi$. The spectrum $X(e^{j\omega})$ is the DTFT of the windowed signal $\{x_w(n)\}$. The periodicity is easily shown by introducing $\omega + 2\pi$ in place of ω in (4.46) and remembering that $\exp(j2\pi) = 1$. The values of the periodogram at discrete points in the frequency domain are located at $\{\omega_k = 2\pi k/N\}_0^{N-1}$ and are found by

$$\hat{S}_x(k) \triangleq \hat{S}_x(e^{j2\pi k/N}) = \frac{1}{N}\left|X(\omega_k)\right|^2 \quad k = 0, 1,\ldots, N-1 \qquad \omega_k = \frac{2\pi}{N}k \tag{4.48}$$

where:

$X(k)$ is the N-point DFT of the windowed signal $x_w(n)$

It is recommended that we multiply the random signal with an appropriate window for a smoother PSD. The following example and the Book m-file explain the use of the windows. Some of the MATLAB window functions are given below. See also Appendix 1 at the end of the book for more types of windows.

MATLAB functions for several types of windows are as follows:

@blackmanharris—Minimum 4-term Blackman–Harris window
@gausswin—Gaussian window
@hamming—Hamming window
@rectwin—Rectangular window

An example of MATLAB program using window functions is given as follows:

```
N = 64;
w = window(@blackmanharris,N);
w1 = window(@hamming,N);
w2 = window(@gausswin,N,2.5);
plot(1:N,[w,w1,w2]); axis([1 N 0 1]);
legend('Blackman-Harris','Hamming','Gaussian');
```

Example 4.7.1

Plot the PSD of the signal $y(n)$ that is produced by the filter $y(n) - 0.9y(n-1) + 0.2y(n-2) = v(n)$, where $v(n) = 0.8 * \text{randn}(n)$. Use a rectangular and a Hamming window with $N = 128$.

Solution: The results are shown in Figure 4.9 using the following Book m-file.

Book m-File: ex4_7_1

```
%Book m-file: ex4_7_1;
y(1) = 0;y(2) = 0;
for n = 3:1024
     y(n) = 0.9*y(n-1)-0.2*y(n-2)+0.8*randn;
end;
N = 128;
xw = y(1,50:50+N-1).*window(@rectwin,N)';%the window
     %is given as a column vector, here is
     %a rectangular window;
psd = (abs(fft(xw,N))).^2/N;
xw1 = y(1,1:N).*window(@hamming,N)';
psd1 = (abs(fft(xw1,N))).^2/N;
sx128 = psd(1,1:N);
sx1128 = psd1(1,1:N);
om = 0:2*pi/N:2*pi-(2*pi/N);
subplot(2,1,1);stem(om,sx128,'k');
xlabel('k');ylabel('PSD');
subplot(2,1,2);stem(om,sx1128,'k');
xlabel('k');ylabel('PSD');
```

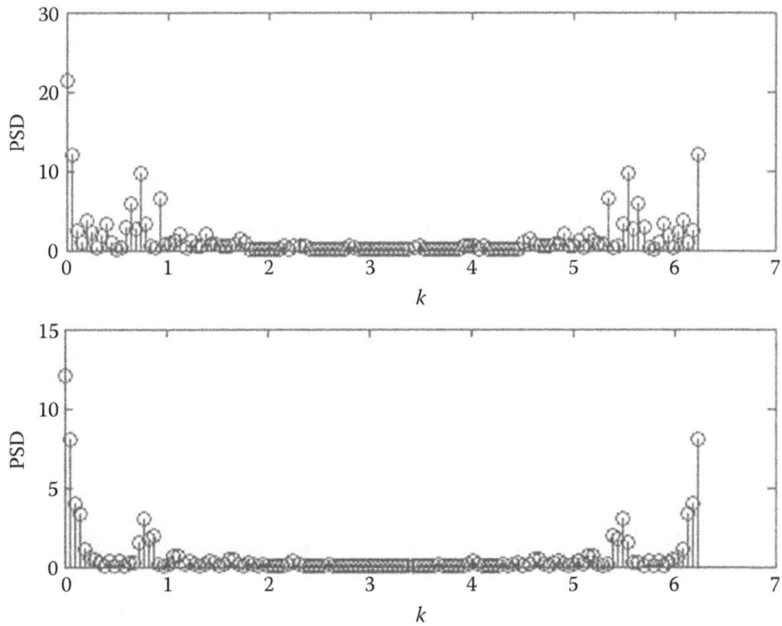

FIGURE 4.9

The reader will observe that the Hamming window (and any other one) produces a smoother PSD than the rectangular one. We must point out that the results showing in the figure is only for one realization. This means that any time we run the program we will observe different PSDs. We can also proceed to create a number of these realizations and then average them.

■

4.7.2 Correlogram

The **correlogram spectral estimator** is given by the formula:

$$\hat{S}_r(e^{j\omega}) = \sum_{m=-(N-1)}^{N-1} \hat{r}_{xx}(m)e^{-j\omega m} \tag{4.49}$$

where:

$\hat{r}_{xx}(m)$ is the estimate of the biased autocorrelation (assumed mean value of $\{x(n)\}$ to be 0) given by (4.11)

It can be shown that the correlogram spectral estimator, evaluated using the standard biased autocorrelation estimates, coincides with that of the periodogram spectral estimator. As in (4.46), the correlogram is a periodic function of ω with period 2π.

4.7.3 Computation of Periodogram and Correlogram Using FFT

Since both functions are continuous with respect to frequency (DTFT), we can sample the frequency as follows:

$$\omega_k = \frac{2\pi}{N}k \qquad k = 0, 1, 2, \ldots, N-1 \tag{4.50}$$

Introducing the above equation in (4.46) and (4.49), we obtain

$$\begin{aligned} &\hat{S}_x(e^{j\omega_k}) = \frac{1}{N}\left|X(e^{j\omega_k})\right|^2; \quad \hat{S}_r(e^{j\omega_k}) = \sum_{m=-(N-1)}^{N-1} \hat{r}_{xx}(m)e^{-j\frac{2\pi}{N}km} \qquad 0 \le k \le N-1 \\ &X(e^{j\omega_k}) = \sum_{n=0}^{N-1} x(n)e^{-j\frac{2\pi}{N}nk} = \sum_{n=0}^{N-1} x(n)W^{nk} \qquad W = e^{-j2\pi/N} \qquad 0 \le k \le N-1 \end{aligned} \tag{4.51}$$

The most efficient way to find the DFT using fast Fourier transform (FFT) is to set $N = 2^r$ for some integer r. The following two Book m-functions give the windowed periodogram and correlogram.

Book m-Function for Windowed Periodogram: [s,as,phs] = lmsperiodogram (x,w,L)

```
        %Book function: [s,as,phs] = lmsperiodogram(x,w,L);
function[s,as,phs] = lmsperiodogram(x,w,L)
        %w = window(@name,length(x)),w is in column form
        %(see also Appendix 1
        %name = hamming,kaiser,hann,rectwin,
```

```
        %bartlett,tukeywin,blackman,gausswin,nattallwin,
        %triang,blackmanharris);
        %L = desired number of points (bins) of the spectrum;
        %x = data in row form;s = complex form of its DFT;
xw = x.*w';
for m = 1:L
        n = 1:length(x);
        s(m) = sum(xw.*exp(-j*(m-1)*(2*pi/L)*n));
end;%the for loom finds the set of complex numbers
        %of the summation for each m;
as = ((abs(s)).^2/length(x))/norm(w);
        %as = amplitude value of the periodogram;
phs = (atan(imag(s)./real(s))/length(x))/norm(w);
        %phase of the periodogram;
wb = 0:2*pi/L:2*pi-(2*pi/L);stem(wb,as,'k');
xlabel('Frequency bins, 2\pi/L');ylabel('Magnitude');
```

Book m-Function for Windowed Correlogram: [s,as,ps] = lms_correlogram (x,w,L)

```
        %Book m-function: [s,as,ps] =
        %lms_correlogram(x,w,lg,L);
function[s,as,ps] = lms_correlogram(x,w,lg,L)
        %x = data with mean zero;w = window(@name,length(2*lg)),
        %L = desired number of spectral points;
        %lg = lag number<<N;rc = symmetric autocorrelation
        %function;
r = lms_sample_biased_autoc(x,lg);
rc = [fliplr(r(1,2:lg)) r 0];
rcw = rc.*w';
for m = 1:L
        n = -lg+1:lg;
        s(m) = sum(rcw.*exp(-j*(m-1)*(2*pi/L)*n));
end;
as = (abs(s).^2)/norm(w)/L;%amplitude spectrum;
ps = (atan(imag(s))/real(s))/norm(w);%phase spectrum;
stem(0:2*pi/L:2*pi-(2*pi/L),as,'k');xlabel('\omega(k)');
ylabel('Correlogram');
```

To plot, for example, `as` or `ps`, we can use the command: `plot (0:2*pi/L:2*pi-(2*pi/L),as)`.

Note: In case the reader does not want to apply window, we write `w = window(@rectwin,length(2*lg)')` or `w = rectwin(2*lg)'`.

4.7.4 General Remarks on the Periodogram

The general remarks on the periodogram are as follows:

1. The variance of the periodogram does not tend to 0 as $N \to \infty$. This indicates that the periodogram is an **inconsistent** estimator; that is, its distribution does not tend to cluster more closely around the true spectrum as N increases.

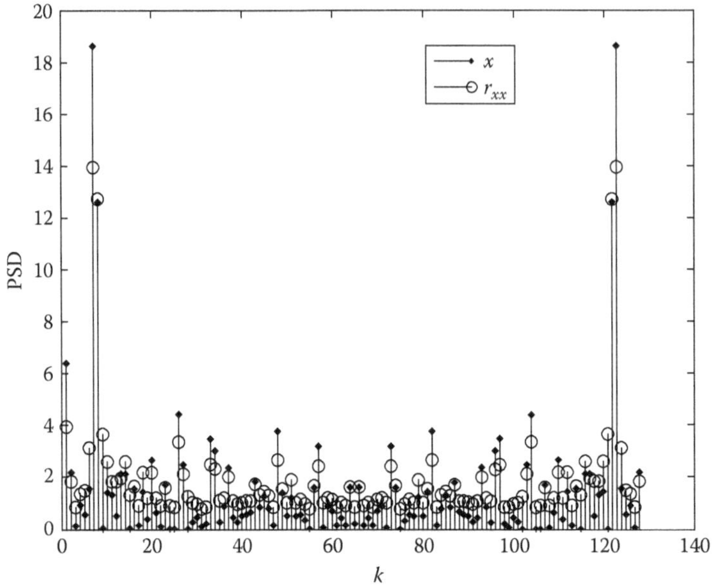

FIGURE 4.10

2. To reduce the variance and, thus, produce smoother spectral estimator, we must obtain
   a. The average contiguous values of the periodogram.
   b. The average periodogram obtained from multiple data segments.
3. The effect of the side lobes of the windows on the estimated spectrum consists of transferring power from strong bands to less strong bands or bands with no power. This process is known as the **leakage** problem.

Figure 4.10 shows a plot of both the periodogram and the correlogram.

Example 4.7.2

It is instructive to obtain the average of a number of realizations of a periodogram.

Solution: The signal is the sum of two sinusoids and a WN of unit variance. Using the Book m-file given below, the results are shown in Figure 4.11. The top plot shows 50 realizations of the signal, the second shows 50 realizations of the periodograms, the third shows the average PSD of all 50 PSDs, and the last one shows just the PSD of one of the realizations.

Book m-File: ex4_7_2

```
%Book m-file: ex4_7_2;
for m = 1:60
     n = 0:127;
     x(m,:) = sin(0.35*pi*n)+sin(0.37*pi*n)...
     +3.5*(rand(1,128)-0.5);
     subplot(4,1,1);plot(x(m,1:64),'k');xlabel('n');
```

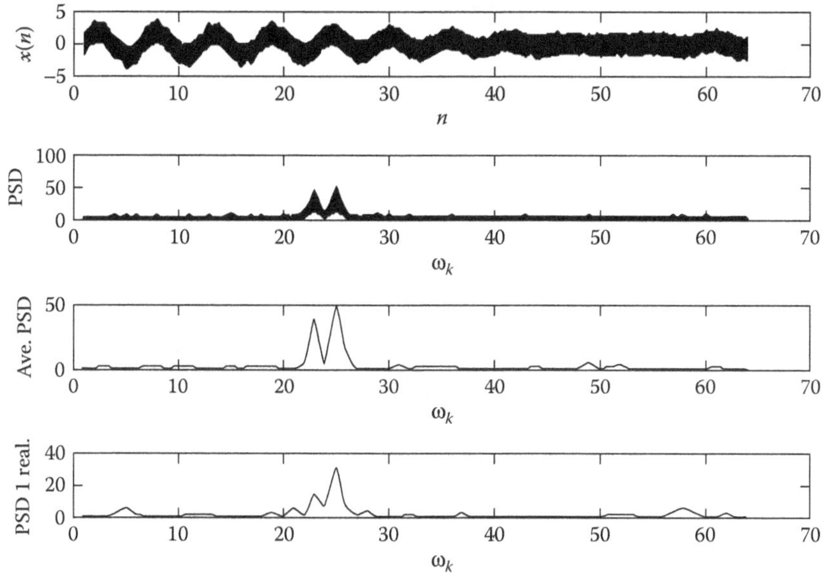

FIGURE 4.11

```
        ylabel('x(n)');
        hold on;
        x(m+1,:) = sin(0.35*pi*n)+sin(0.37*pi*n)...
        +3.5*(rand(1,128)-0.5);
        ax(m,:) = abs(fft(x(m,:),128)).^2/128;
        ax(m+1,:) = abs(fft(x(m+1,:),128)).^2/128;
        ax(m+2,:) = (ax(m,:)+ax(m+1,:))/2;
        subplot(4,1,2);plot(ax(m,1:64),'k');xlabel('\omega_k');
        ylabel('PSD');
        hold on;
end;
subplot(4,1,3);plot(ax(m,1:64),'k');xlabel('\omega_k');
ylabel('Ave. PSD');
subplot(4,1,4);plot(ax(13,1:64),'k');xlabel('\omega_k');
ylabel('PSD 1 real.');
```

■

4.7.4.1 Windowed Periodogram

We can define a window periodogram as the DTFT of the time function $\{x(n)w(n)\}$ multiplied by a window, where $w(n)$ is any desired window function. Hence, the periodogram with temporal window is given by

$$\hat{S}_x(e^{j\omega}) = \frac{1}{NP_w}\left|\sum_{n=0}^{N-1} x(n)w(n)e^{-j\omega n}\right|^2 \qquad P_w = \frac{1}{N}\sum_{n=0}^{N-1}|w(n)|^2 \tag{4.52}$$

where:

P_w is the average power of the window

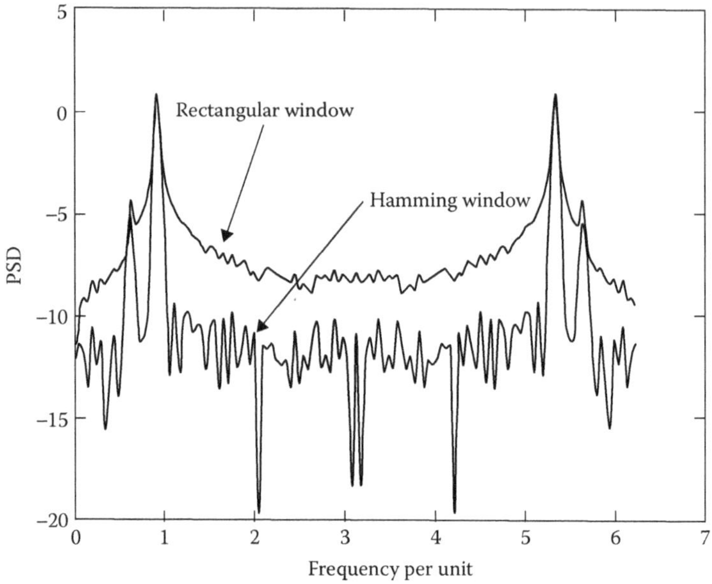

FIGURE 4.12

Example 4.7.3

Apply the rectangular and Hamming windows to the signal given below and compare the resulting spectra:

$$x(n) = \{0.06\sin(0.2\pi n) + \sin(0.3\pi n) + 0.05[\text{rand}(1,128) - 0.5]\}w(n) \qquad 0 \le n \le 127$$

Solution: The window provides a trade-off between spectra resolution (main lobe) and spectral masking (side lobe amplitude). In this case, the Hamming window has a side lobe at about −45 dB compared with the rectangular window having a side lobe at about −13 dB. Figure 4.12 shows that although the side lobe of the rectangular window just about obscures the 0.2π frequency, the Hamming window clearly resolves it.

■

It can be shown that that the amount of smoothing in the periodogram is determined by the type of the window that is applied to the data. Selecting the type of window, we must weigh our intention between resolution and smoothing.

The following Book m-function calculates the temporal windowed data with the ability to introduce any one of the several windows that are given in the function without the obligation to introduce it in the Command window. This helps if one wants to repeat the simulation using different windows.

Book m-Function That Produces Windowed Periodograms

```
      %Book m-function: [s,as,ps] = lmsperiodogram(x,win,L);
function[s,as,ps] = lmsperiodogramwin(x,win,L)
      %window names = hamming,kaiser,hann,rectwin,
      %bartlett,tukeywin,blackman,gausswin,nattallwin,
      %triang,blackmanharris;
```

```
        %L = desired number of points (bins) of the spectrum;
        %x = data in row form;s = complex form of the DFT;
if (win = =2) w = rectwin(length(x));
elseif (win = =3) w = hamming(length(x));
elseif (win = =4) w = bartlett(length(x));
elseif (win = =5) w = tukeywin(length(x));
elseif (win = =6) w = blackman(length(x));
elseif (win = =7) w = triang(length(x));
elseif (win = =8) w = blackmanharris(length(x));
end;
xw = x.*w';
for m = 1:L
        n = 1:length(x);
        s(m) = sum(xw.*exp(-j*(m-1)*(2*pi/L)*n));
end;
as = ((abs(s)).^2/length(x))/norm(w);%as = amplitude spectral
        density;
ps = (atan(imag(s)./real(s))/length(x))/norm(w);%ps = phase
        spectrum;

        %To plot as or ps we can use the command:
        %plot(0:2*pi/L:2*pi-(2*pi/L),as);
```

4.7.5 Proposed Book Modified Method for Better Frequency Resolution

4.7.5.1 Using Transformation of the rv's

One of the difficulties is that if the sequence is short (N small number), we may not be able to sufficiently resolve frequencies being close together. Since we have at hand one realization and we cannot extract another one from the population in the probability space, we propose to create a **pseudosequence** from the data at hand by linear transformation of the rv's. Next, we overlap (about 25%) these two series. These series can also be multiplied by a window and then processed.

Note: The proposed Book Modified method is based on the linear transformation of rv's.

Example 4.7.4

Compare the spectrum based on the periodogram and the proposed modified one.

Solution: Figure 4.13 shows the resolution capabilities of the sequence that is made up from the original sequence as follows: $y = 0.2*[x\text{zeros}(1,48)] + 0.2*[\text{zeros}(1,48)x]$. The original signal, a 64-term series, was $x = \sin(0.3*\text{pi}*n) + \sin(0.32*\text{pi}*n) + 0.2*\text{randn}(1,64)$. The following Book m-function produces the figure.

■

Book m-Function Linearly Modified Signal: [ax,ay,w,y] = lmslinear_modified_periodogram(x)

```
function[ax,ay,w,y] = lmslinear_modified_periodogram(x)
        %Book MATLAB function:
```

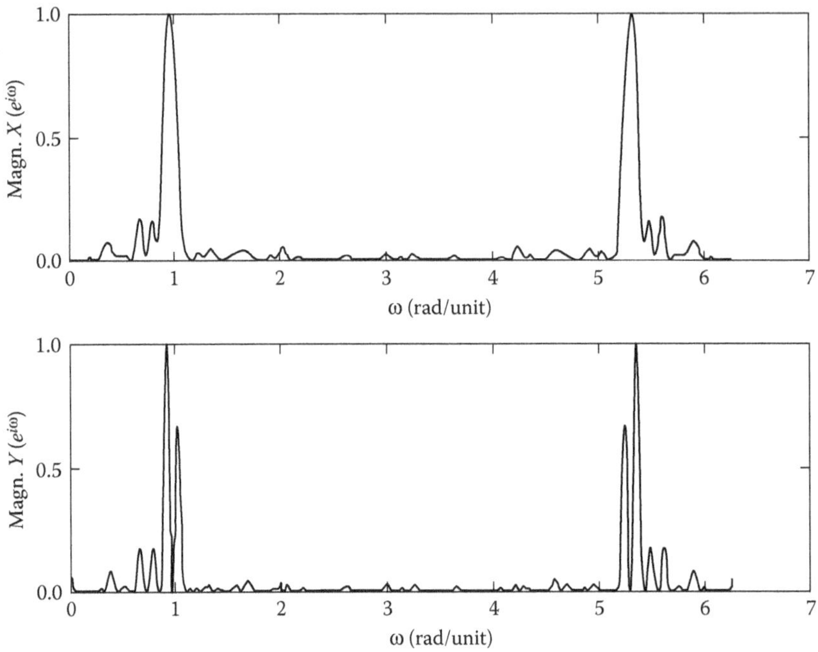

FIGURE 4.13

```
        %[ax,ay,w]: lmslinear_modified_periodogram(x);
[sx,ax,px] = lmsperiodogramwin(x,2,512);
y1 = [x zeros(1,48)]+0.1*rand(1,112);
y2 = [zeros(1,48) x]+0.1*rand(1,112);
y = 0.2*y1+y2*0.2;
[sy,ay,py] = lmsperiodogramwin(y,2,512);
w = 0:2*pi/512:2*pi-(2*pi/512);
        %2 implies rectangular window, see Book
        %MATLAB function lmsperiodogramwin
        %for other windows;to plot we must
        %write in the command window:
        %plot(w,20*log10(ax/max(ax)),'k') and
        %similar for the ay;
```

■

4.7.5.2 Blackman–Tukey Method

Because the correlation function at its extreme lag values is not reliable due to the small overlapping of the correlation process, it is recommended to use lag values of about 30%–40% of the total length of the data. The Blackman–Tukey (BT) estimator is a windowed correlogram and is given by

$$\hat{S}_{\mathrm{BT}}(e^{j\omega}) = \sum_{m=-(L-1)}^{L-1} \hat{r}_{xx}(m)w(m)e^{-j\omega m} \tag{4.53}$$

where:

$w(m)$ is the window with 0 value for $|m| > L-1$ and $L << N$

The above equation can also be written in the form:

$$\begin{aligned}\hat{S}_{\text{BT}}(e^{j\omega}) &= \sum_{m=-\infty}^{\infty} \hat{r}_{xx}(m)w(m)e^{-j\omega m} \\ &= \hat{S}_r(e^{j\omega}) * W(e^{j\omega}) = \frac{1}{2\pi}\int_{-\pi}^{\pi} \hat{S}_r(e^{j\tau})W(e^{j(\omega-\tau)})d\tau\end{aligned} \tag{4.54}$$

where we applied the DTFT frequency convolution property (the DTFT of the multiplication of two functions is equal to the convolution of their Fourier transforms). Since windows have a dominant and relatively strong main lob, the BT estimator corresponds to a "locally" weighting average of the periodogram. Although the convolution smoothes the periodogram, it reduces the resolution at the same time. It is expected that the smaller the L, the larger the reduction in variance and the lower the resolution. It turns out that the resolution of this spectral estimator is on the order of $1/L$, whereas its variance is on the order of L/N.

For convenience, we give again some of the most common windows below. For the Kaiser window, the parameter β trades the main lobe width for the side-lobe leakage; $\beta = 0$ corresponds to a rectangular window, and $\beta > 0$ produces lower side lobe at the expense of a broader main lobe.

1. Rectangular window

$$w(n) = 1 \quad n = 0, 1, 2, \cdots, L-1$$

2. Bartlett (triangle) window

$$w(n) = \begin{cases} \dfrac{n}{L/2} & n = 0, 1, \cdots, L/2 \\ \dfrac{L-n}{L/2} & n = \dfrac{L}{2}+1, \cdots, L-1 \end{cases}$$

3. Hann window

$$w(n) = 0.5\left[1-\cos\left(\frac{2n}{L}\pi\right)\right] \quad n = 0, 1, 2, \cdots, L-1$$

4. Hamming window

$$w(n) = 0.54 - 0.46\cos\left(\frac{2\pi}{L}n\right) \quad n = 0, 1, 2, \cdots, L-1$$

5. Blackman window

$$\begin{aligned} w(n) = 0.42 + 0.5\cos\left[\frac{2\pi}{L}\left(n-\frac{L}{2}\right)\right] \\ +0.08\cos\left[\frac{2\pi}{L}2\left(n-\frac{L}{2}\right)\right] \quad n = 1, 2, \cdots, L-1 \end{aligned}$$

6. Kaiser window

$$w(n)=\frac{I_0\left[\beta\sqrt{1.0-\left(\frac{n}{L/2}\right)^2}\right]}{I_0(\beta)}\qquad -(L-1)\le n\le L-1$$

$$I_0(x)=\sum_{k=0}^{\infty}\left[\frac{\left(\frac{x}{2}\right)^k}{k!}\right]^2 = \text{Zero-order modified Bessel function}$$

The windows $w(k)=0$ for $|k|\ge L$ and $w(k)=w(-k)$, and the equations are valid for $0\le k\le L-1$.

The following Book m-function produces the spectrum under the BT method. Figure 4.14 shows the effect of averaging of different lengths of the autocorrelation function that was proposed in this chapter. The signal was produced as follows: $n=0:127$, $x=\sin(0.3\pi n)+\sin(0.33\pi n)+\text{randn}(1,128)$.

Book m-Function: [rx,asn,ps] = lmsBTperiodogram(x,win,L,lg)

```
function[rx,asn,ps] = lmsBTperiodogram(x,win,L,lg)
        %[rx,asn,ps] = lmsBTperiodogram(x,win,L,lg);
        %window names = hamming,kaiser,hann,rectwin,
```

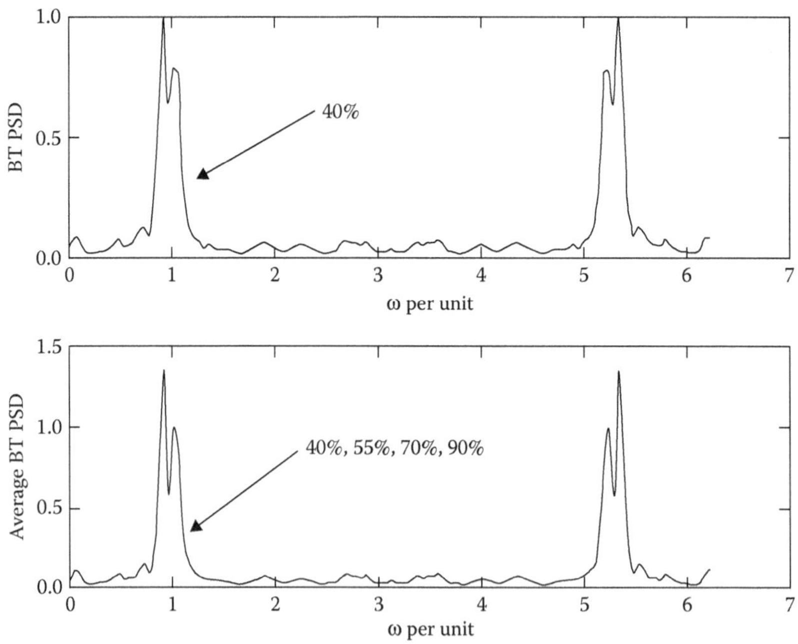

FIGURE 4.14

```
        %bartlett,tukeywin,blackman,gausswin,nattallwin,triang,
        %blackmanharris;lg = lag number;
        %L = desired number of points (bins) of the spectrum;
        %x = data in row form;s = complex DFT data;
rx = lms_sample_biased_autoc(x,lg);
if (win = =2) w = rectwin(2*length(rx));
elseif (win = =3) w = hamming(2*length(rx));
elseif (win = =4) w = bartlett(2*length(rx));
elseif (win = =5) w = tukeywin(2*length(rx));
elseif (win = =6) w = blackman(2*length(rx));
elseif (win = =7) w = triang(2*length(rx));
elseif (win = =8) w = blackmanharris(2*length(rx));
end;
wh = w(length(rx):2*length(rx)-1,1)';
rxw = (rx.*wh);
for m = 1:L
        n = 1:length(rx);
        s(m) = sum(rxw.*exp(-j*(m-1)*(2*pi/L)*n));
end;
as = (abs(s)).^2/norm(w);%as = amplitude spectral density;
asn = as/max(abs(as));
ps = atan((imag(s)./(real(s)+eps))/((length(x))/...
(norm(w)+eps)));
        %ps = phase spectrum;
        %To plot as or ps we can use the command:
        %plot(0:2*pi/L:2*pi-(2*pi/L),as);
```

Another Book proposed method is based on linearly extending the short windowed signal. The following Book m-file is used to produce Figure 4.15.

Book m-File: fig4_15

```
%Book m-file: fig4_15;
n = 0:63;
x = sin(0.3*pi*n)+sin(0.32*pi*n)+0.5*randn(1,64);
y = [0.4*x 0.2*x+0.05].*hamming(128)';
[wh,rx,rxw,asnx,psx] = lmsBTperiodogram(x,2,512,64);
[why,ry,ryw,asny,psy] = lmsBTperiodogram(y,2,512,64);
om = 0:2*pi/512:2*pi-(2*pi/512);
subplot(2,1,1);plot(om,asnx,'k');xlabel('\omega per unit');
ylabel('BT PSD');
subplot(2,1,2);plot(om,asny,'k');xlabel('\omega per unit');
ylabel('BT Modif. PSD');
```

Further modification can be achieved by the proposed BT method using the combination of the rv's transformation and averaging. The Book m-function is given as follows:

Book m-Function for Modified BT method Using rv's Transformation and Averaging: [rz,az,pz] = lmsav_modifiedBTpsd(x,win,avn,L,lg)

```
function[sz,az,pz] = lmsav_modifiedBTpsd(x,win,avn,L,lg)
        %win = 2 implies rectangular, see the function
```

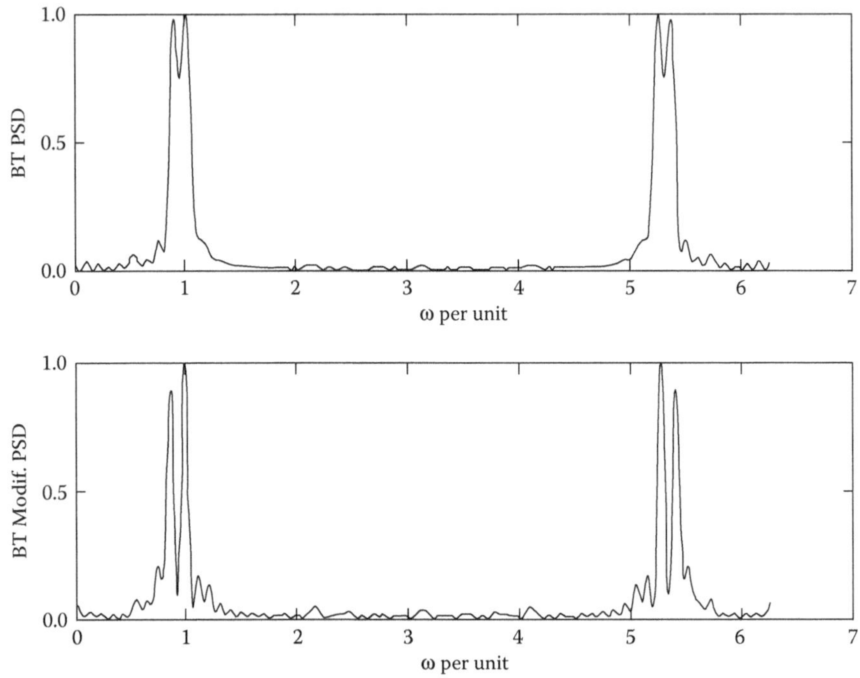

FIGURE 4.15

```
        %lmsBTperiodogram
        %to identify the appropriate numbers for other windows;
        %avn = number of ensemble sequences to be averaged;L = number
        %of frequence bins;lg = lag number for the autocorrelation;
for m = 1:avn
        z1(m,:) = [x*rand*0.5 x*rand*0.2+...
        0.05*(rand-0.5)].*hamming(2*length(x))';
end;
z = sum(z1,1);
[rz,az,pz] = lmsBTperiodogram(z,win,L,lg);
```

The top of Figure 4.16 is the BT periodogram with Hamming window. The data were as follows: $x = \sin(0.3\pi n) + \sin(0.33\pi n) + 0.8\text{randn}(1,128)$, $\text{lg} = 64$, $L = 256$, Hamming window, and number of averaging 40. The bottom figure is the proposed modified average BT periodogram. The reader will observe that by increasing the resolution the variance also increases.

4.7.6 Bartlett Periodogram

The variance of the periodogram presupposes that an ensemble averaging for variance decreases. However, for a single realization, the variance does not decrease as $N \to \infty$ (**inconsistent estimator**), and hence, there is a need for other approaches to

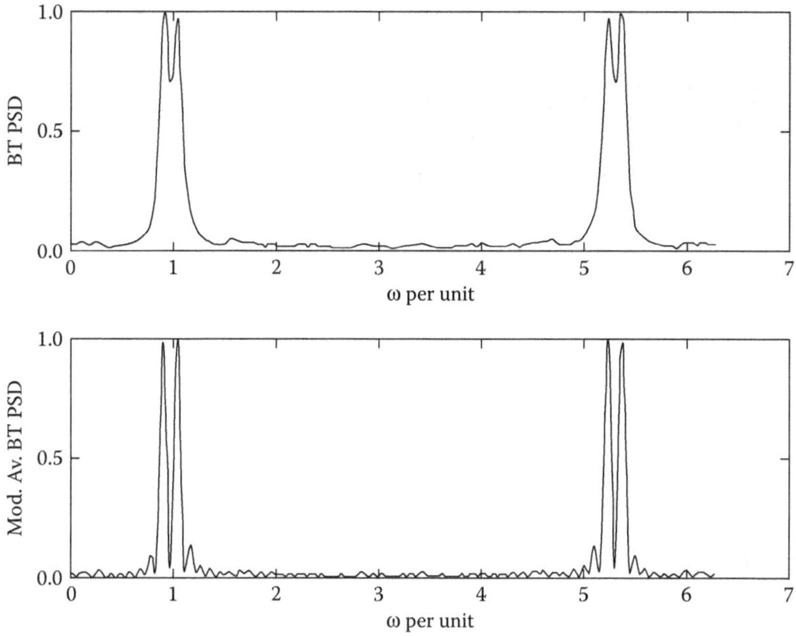

FIGURE 4.16

reduce the variance. We can improve the statistical properties of the periodogram by replacing the expectation operator with averaging a set of periodograms. Figure 4.17 shows the ensemble averaging effect on the variance using 10 realizations of the PSD only. The variance of one realization is about 55 and the variance of the ensemble is about 2.5. The number of realizations in this case was 30. The following Book m-file produces Figure 4.17.

Book m-File: fig4_17

```
%Book m-file: fig4_17;
N = 256;
for m = 1:30
      x(m,1:N) = randn(1,N);
end;
w = 0:2*pi/N:2*pi-(2*pi/N);
subplot(2,1,1);plot(w,20*log10(abs(fft...
(x(10,:),N))),'k');
xlabel('\omega per unit');ylabel('PSD');
subplot(2,1,2);plot(w,20*log10(abs(fft...
(sum(x,1),N)))/30,'k');
      %x is a matrix and y = sum(x,1) produces
      %a vector y with the summation of all the
      %rows of x, if we had used sum(x,2) we could
      %create a column vector with the summation
      %of all the columns of x;
xlabel('\omega per unit');ylabel('Ave. PSD');
```

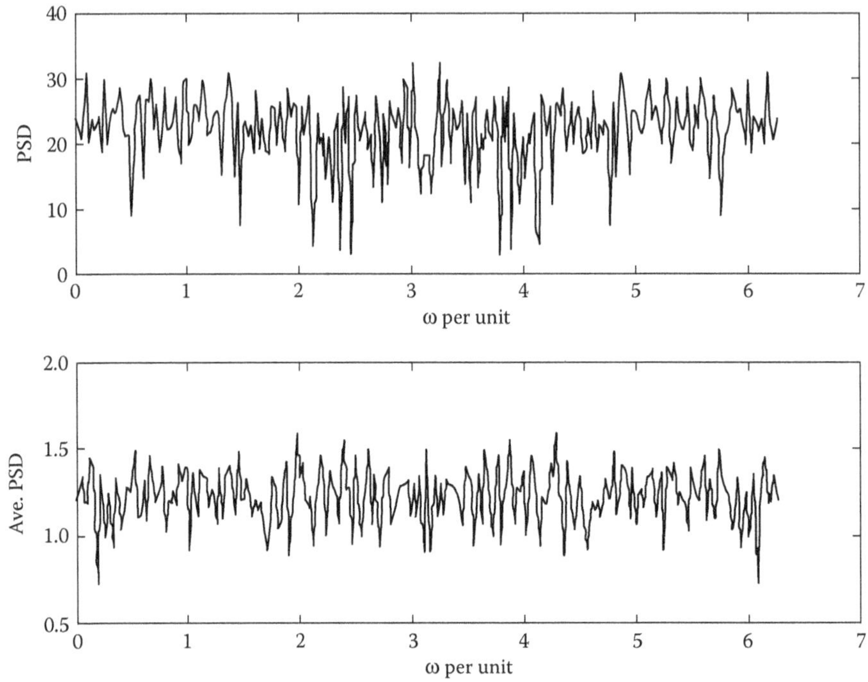

FIGURE 4.17

Bartlett proposed to split the data $\{x(n)\}$ into K nonoverlapping segments, which is assumed to be statistically independent. This condition presupposes that the autocorrelation of each segment decays much faster than the length of the segments. The N samples of the data signal are divided into K nonoverlapping segments of M samples each such that $KM \le N$. The ith segment will then consist of the samples:

$$x^{(i)}(n) = x(iM + n) \quad n = 0, 1, \ldots, M - 1$$
$$i = \text{segment number} = 0, 1, \ldots, K - 1 \tag{4.55}$$

Each of the segments has the spectrum

$$\hat{S}_x^{(i)}(e^{j\omega}) = \frac{1}{M}\left|\sum_{n=0}^{M-1} x(iM + n)e^{-j\omega n}\right|^2 \tag{4.56}$$

and hence, the Bartlett average periodogram is given by

$$\hat{S}_{xB} = \frac{1}{K}\sum_{i=0}^{K-1} \hat{S}_x^{(i)}(e^{j\omega}) \tag{4.57}$$

The diagrammatic presentation of the above Bartlett equation is shown in Figure 4.18. Since the power spectrum is due to the reduced number of terms, $M << N$, compared to the total available data, the resolution reduces from the order of $1/N$ to the order

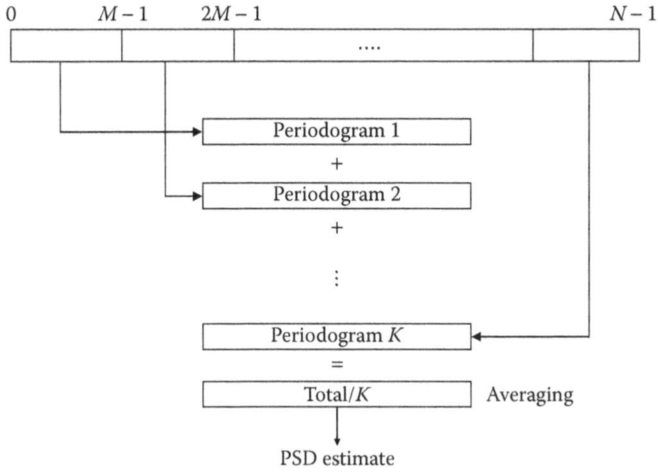

FIGURE 4.18

of $1/M$. This implies that the resolution is reduced by $N/M = K$ compared to the resolution of the original data set. It can be shown that the variance is reduced by the same amount.

The following Book m-function produces the Bartlett periodogram with any desired window:

Book m-Function: [s,as,ps] = lmsbartlettpsd(x,k,win,L)

```
function[s,as,ps] = lmsbartlettpsd(x,k,win,L);
       %x = data;k = number of sections;
       %L = number of points desired in the FT domain;
       %M = number of points in each section; kM< = N = length(x);
       %the number win identifies the type of
       %window;
M = floor(length(x)/k);
s = 0;
ns = 1;
for m = 1:k
       s = s+lmsperiodogramwin(x(1,ns:ns+M-1),win,L);
       ns = ns+M;
end;
as = ((abs(s)/k).^2)/length(x);
ps = (atan(imag(s/k)./real(s/k)))/length(x);
```

Figure 4.19 shows the Bartlett PSD for two windows: the rectangular and the Hamming. We observe that the PSD using the Hamming window is smoother (smaller variance) than that using the rectangular window. The data used were as follows: $n = 0:127$, win = 2 and 3, $L = 256$ (number of bins in the frequency domain), $k = 2$ (the length 128 of the signal was split into two equal sections with 64 points each) and the signal was $x = \sin(0.3\pi n) + \sin(0.315\pi n) + 0.5\text{randn}(1,128)$.

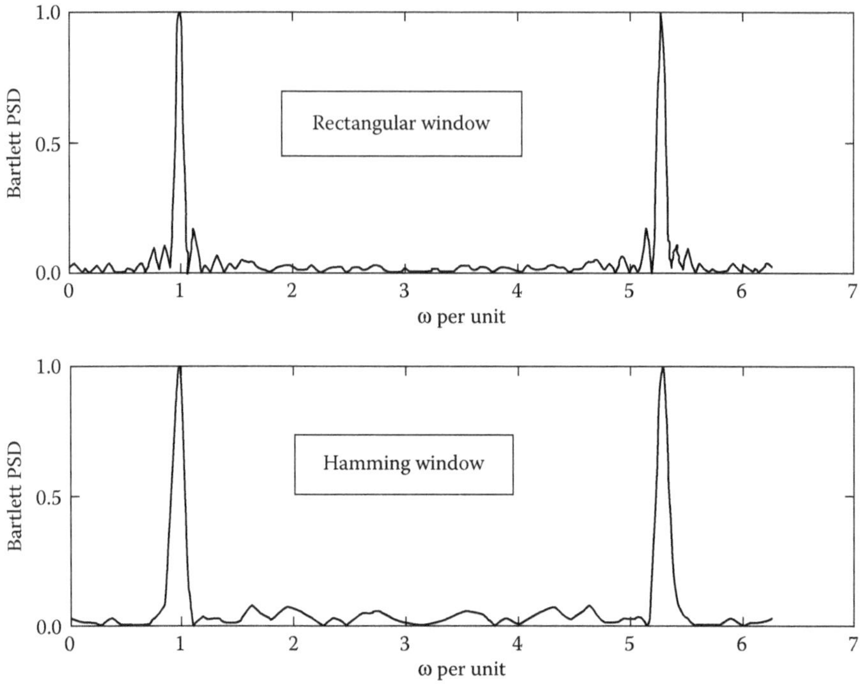

FIGURE 4.19

We propose three different Book m-functions that increase the resolution of the desired Bartlett-type spectrum as follows:

Book m-Function for Modified Bartlett Spectrum-Modified No. 1

```
function[s,as,ps,apsd] = lms_aver_mod_bartlettpsd1(x,k,win,L,R)
      %x = data; k = number of sections; L = number of points
      %desired in the FT domain; M = number of points in each
      %section;kM< = N = length(x);R = number of realizations;
      %win = 2 for rectwin, 3 for hamming etc.,
      %see sspperiodogramwin function;
      %s, as and ps are RxL amtrices;apsd is
      %the average spectrum of the R realizations;
      %if desired we can plot the amplitude
      %spectrum of one realization:
      %plot(as(5,1:512));
for r = 1:R
      xr(r,:) = ([x zeros(1,floor(0.8*length(x)))]+...
                [zeros(1,floor(0.8*length(x))) x...
                +0.1*rand(1,length(x))])*.2;
                [s(r,:),as(r,:),ps(r,:)]=lmsbartlettpsd
                (xr,k,win,L);
end;
apsd = sum(as,1)/R;%as is matrix and 1 means add the rows;
```

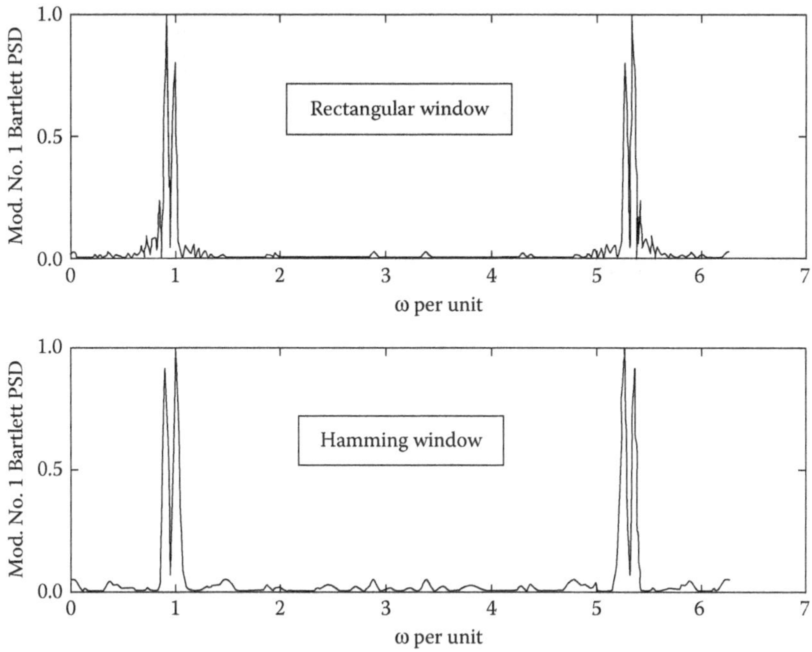

FIGURE 4.20

Figure 4.20 shows the result of the proposed Book m-file No. 1 given above. It shows that the proposed method has higher resolution capabilities than the Bartlett proposed approach.

Book m-Function for Modified Bartlett Spectrum-Modified No. 2

```
function[s,as,ps,apsd] =...
lms_aver_mod_bartlettpsd2(x,k,win,L,R)
      %x = data; k = number of sections; L = number of points
      %desired in the FT domain; M = number of points in each
      %section;kM< = N = length(x);R = number of realizations;
for r = 1:R
      x1(r,:) = [x*0.5+0.05*randn(1,length(x))...
      x*0.5+0.05*randn(1,length(x))];
      [s(r,:),as(r,:),ps(r,:)] = lmsbartlettpsd(x1,k,win,L);
end;
apsd = sum(as,1)/R;
```

Figure 4.21 shows the results of the No. 2 proposed modification. The input data were identical to the above two figures. We find similar results using the modified No. 3 given below.

Book m-Function for Modified Bartlett Spectrum-Modified No. 3

```
function[s,as,ps,apsd] =...
lms_aver_mod_bartlettpsd3(x,k,win,L,R)
```

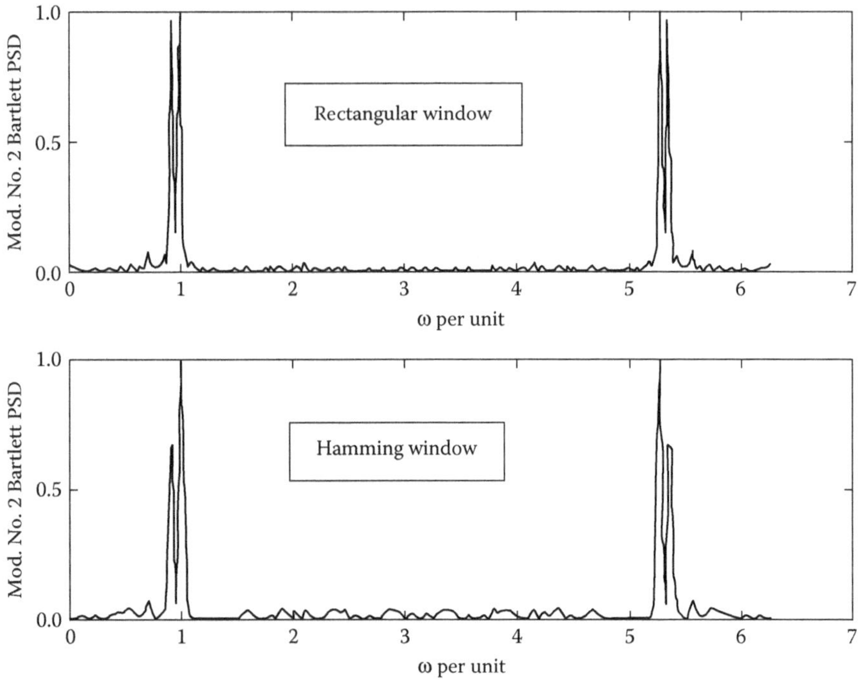

FIGURE 4.21

```
        %x = data; k = number of sections;
        %L = number of points
        %desired in the FT domain;
        %M = number of points in each
        %section;kM< = N = length(x);
        %R = number of realizations;
for m = 1:R
        x1(m,:) = [x 0.2*x+0.1*randn(1,length(x))];
        [s(m,:),as(m,:),ps(m,:)] =...
        lmsbartlettpsd(x1,k,win,L);
end;
apsd = sum(as,1);
```

4.7.7 The Welch Method

Welch proposed the modifications to Bartlett method as follows: Data segments are allowed to overlap and each segment is windowed prior to computing the periodogram. In most practical applications, we create smaller sections as follows:

$$x_i(n) = x(iD+n)w(n) \qquad 0 \le n \le M-1 \qquad 0 \le i \le K-1 \tag{4.58}$$

where:

$w(n)$ is the window of length M
D is an offset distance
K is the number of sections that the sequence $\{x(n)\}$ is divided into

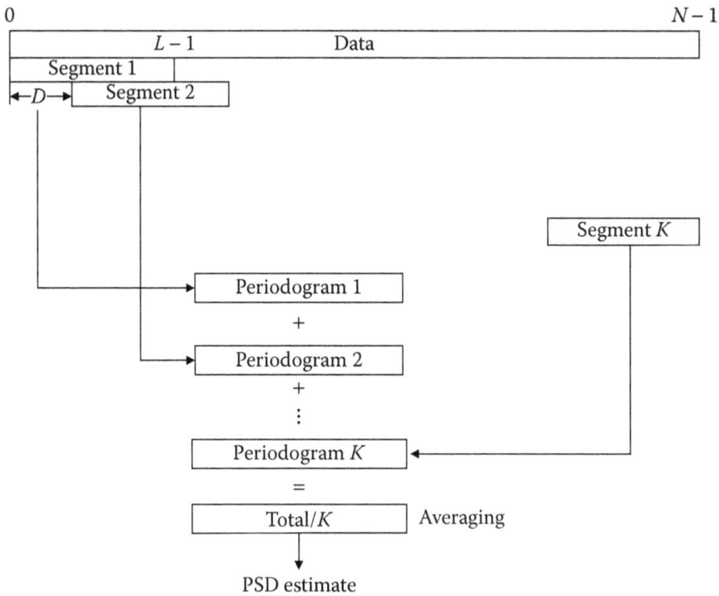

FIGURE 4.22

Pictorially, the Welch method is shown in Figure 4.22. The ith periodogram is given by

$$\hat{S}_{xi}(e^{j\omega}) = \frac{1}{M}\left|\sum_{n=0}^{M-1} x_i(n)e^{-j\omega n}\right|^2 \tag{4.59}$$

Therefore, the average Welch periodogram is given by

$$\hat{S}_x(e^{j\omega}) = \frac{1}{K}\sum_{i=0}^{K-1} \hat{S}_{xi}(e^{j\omega}) \tag{4.60}$$

If $D = M$, then the segments do not overlap and the result is equivalent to the Bartlett method with the exception that the segments are windowed.

The MATLAB program for the Welch PSD is given as follows:

```
ps = spectrum.welch(window,segmentlength,overlappercent)
        %window = 'hamming', we could have put rectangular etc;
        %segmentlength is anumber power of two,
        %e.g. if length(x) = 128
        %we put for segmentlength = 64 or 32 etc;
        %we write overlappercent = 50 for 50%, 30 for 30% etc;
hpsd = psd(ps,x,'nfft',1024);%the DFT with 1024 bins, we can
        %use 2048 etc; 'nfft' is written as is given;
plot(hpsd);%spectrum.welch() and psd() are MATLAB functions
        %and must be used as shown in the example below;
```

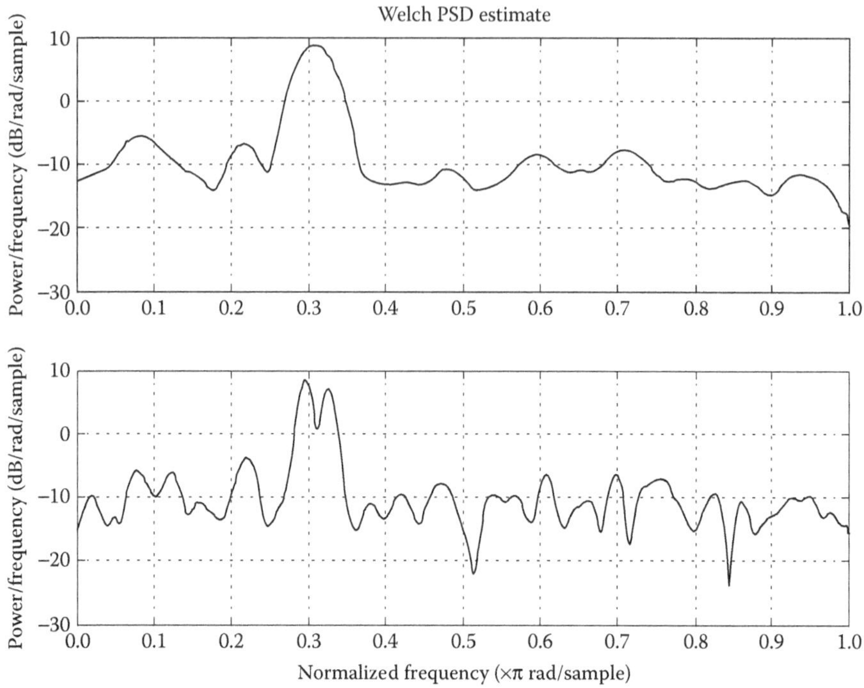

FIGURE 4.23

Example 4.7.5

Find the Welch spectrum for two lengths of the sections: 64 and 128. Create a signal with 256 elements, hamming window, and overlap 30%. Figure 4.23 shows the results.

Solution: In the Command window, we write

```
>>n = 0:255;
>>x = sin(0.30*π*n)+sin(0.30*π*n)+0.5*randn(1,256);
>>ps = spectrum.welch('hamming',64,30);
>>hpsd = psd(ps,x,'nfft',1024);
>>subplot(2,1,1);
>>plot(hpsd);
```

By changing segmentlength = 128, we proceed the same way and find the PSD given in the bottom of Figure 4.23. The top figure corresponds to a length of 64.

■

Book m-Function for Welch PSD: [s,ast,ps,K] = lmswelchpsd(x,win,frac, overlap,L)

```
function[s,ast,ps,K] = lmswelchpsd(x,win,frac,overlap,L)
        %x = data; M = section length; L = number of samples
        %desired in the frequency domain; win = 2 means rectwin,
```

```
        %number 3 means hamming window; see lmsbartlettpsd
        %function to add more windows if you desire; frac
        %defines the number of data for each section, depending
        %on the data length it is recommended the number
        %to vary between 0.2 and 0.5; overlap defines the
        %overlapping of the sections, it is recommended the
        %overlap to vary from 1 (no overlap) to 0.2
        %which means a 20% overlap; M<<N = length(x);
if (win = =2) w = rectwin(floor(frac*length(x)));
elseif (win = =3) w = hamming(frac*length(x));
end;
N = length(x);
M = floor(frac*length(x));
K = floor(floor((N-M+floor(overlap*M)))/floor(overlap*M));
        %K = number of processings;
s = 0;as = 0;
for i = 1:K
        s = s+fft(x(1,(i-1)*(floor(overlap*M))+...
        1:(i-1)*floor(overlap*M)+M).*w',L);
        as = as+abs(s);
end;
ast = as/(M*K);%as = amplitude spectral density;
ps = atan(imag(s)./real(s))/(M*K);%phase spectral density;
```

The data used for Figure 4.24 were as follows: rectangular window, 2; FFT length 512; $x = \sin(0.3\pi n) + \sin(0.32\pi n) + 0.5\mathrm{randn}(1,256)$; $n = 1,2,\ldots,256$. The following additional inputs were used for Figure 4.24a: length of sections is 0.25 or equivalently 64 elements and overlap is 0.25 or equivalently 16 elements overlapping. The following additional inputs were used for Figure 4.24b: length of sections is 0.50 or equivalently 128 elements and overlap is 0.25 or equivalently 32 elements. The following additional inputs were used for Figure 4.24c: length of sections is 0.25 or equivalently 64 elements and overlap is 0.50 or equivalently 32 elements. The following additional inputs were used for Figure 4.24d: length of sections is 0.50 or equivalently 128 elements and overlap is 0.50 or equivalently 64 elements.

4.7.8 Proposed Modified Welch Methods

4.7.8.1 Modified Method Using Different Types of Overlapping

It is evident from Figure 4.24 that if the lengths of the sections are not long enough and the overlaps are not enough, frequencies close together cannot be differentiated and the variance is large. Therefore, we propose a procedure defined as the **symmetric modified Welch method** (SMWM), and its implementation is shown in Figure 4.25. Windowing of the segments can also be incorporated. This approach and the rest of the proposed schemes have the advantage of progressively incorporating longer and longer segments of the data and thus introducing better and better resolution. In addition, due to the averaging process, the variance decreases and smoother periodograms are obtained but not as smooth as the Welch method. It is

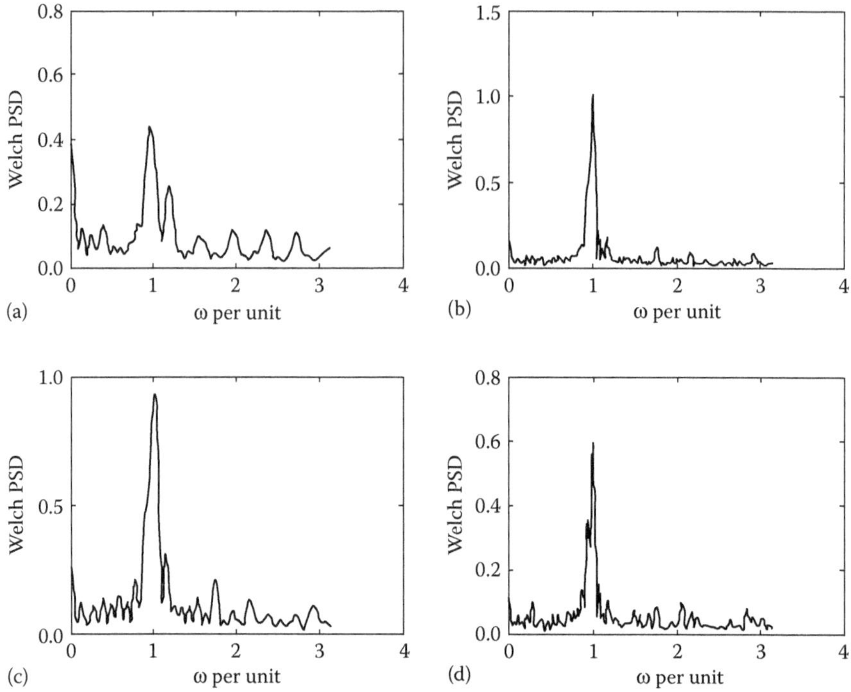

FIGURE 4.24

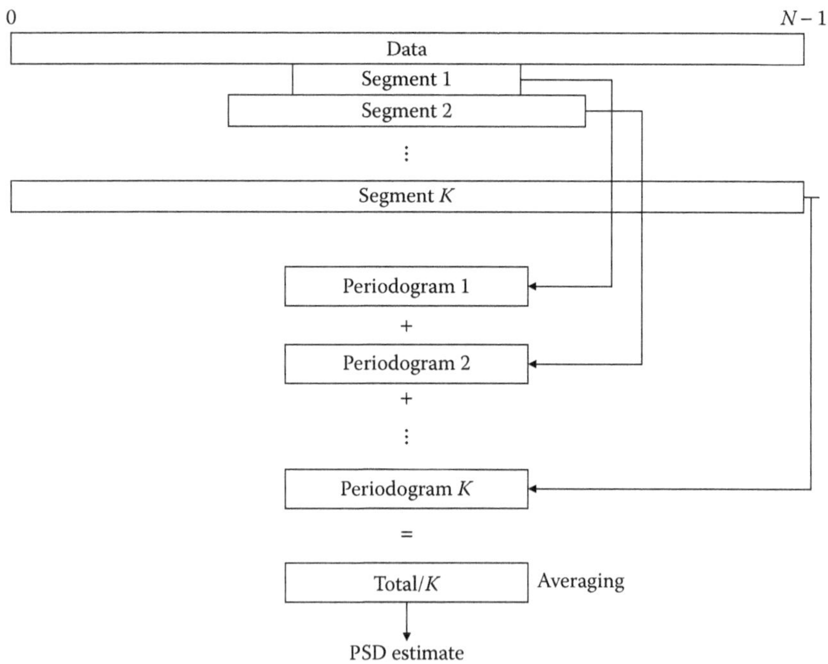

FIGURE 4.25

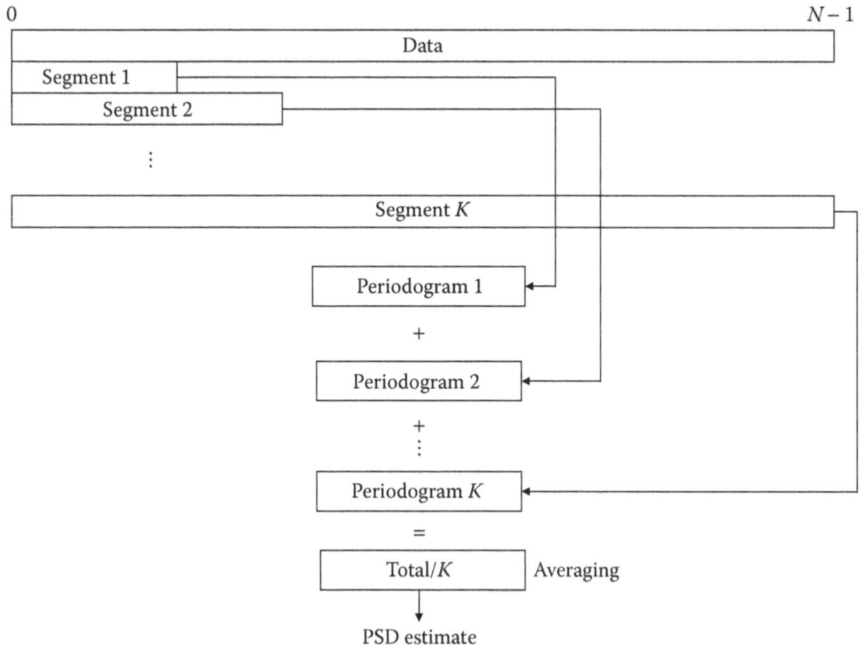

FIGURE 4.26

up to the reader to decide between smoothness of the periodogram and resolution of frequencies. Figure 4.26 shows another proposed method defined as the **asymmetric modified Welch method** (AMWM). Figure 4.27 shows another suggested approach for better resolution and reduced variance. The procedure is based on the method of prediction and averaging. This proposed method is defined as the **symmetric prediction modified Welch method** (SPMWM). This procedure can be used in all the other forms, for example, nonsymmetric. The above methods can also be used for **spectral estimation** if we substitute the word **periodogram** for the word **correlogram**.

4.7.8.2 Modified Welch Method Using Transformation of rv's

Figure 4.28a shows the result of the Welch method for the signal $x = \sin(0.3\pi n) + \sin(0.33\pi n) + 0.2\text{randn}(1,128)$, $n = 0, 1, \ldots, 127$. The signal was split into two sections with 50% overlap. For Figure 4.28b, the following transformed signal was used by incorporating a linear transformation of the rv's: $z = [x \quad 0.1 * x + 0.05 * \text{randn}(1,128)]$ For this signal, the Welch method was used splitting the series into two parts and using 50% overlapping. The figure was obtained using the following Book MATLAB program:

```
>>n = 0:127;
>>x = sin(0.30*pi*n)+sin(0.33*pi*n)+0.2*randn(1,128);
>>[sx,as,ps,K] = lmswelchpsd(x,2,0.5,0.5,512);
```

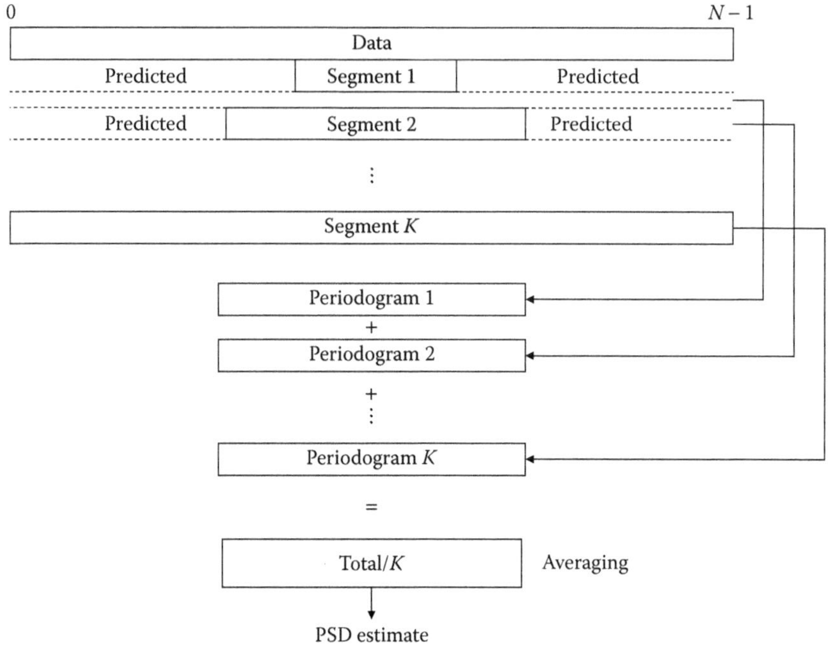

FIGURE 4.27

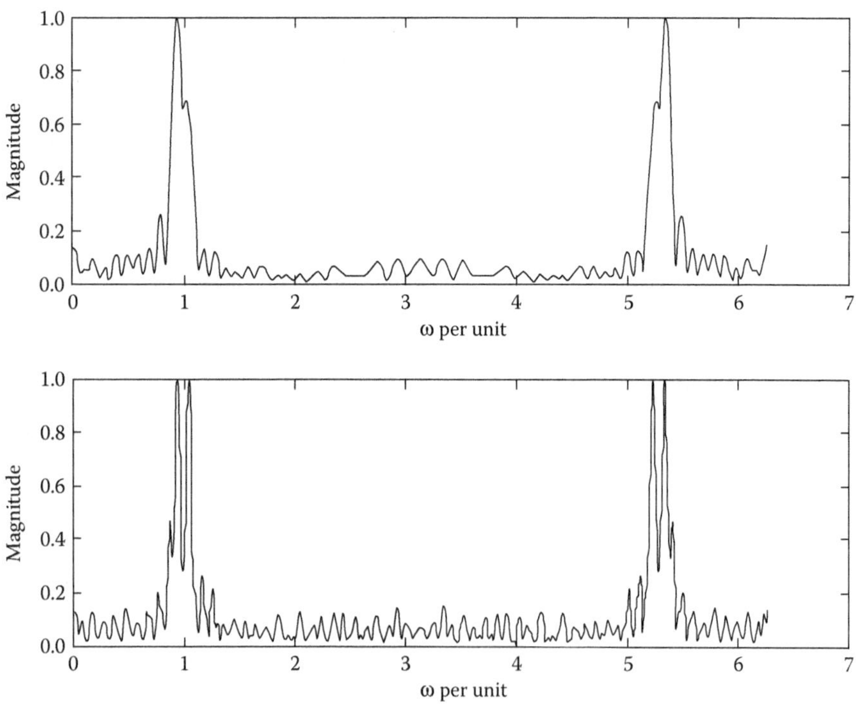

FIGURE 4.28

```
        %rectangular
        %window;
>>om = 0:2*pi/512:2*pi-(2*pi/512);
>>z = [x 0.1*x+0.05*randn(1,128)];
>>[sz,asz,psz,Kz] = lmswelchpsd(z,2,0.5,0.5,512);
        %rectangular
         %window;
>>subplot(2,1,1);plot(om,as/max(as),'k');
>>subplot(2,1,2);plot(om,asz/max(asz),'k');
>>xlabel('\omega per unit');ylabel('Magnitude');
```

PROBLEMS

4.1.1 Verify that the sample mean is equal to the ensemble mean.

4.1.2 If two random processes $x(n)$ and $y(n)$ with zero mean value are uncorrelated, find the autocorrelation of their sum.

4.1.3 Find the mean and autocorrelation function using the frequency interpretation approach. Use 10 and 1000 realizations.

4.2.1 Verify the properties 1, 4, 2, and 6 of a WSS process (see Table 4.1).

4.2.2 If $\boldsymbol{R}_x$ is an autocorrelation matrix, then show that $\boldsymbol{a}^T \boldsymbol{R}_x \boldsymbol{a} \geq 0$ for any vector $\boldsymbol{a}$.

4.2.3 Assume a discrete sine wave (vector) $s(n) = \sin(0.2\pi n)$ with $n = 0:39$, a vector $v(n) = \text{randn}(1,40)$ with statistically independent elements and normally distributed, and the sum of the above is $x(n) = s(n) + v(n)$. (a) Find the biased autocorrelation function up to lag 20, (b) find the FFT of the sine wave, and (c) find the FFT of the autocorrelation function. Plot the functions $s(n)$, $v(n)$, $s(n)+v(n)$, and $r_{xx}(k)$, and their FFTs (their spectra).

4.2.4 Verify (4.19).

4.3.1 Verify (4.25).

4.3.2 Find the joint pdf of a sequence of WGN with n elements, each one having a zero mean value and the same variance.

4.3.3 If an rv x is $N(m_x, \sigma^2)$ with variance greater than 0, then the rv $w = (x - m_x)/\sigma$ is $N(0,1)$.

4.3.4 Show that the sample mean is equal to the population mean.

4.4.1 Verify (4.33).

4.4.2 Determine the PSD of a zero-mean WSS process $\{x(n)\}$ with $r_{xx}(k) = a^{|k|}, -1 < a < 1$.

4.5.1 Verify (4.36).

4.6.1 Verify (4.42).

4.7.1 Find the periodogram for a white Gaussian (WG) signal with $N = 128$ and $N = 1024$. Verify that the variance (variability) does not decrease with increasing the number of the sequence.

HINTS–SOLUTIONS–SUGGESTIONS

4.1.1

$$E\{\hat{m}\} = E\left\{\frac{1}{N}\sum_{n=1}^{N} x(n)\right\} = \frac{1}{N}\sum_{n=1}^{N} E\{x(n)\} = \frac{1}{N}\sum_{n=1}^{N} m = \frac{Nm}{N} = m$$

4.1.2

$$r_{zz}(k,l) = E\{[x(k)+y(k)][x(l)+y(l)]\}$$

$$= E\{x(k)x(l)\} + E\{y(k)\}E\{x(l)\} + E\{x(k)\}E\{y(l)\} + E\{y(k)y(l)\}$$

$$= r_{xx}(k,l) + r_{yy}(k,l)$$

4.1.3 The Book MATLAB program is given as follows:

```
function[mx,rx] = lms_mean_autoc_ensemble(M,N)
      %N = number of data whichever distribution is
      %desired; M = number of realizations;
x = randn(M,N);%randn = is a MATLAB function producing
      %zero mean normal distribution; x = MxN
      %matrix;sum(x,1) = MATLAB function that
      %sums all the rows of the mareix x;
      %sum(x,2) sums all the columns;
mx = sum(x,1)/M;
for i = 1:N
      rx(i) = sum(x(:,1).*x(:,i))/M;
end;

>>[mx,rx] = lms_mean_autoc_ensemble(10,50);
>>[mx1,rx1] = lms_mean_autoc_ensemble(1000,50);
>>subplot(2,2,1);plot(mx,'k');
>>xlabel('Mean of 10 realizations');
>>subplot(2,2,2);plot(mx1,'k');
>>xlabel('Mean of 1000 realizations');
>>subplot(2,2,3);plot(rx,'k');
>>xlabel('Autoc. of 10 realizations');
>>subplot(2,2,4);plot(rx1,'k');
>>xlabel('Autoc. of 1000 realizations');
```

Figure of P4.1.3 shows the results.

4.2.1

a. $E\{x(n+l)\} = E\{x(m+l)\} \Rightarrow$ Mean must be constant

b. $r_{xx}(l) = E\{x(n+l)x(n)\} = E\{x(n)x(n+l)\} = r_{xx}(n-n-l) = r_{xx}(-l)$

c. $$r_{xx}(m,n) = \int_{-\infty}^{\infty}\int_{-\infty}^{\infty} x(m)x(n)f(x(m+l)x(n+l))dx(m)dx(n) = r_{xx}(m+l,n+l)$$

$$= r_{xx}(m+l-n-l) = r_{xx}(m-n)$$

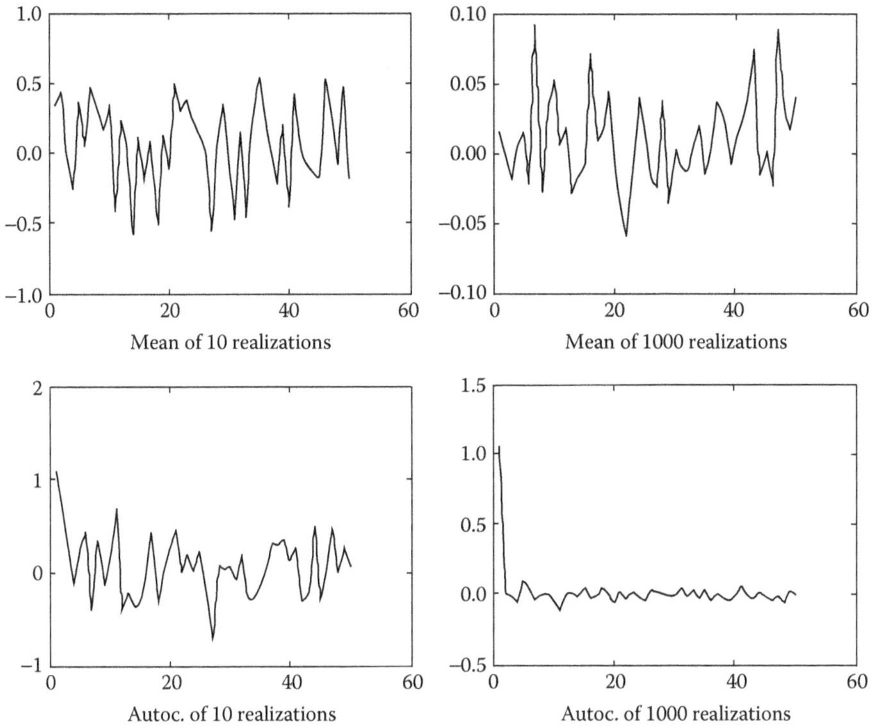

FIGURE P4.1.3

d. $E\{[x(n+l)-x(n)]^2\} = E\{x^2(n+l)+x^2(n)-2x(n+l)x(n)\}$

$$= r_{xx}(0)+r_{xx}(0)-2r_{xx}(l)$$

$$= 2r_{xx}(0)-2r_{xx}(l) \geq 0 \text{ or } r_{xx}(0) \geq r_{xx}(l)$$

4.2.2

$$\boldsymbol{a}^T\boldsymbol{R}\boldsymbol{a} = \boldsymbol{a}^T E\{\boldsymbol{x}\boldsymbol{x}^T\}\boldsymbol{a} = E\{\boldsymbol{a}^T(\boldsymbol{x}\boldsymbol{x}^T)\boldsymbol{a}\} = E\{(\boldsymbol{a}^T\boldsymbol{x})(\boldsymbol{x}^T\boldsymbol{a})\} = E\{(\boldsymbol{a}^T\boldsymbol{x})^2\} \geq 0,$$

$$\boldsymbol{a}^T\boldsymbol{x} = \boldsymbol{x}^T\boldsymbol{a} = \text{a number; } \boldsymbol{R} = \begin{bmatrix} 1 & 0 \\ 0 & 1 \end{bmatrix}, \boldsymbol{a} = \begin{bmatrix} 0.1 \\ -0.2 \end{bmatrix},$$

$$\boldsymbol{a}^T\boldsymbol{R}\boldsymbol{a} = [0.1 \ -0.2]\begin{bmatrix} 1 & 0 \\ 0 & 1 \end{bmatrix}\begin{bmatrix} 0.1 \\ -0.2 \end{bmatrix} = \begin{bmatrix} 0.1 & -0.2 \end{bmatrix}\begin{bmatrix} 0.1 \\ -0.2 \end{bmatrix} = 0.1^2 + (-0.2)^2 > 0$$

4.2.3 The Book m-file is given as follows:

```
%prob4_2_3 Book m-file;
n = 0:39;s = sin(0.2*pi*n);v = randn(1,40);
```

```
x = s+v;
rxx = lms_sample_biased_autoc(x,20);
fts = fft(s,40);ftrxx = fft(rxx,40);
subplot(3,2,1);stem(s,'k');xlabel('n');ylabel('s(n)');
subplot(3,2,2);stem(v,'k');xlabel('n');ylabel('v(n)');
subplot(3,2,3);stem(x,'k');xlabel('n');ylabel('x(n) =
      s(n)+v(n)');
subplot(3,2,4);stem(rxx,'k');xlabel('k');ylabel
      ('r_{xx}');
subplot(3,2,5);stem(abs(fts),'k');xlabel('Freq. bins');
ylabel('FT of s(n)');
subplot(3,2,6);stem(abs(ftrxx),'k');xlabel('Freq. bins');
ylabel('FT of r_{xx}');
```

Figure of P4.2.3 shows the results.

4.2.4

$$\boldsymbol{C}_x = E\{(\boldsymbol{x}-\boldsymbol{m}_x)(\boldsymbol{x}-\boldsymbol{m}_x)^T\} = E\{(\boldsymbol{x}-\boldsymbol{m}_x)(\boldsymbol{x}^T-\boldsymbol{m}_x^T)\}$$

$$= E\{\boldsymbol{x}\boldsymbol{x}^T - \boldsymbol{m}_x\boldsymbol{x}^T - \boldsymbol{x}\boldsymbol{m}_x^T + \boldsymbol{m}_x\boldsymbol{x}^T\}$$

$$= E\{\boldsymbol{x}\boldsymbol{x}^T - \boldsymbol{m}_x\boldsymbol{x}^T - \boldsymbol{m}_x\boldsymbol{x}^T + \boldsymbol{m}_x\boldsymbol{x}^T\} = E\{\boldsymbol{x}\boldsymbol{x}^T\} - \boldsymbol{m}_x E\{\boldsymbol{x}^T\} = \boldsymbol{R}_x - \boldsymbol{m}_x\boldsymbol{m}_x^T$$

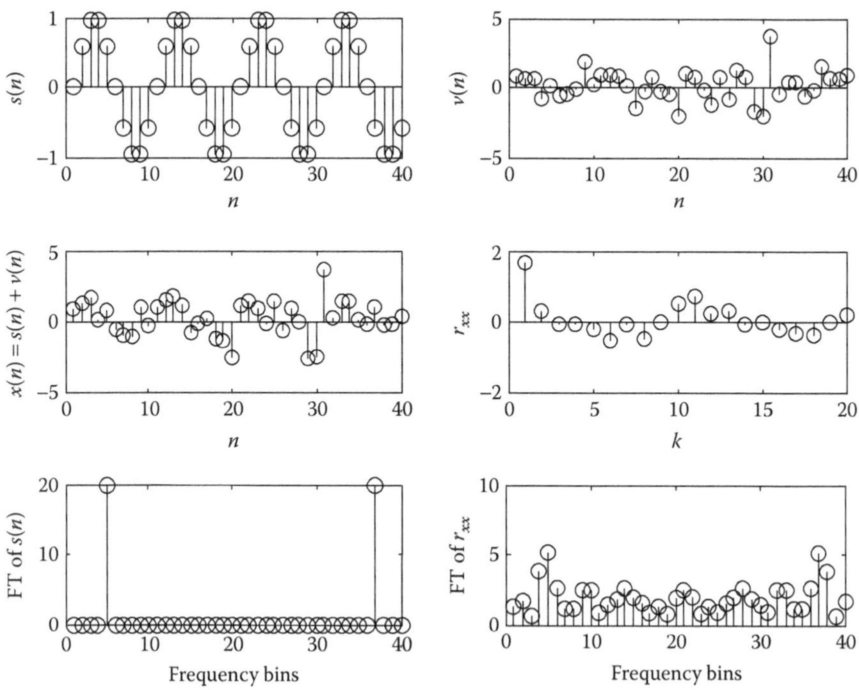

FIGURE P4.2.3

4.3.1

$$r_{xx}(m-n)=\frac{1}{2\pi\sigma_{x(m)}\sigma_{x(n)}}\int\int x(m)x(n)\exp\left[-\frac{x^2(m)}{2\sigma^2_{x(m)}}-\frac{x^2(n)}{2\sigma^2_{x(n)}}\right]$$

$$=\frac{1}{2\pi\sigma_{x(m)}\sigma_{x(n)}}\left\{\int x(m)\exp\left[-\frac{x^2(m)}{2\sigma^2_{x(m)}}\right]\int x(n)\exp\left[-\frac{x^2(n)}{2\sigma^2_{x(n)}}\right]\right\}$$

$$=\frac{1}{2\pi\sigma_{x(m)}\sigma_{x(n)}}0\times 0=0$$

4.3.2 The joint pdf is

$$f(x(1), x(2),\cdots x(n))=f_1(x(1))f_2(x(2))\cdots f_n(x(n))$$

$$=\frac{1}{(2\pi)^{n/2}\sigma_x^2}\exp\left[-\frac{1}{2\sigma_x^2}\sum_{k=1}^{n}x^2(k)\right]$$

4.3.3

$$F(w)=\text{cdf}=\text{pr}\left\{\frac{x-m_x}{\sigma}\le w\right\}=\text{pr}\{x\le w\sigma+m_x\}$$

$$=\int_{-\infty}^{w\sigma+m_x}\frac{1}{\sigma\sqrt{2\pi}}\exp\left[-\frac{(x-m_x)}{2\sigma^2}\right]dx$$

If we change the variable of integration by setting $y=(x-m_x)/\sigma_x$, then $F(w)=\int_{-\infty}^{w}(1/\sqrt{2\pi})\exp[-(y^2/2)]dy$. But $f(w)=dF(w)/dw$, and hence, $f(w)$ is equal to the indegrand which is the desired solution.

4.3.4

$$E\{\hat{m}\}=E\left\{\frac{1}{N}\sum_{k=1}^{N}x(k)\right\}=\frac{1}{N}\sum_{k=1}^{N}E\{x(k)\}=\frac{1}{N}\sum_{k=1}^{N}m=\frac{Nm}{m}=m$$

4.4.1

$$\int_{-\pi}^{\pi}S_x(e^{j\omega})e^{j\omega m}d\omega=\sum_{k=-\infty}^{\infty}\int_{-\pi}^{\pi}r_{xx}(k)e^{j\omega(m-k)}d\omega=r_{xx}(m)2\pi$$

4.4.2

$$S_x(e^{j\omega})=\sum_{k=-\infty}^{\infty}a^{|k|}e^{-j\omega k}=\sum_{k=-\infty}^{0}a^{|k|}e^{-j\omega k}-1+\sum_{k=0}^{\infty}a^{|k|}e^{-j\omega k}=\sum_{k=0}^{\infty}a^{k}e^{j\omega k}+\sum_{k=0}^{\infty}a^{k}e^{-j\omega k}$$

$$-1=\frac{1}{1-ae^{j\omega}}+\frac{1}{1-ae^{-j\omega}}-1=\frac{1-a^2}{1+a^2-2a\cos\omega}\qquad -1<a<1$$

The PSD is a real-valued, even, and nonnegative function of ω.

4.5.1

$$y(n)=\sum_{m=-\infty}^{\infty}h(m)x(n-m) \qquad (1)$$

$$E\{x(n+l)y(n)\}\triangleq r_{xy}(l)=\sum_{m=-\infty}^{\infty}h(m)E\{x(n+l)x(n-m)\}$$

$$=\sum_{m=-\infty}^{\infty}h(m)r_{xx}(l+m)=\sum_{k=-\infty}^{\infty}h(-k)r_{xx}(l-k)=h(-l)*r_{xx}(l) \qquad (2)$$

similarly, we get $r_{yx}(l)=h(l)*r_{xx}(l)$ from (1)

$$E\{y(n)y(n-l)\}\triangleq r_{yy}(l)=\sum_{m=-\infty}^{\infty}h(m)E\{x(n-m)y(n-l)\}$$

$$=\sum_{m=-\infty}^{\infty}h(m)r_{xy}(l-m)=h(l)*r_{xy}(l) \qquad (3)$$

We get $r_{yy}(l)=h(l)*h(-l)*r_{xx}(l)$ or $r_{yy}(k)$

$$=\sum_{l=-\infty}^{\infty}\sum_{m=-\infty}^{\infty}h(l)r_{xx}(m-l+k)h(m) \text{ from (2) and (3)}$$

4.6.1

$$H(z)=\frac{Y(z)}{V(z)}=\frac{b(0)}{1+\sum_{l=0}^{p}a(l)z^{-l}}\Rightarrow y(n)+\sum_{l=1}^{p}a(l)y(n-l)=\sum_{l=0}^{q}b(l)v(n-l) \text{ or}$$

$$E\{y(n)y(n-k)\}+\sum_{l=1}^{p}a(l)E\{y(n-l)y(n-k)\}=\sum_{l=0}^{q}b(l)E\{v(n-l)y(n-k)\}$$

$$\Rightarrow r_{yy}(k)+\sum_{l=1}^{p}a(l)r_{yy}(k-l)=\sum_{l=0}^{q}b(l)r_{vy}(k-l) \qquad (1)$$

$$y(n)=h(n)*v(n)=\sum_{m=-\infty}^{\infty}v(m)h(n-m) \qquad (2)$$

From (2) we get

$$E\{v(n-l)y(n-k)\} = \sum_{m=-\infty}^{\infty} E\{v(n-l)v(m)\}h(n-k-m)$$

$$= \sigma_v^2\delta(n-l-m)h(n-k-m) = \sigma_v^2 h(l-k) \quad (3)$$

Setting (3) in (1) we get

$$r_{yy}(k) + \sum_{l=1}^{p} a(l) r_{yy}(k-l) = \sum_{l=0}^{q} b(l)\sigma_v^2 h(l-k) \quad (4)$$

$$H(z) = \frac{Y(z)}{V(z)} = b(0)\frac{1}{1+\sum_{l=0}^{p} a(l)z^{-l}} \Rightarrow H(z) = b(0)[1 - a(1)z^{-1} - \cdots]$$

$$\Rightarrow h(n) = b(0)[\delta(k) - a(1)\delta(k-1) - \cdots] \Rightarrow h(0) = b(0)\delta(k) k \geq 0 \quad (5)$$

$$\Rightarrow (4) \quad r_{yy}(k) + \sum_{l=1}^{p} a(l) r_{yy}(k-l) = b(0)^2 \sigma_v^2 \delta(k) \quad k \geq 0$$

5 The Wiener Filter

5.1 INTRODUCTION

In this chapter, we shall develop two basic important principles used extensively in signal processing, communications, system identification, denoising signals, and so on. The least-squares (LS) principle, proposed first by Gauss when he was 18 years old, is widely applicable to the design of digital signal processing systems. We shall first describe the use of the LS approach to modeling, interference canceling as well as the cases involving prediction. Next, we shall study the celebrated Wiener filter, which was developed during World War II.

5.2 THE LS TECHNIQUE

The principle of LS is used in this chapter for signals of one variable, although the concept applies equally well for signals with more than one variable. Furthermore, we study discrete signals, which can always be derived from continuous signals if the sampling frequency is sufficient high enough so that aliasing does not take place.

Let us, for example, have a discrete signal that is represented in its vector form as follows: $\boldsymbol{x} = [x(1) \quad x(2) \quad \cdots \quad x(N)]$. We would like, next, to approximate the function with a polynomial of the form:

$$\hat{x}(nT) = c_1 + c_2 \times (nT) + c_3 \times (nT)^2 + \cdots + c_M \times (nT)^{M-1} \qquad T = \text{sampling time} \tag{5.1}$$

Since the approximating polynomial is a linear function of the unknown c_i's, we are dealing with **linear LS approximation**. The difference between the exact and the approximate function is the error. We can define a **cost function *J***, which is equal to the sum of the difference of errors squared, the total square error. Hence, we write

$$J = \sum_{n=1}^{N} [x(n) - \hat{x}(n)]^2 \tag{5.2}$$

Example 5.2.1

Let $x(n) = 2\sin(0.05\pi n) + \text{randn}(1,20)$, with $T = 1$, be the signal to be approximated. Let the signal $\hat{x}(n) = c_1 + c_2 n + c_3 n^2$, the sampling time $T = 1$, approximate the original signal. Find the unknown constants c_i's and plot the results.

Solution: The cost function for this case is

$$J = \sum_{n=1}^{20} [x(n) - \hat{x}(n)]^2 = \sum_{n=1}^{20} [x(n) - c_1 - c_2 n - c_3 n^2]^2$$

The partial derivatives with respect to c_i's are

$$\frac{\partial J}{\partial c_1} = -2\sum_{n=1}^{20}[x(n) - c_1 - c_2 n - c_3 n^2] = 0 \quad \text{or}$$

$$\sum_{n=1}^{20} x(n) - c_1\sum_{n=1}^{20} 1 - c_2\sum_{n=1}^{20} n - c_3\sum_{n=1}^{20} n^2 = 0$$

$$\frac{\partial J}{\partial c_2} = -2\sum_{n=1}^{20}[x(n) - c_1 - c_2 n - c_3 n^2]n = 0 \quad \text{or}$$

$$\sum_{n=1}^{20} x(n)n - c_1\sum_{n=1}^{20} n - c_2\sum_{n=1}^{20} n^2 - c_3\sum_{n=1}^{20} n^3 = 0$$

$$\frac{\partial J}{\partial c_3} = -2\sum_{n=1}^{20}[x(n) - c_1 - c_2 n - c_3 n^2]n^2 = 0 \quad \text{or}$$

$$\sum_{n=1}^{20} x(n)n^2 - c_1\sum_{n=1}^{20} n^2 - c_2\sum_{n=1}^{20} n^3 - c_3\sum_{n=1}^{20} n^4 = 0$$

Using MATLAB, we obtain the following system:

$$20c_1 + 210c_2 + 2870c_3 = 27.7347$$

$$210c_1 + 2870c_2 + 44100c_3 = 296.1023$$

$$2870c_1 + 44100c_2 + 722666c_3 = 3499.9838$$

The values of the unknown c_i's are found by using the expression:

$$\begin{bmatrix} c_1 \\ c_2 \\ c_3 \end{bmatrix} = \begin{bmatrix} 20 & 210 & 2370 \\ 210 & 2870 & 44100 \\ 2870 & 44100 & 722666 \end{bmatrix}^{-1} \begin{bmatrix} 27.7347 \\ 296.1023 \\ 3499.9838 \end{bmatrix} = \begin{bmatrix} -1.2457 \\ 0.7042 \\ -0.0332 \end{bmatrix}$$

Therefore, the estimate curve is given by

$$\hat{x}(n) = -1.2457 + 0.7042n - 0.0332n^2$$

Figure 5.1 shows the original signal $\{x(n)\}$, the approximate one $\{\hat{x}(n)\}$, based on the LS approach, and the undisturbed signal $\{2\sin(0.05\pi n)\}$.

■

5.2.1 Linear LS

We proceed to generalize the linear LS technique by a set of known vectors $\{h(n)\}$, $n = 0, 1, \ldots, M$ with $M < N$. $\boldsymbol{H}$ is a matrix having inverse. Therefore, we write

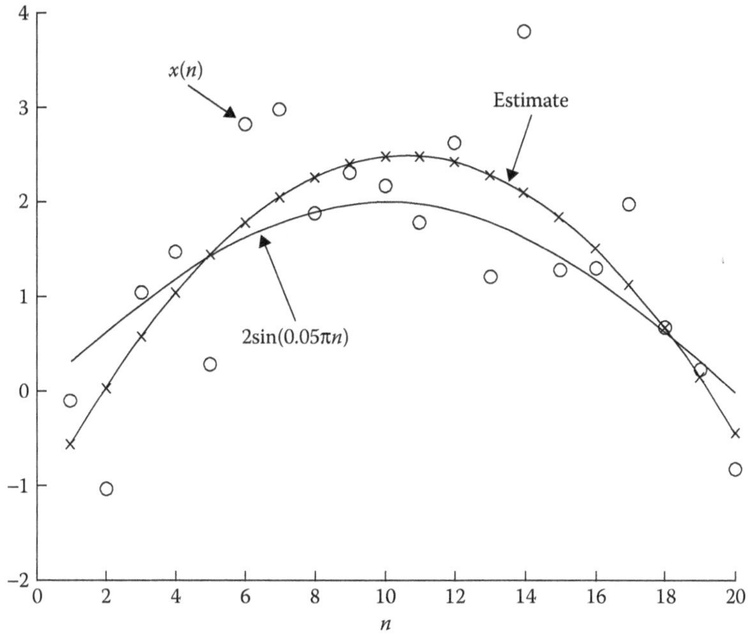

FIGURE 5.1

$$J(\boldsymbol{c})=\sum_{n=0}^{N-1}[x(n)-\hat{x}(n)]^2=\sum_{n=0}^{N-1}[x(n)-\boldsymbol{c}^T\boldsymbol{h}(n)]^2$$
$$=\sum_{n=0}^{N-1}\{x(n)-[c_1h_1(n)+\cdots+c_Mh_M(n)]\}^2=(\boldsymbol{x}-\boldsymbol{Hc})^T(\boldsymbol{x}-\boldsymbol{Hc}) \tag{5.3}$$

Next, to verify (5.3), we proceed as follows:

$$J(\boldsymbol{c})=\left(\begin{bmatrix}x(0)\\x(1)\end{bmatrix}-\begin{bmatrix}h_1(0)&h_2(0)\\h_1(1)&h_2(1)\end{bmatrix}\begin{bmatrix}c_1\\c_2\end{bmatrix}\right)^T\left(\begin{bmatrix}x(0)\\x(1)\end{bmatrix}-\begin{bmatrix}h_1(0)&h_2(0)\\h_1(1)&h_2(1)\end{bmatrix}\begin{bmatrix}c_1\\c_2\end{bmatrix}\right)$$
$$=\left([x(0)\;\;x(1)]-[c_1\;\;c_2]\begin{bmatrix}h_1(0)&h_1(1)\\h_2(0)&h_2(1)\end{bmatrix}\right)\begin{pmatrix}x(0)-h_1(0)c_1-h_2(0)c_2\\x(1)-h_1(1)c_1-h_2(1)c_2\end{pmatrix}$$
$$=[x(0)-c_1h_1(0)-c_2h_2(0)\quad x(1)-c_1h_1(1)-c_2h_2(1)]\begin{pmatrix}x(0)-h_1(0)c_1-h_2(0)c_2\\x(1)-h_1(1)c_1-h_2(1)c_2\end{pmatrix}$$

The last two expressions are identical. The reader should observe that the following matrix relationship was used:

$$(\boldsymbol{x}-\boldsymbol{Hc})^T=(\boldsymbol{x}^T-\boldsymbol{c}^T\boldsymbol{H}^T) \tag{5.4}$$

where:

The exponent T stands for transpose of a matrix

To find the unknowns c_i, differentiate the cost function of (5.3) with respect to each c_i and then set the developed equations equal to zero. Therefore, we have a system with c_i, the unknowns that are determined by solving the system. The following example will elucidate the procedure.

Example 5.2.2

Let a signal be a constant, $s(n) = A$, and let the received signal be given by $x(n) = 5 + \text{randn}(1,10)$. Find A.

Solution: According to the LS approach, we can estimate the constant 5 by minimizing the cost function. Taking the derivative J with respect to A and setting the results equal to zero, we obtain

$$J(A)=\sum_{n=0}^{N-1}[A-x(n)]^2;\ \frac{\partial J(A)}{\partial A}=\sum_{n=0}^{N-1}[2A-2x(n)]=0 \text{ or}$$

$$A\sum_{n=0}^{N-1}1=\sum_{n=0}^{N-1}x(n) \text{ or } A=\frac{1}{N}\sum_{n=0}^{N-1}x(n)$$

Using MATLAB, we obtain $\hat{A}=\text{sum}(x)/10=5.1794$. The minimum $J(A)$ is given by

$$J_{\min}=\sum_{n=0}^{N-1}[x(n)-\hat{A}]^2$$

and for this case

$$J_{\min}=\sum_{n=1}^{10}[x(n)-5.1794]^2=1.3803$$

■

Example 5.2.3

Find the amplitude constants of the signal

$$s(n)=A\sin(0.1\pi n)+B\sin(0.4\pi n) \qquad n=0,1,\ldots,N-1$$

if the received signal is $x(n)=s(n)+\text{randn}(1,N)$ and their exact values are $A=0.2$ and $B=5.2$.

Solution: For this case, and for $n = 0, 1, 2, 3$ and $N = 4$, we obtain

$$J(A,B) = \sum_{n=0}^{3}[x(n) - h_1(n)A - h_2(n)B]^2$$

$$h_1(n) = \sin(0.1\pi n) \quad h_2(n) = \sin(0.4\pi n)$$

If we differentiate $J(A,B)$ first with respect to A and next with respect to B, and then equate the results to zero, we obtain

$$\begin{bmatrix} A \\ B \end{bmatrix} = \begin{bmatrix} \sum_{n=0}^{3} h_1^2(n) & \sum_{n=0}^{3} h_1(n)h_2(n) \\ \sum_{n=0}^{3} h_1(n)h_2(n) & \sum_{n=0}^{3} h_2^2(n) \end{bmatrix}^{-1} \begin{bmatrix} \sum_{n=0}^{3} x(n)h_1(n) \\ \sum_{n=0}^{3} x(n)h_2(n) \end{bmatrix}$$

The following MATLAB program was used to find $\hat{A} = 0.5372$, $\hat{B} = 4.7868$ from the vector ab:

```
>>n = 0:3;x = 0.2*sin(0.1*pi*n)+5.2*sin(0.4*pi*n)...
>>+0.5*randn(1,4);
>>h1 = sin(0.1*pi*n);h2 = sin(0.4*pi*n);
>>ab = inv([sum(h1.^2) sum(h1.*h2);sum(h1.*h2)...
>>sum(h2.^2)])*[sum(x.*h1);sum(x.*h2)];
```

■

5.2.2 LS Formulation

We consider a linear adaptive filter with coefficients at time n:

$\boldsymbol{w}(n) = [w_1(n) \quad w_2(n) \quad \dots \quad w_M(n)]^T$ a measured real-valued input vector

$\boldsymbol{x}(n) = [x_1(n) \quad x_2(n) \quad \dots \quad x_M(n)]^T$ and a measured desired response $d(n)$

Note that no structure has been specified for the input vector $\boldsymbol{x}(n)$, and therefore, it can be considered as the successive samples of a particular process or as a snapshot of M detectors as shown in Figure 5.2. Hence, the problem is to estimate the desired response $d(n)$ using the linear combination:

$$y(n) = \boldsymbol{w}^T(n)\boldsymbol{x}(n) = \sum_{k=1}^{M} w_k(n)x_k(n) \qquad n = 1, 2, \dots, N \tag{5.5}$$

The above equation can be represented by a linear combiner as shown in Figure 5.3. The combiner error is defined by the relation:

$$e(n) = d(n) - y(n) = d(n) - \boldsymbol{w}^T(n)\boldsymbol{x}(n) \tag{5.6}$$

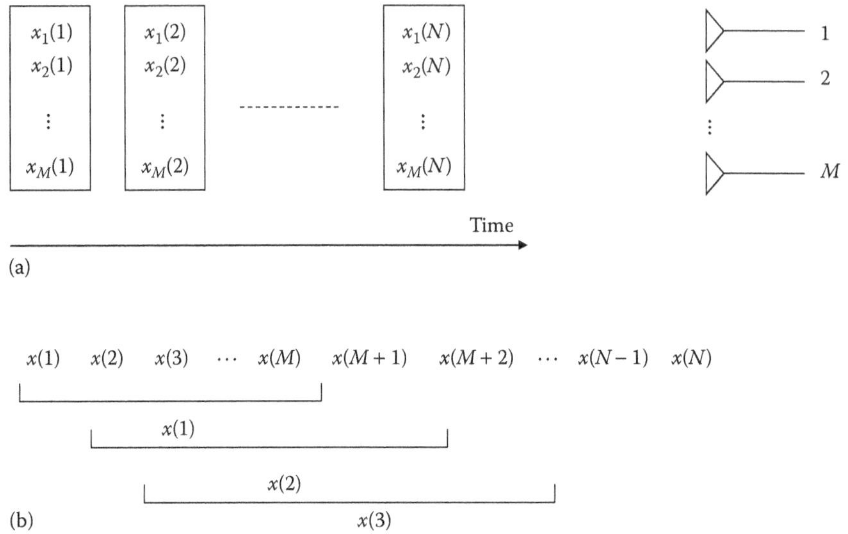

FIGURE 5.2

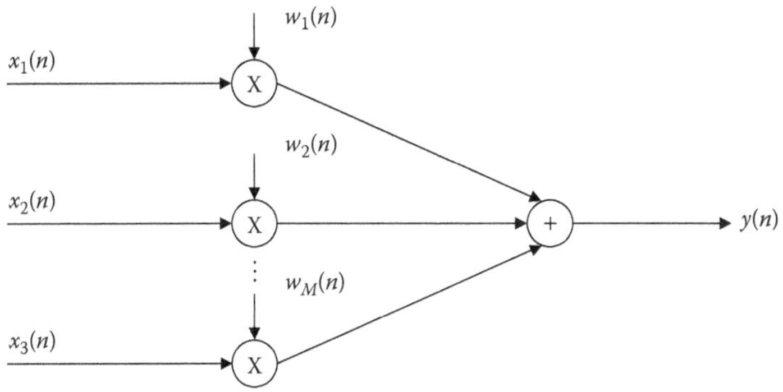

FIGURE 5.3

The coefficients of the adaptive filter are found by minimizing the sum of the squares of the error (LS):

$$J \equiv E_e = \sum_{n=1}^{N} g(n)e^2(n) \tag{5.7}$$

where:

$g(n)$ is a weighting function

Therefore, in the LS method, the filter coefficients are optimized by using all the observations from the time the filter begins until the present time and minimizing the sum of the squared values of the error samples, which are equal to the measured desired signal and the output signal of the filter. The minimization is valid when the filter coefficient vector $\boldsymbol{w}(n)$ is kept constant, $\boldsymbol{w}$, over the measurement time interval $1 \le n \le N$. In statistics, the LS estimation is known as **regression**, $e(n)$ are known as **signals**, and $\boldsymbol{w}$ is the **regression vector**.

We next define the matrix of the observed input samples as follows:

$$\boldsymbol{X}^T = \begin{bmatrix} x_1(1) & x_1(2) & \cdots & x_1(N) \\ x_2(1) & x_2(2) & \cdots & x_2(N) \\ \vdots & \vdots & & \vdots \\ x_M(1) & x_M(2) & & x_M(N) \end{bmatrix} \underset{\text{Snapshots}}{\downarrow} \rightarrow \text{Data records } (M \times N) \tag{5.8}$$

where we assume that $N > M$. This defines an overdetermined LS problem.

For the case in which we have one-dimensional input signal, as shown in Figure 5.2b, the data matrix takes the form:

$$\boldsymbol{X}^T = \begin{bmatrix} x(M) & x(M+1) & \cdots & x(N) \\ x(M-1) & x(M) & \cdots & x(N-1) \\ x(M-2) & x(M-1) & \cdots & x(N-2) \\ \vdots & \vdots & & \vdots \\ x(1) & x(2) & \cdots & x(N-M+1) \end{bmatrix} \tag{5.9}$$

The output $\boldsymbol{y}$, the error $\boldsymbol{e}$, and the data $\boldsymbol{x}_k$ are as follows:

$$\boldsymbol{y} = \boldsymbol{X}\boldsymbol{w} \tag{5.10}$$

$$\boldsymbol{e} = \boldsymbol{d} - \boldsymbol{y} \tag{5.11}$$

where:

$$\boldsymbol{y} = [\,y(1) \quad y(2) \quad \cdots \quad y(N)]^T \equiv \text{Filter output vector } (N \times 1) \tag{5.12}$$

$$\boldsymbol{d} = [\,d(1) \quad d(2) \quad \cdots \quad d(N)]^T \tag{5.13}$$

$$\boldsymbol{e} = [\,e(1) \quad e(2) \quad \cdots \quad e(N)]^T \equiv \text{Error vector } (N \times 1) \tag{5.14}$$

$$\boldsymbol{x} = [\,x_1(n) \quad x_2(n) \quad \cdots \quad x_N(n)]^T \equiv \text{Snapshot } (N \times 1) \tag{5.15}$$

$$\boldsymbol{x}_k = [x_k(1) \quad x_k(2) \quad \cdots \quad x_k(N)]^T \equiv \text{Data vector, } k = 1, 2, \ldots, M \tag{5.16}$$

$$\boldsymbol{w} = [\,w_1 \quad w_2 \quad \cdots \quad w_M\,]^T \equiv \text{Filter coefficients } (M \times 1) \tag{5.17}$$

$$\boldsymbol{X} = [\boldsymbol{x}_1 \quad \boldsymbol{x}_2 \quad \cdots \quad \boldsymbol{x}_M]^T \equiv \text{Matrix } (N \times M) \tag{5.18}$$

In addition, with $g(n) = 1$ for all n, (5.7) takes the form:

$$\begin{aligned} J &= \boldsymbol{e}^T\boldsymbol{e} = (\boldsymbol{d}-\boldsymbol{y})^T(\boldsymbol{d}-\boldsymbol{y}) = (\boldsymbol{d}-\boldsymbol{X}\boldsymbol{w})^T(\boldsymbol{d}-\boldsymbol{X}\boldsymbol{w}) \\ &= \boldsymbol{d}^T\boldsymbol{d} - \boldsymbol{w}^T\boldsymbol{X}^T\boldsymbol{d} - \boldsymbol{d}^T\boldsymbol{X}\boldsymbol{w} + \boldsymbol{w}^T\boldsymbol{X}^T\boldsymbol{X}\boldsymbol{w} \\ &= E_d - \boldsymbol{w}^T\boldsymbol{p} - \boldsymbol{p}^T\boldsymbol{w} + \boldsymbol{w}^T\boldsymbol{R}\boldsymbol{w} = E_d - 2\boldsymbol{p}^T\boldsymbol{w} + \boldsymbol{w}^T\boldsymbol{R}\boldsymbol{w} \end{aligned} \tag{5.19}$$

where:

$$E_d = \boldsymbol{d}^T\boldsymbol{d} = \sum_{n=1}^{N} d(n)d(n) \tag{5.20}$$

$$\boldsymbol{R} = \boldsymbol{X}^T\boldsymbol{X} = \sum_{n=1}^{N} \boldsymbol{x}(n)\boldsymbol{x}^T(n) \qquad (M\times M) \tag{5.21}$$

$$\boldsymbol{p} = \boldsymbol{X}^T\boldsymbol{d} = \sum_{n=1}^{N} \boldsymbol{x}(n)d(n) \qquad (M\times 1) \tag{5.22}$$

$$\boldsymbol{y} = \boldsymbol{X}\boldsymbol{w} = \sum_{k=1}^{M} w_k\boldsymbol{x}_k \qquad (N\times 1) \tag{5.23}$$

The matrix $\boldsymbol{R}$ becomes time average if it is divided by N. In statistics, the scaled form of $\boldsymbol{R}$ is known as the **sample correlation matrix**.

Setting the gradient of J with respect to the vector coefficients w equal to zero, we obtain

$$\boldsymbol{R}\hat{\boldsymbol{w}} = \boldsymbol{p}; \quad \boldsymbol{p}^T = \hat{\boldsymbol{w}}^T\boldsymbol{R}^T = \hat{\boldsymbol{w}}^T\boldsymbol{R} \quad (\boldsymbol{R}\text{ is symmetric}) \tag{5.24}$$

or

$$\hat{\boldsymbol{w}} = \boldsymbol{R}^{-1}\boldsymbol{p} \tag{5.25}$$

Therefore, the minimum sum of squared errors is given by

$$J_{\min} = \boldsymbol{d}^T\boldsymbol{d} - 2\boldsymbol{p}^T\boldsymbol{R}^{-1}\boldsymbol{p} + \boldsymbol{w}^T\boldsymbol{R}\boldsymbol{R}^{-1}\boldsymbol{p} = E_d - \boldsymbol{p}^T\boldsymbol{R}^{-1}\boldsymbol{p} = E_d - \boldsymbol{p}^T\hat{\boldsymbol{w}} \tag{5.26}$$

since $\boldsymbol{R}$ is symmetric.

Example 5.2.4

Let the desired response be $d = [1 \quad 1 \quad 1 \quad 1]$, and the two measured $x_1 = [0.7 \quad 1.4 \quad 0.4 \quad 1.3]^T$ and $x_2 = [1.2 \quad 0.6 \quad 0.5 \quad 1.1]^T$. Then, we obtain

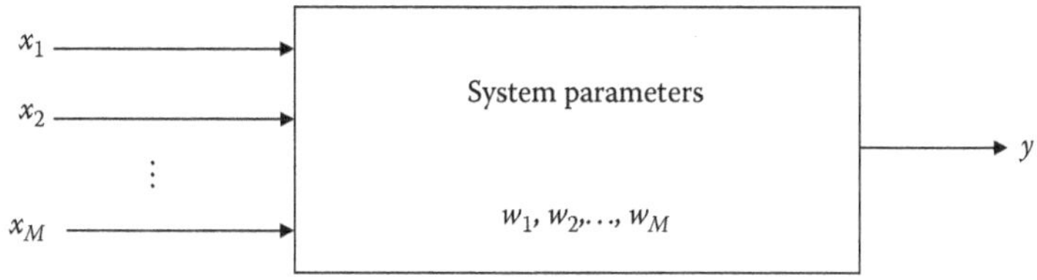

FIGURE 5.4

$$R = X^T X = \begin{bmatrix} 0.7 & 1.4 & 0.4 & 1.3 \\ 1.2 & 0.6 & 0.5 & 1.1 \end{bmatrix} \begin{bmatrix} 0.7 & 1.2 \\ 1.4 & 0.6 \\ 0.4 & 0.5 \\ 1.3 & 1.1 \end{bmatrix} = \begin{bmatrix} 4.30 & 3.31 \\ 3.31 & 3.26 \end{bmatrix}$$

$$p = X^T d = \begin{bmatrix} 3.8 \\ 3.4 \end{bmatrix}, \quad \hat{w} = R^{-1} p = \begin{bmatrix} 0.3704 \\ 0.6669 \end{bmatrix}, \quad J_{\min} = 0.3252$$

$$\hat{y} = X\hat{w} = [1.0595 \quad 0.9187 \quad 0.4816 \quad 1.2150]$$

■

The LS technique is a mathematical procedure that enables us to achieve a best fit of a model to experimental data. In the sense of the M-parameter linear system, shown in Figure 5.4, (5.5) is written in the form:

$$y(n) = w_1 x_1(n) + w_2 x_2(n) + \cdots + w_M x_M(n) \qquad n = 1, 2, \ldots, N \tag{5.27}$$

The above equation takes the following matrix form:

$$y = Xw \tag{5.28}$$

To estimate the M parameters w_i, it is necessary that $N \geq M$. If $N = M$, then we can uniquely solve for w to find

$$\hat{w} = X^{-1} y \tag{5.29}$$

provided that X^{-1} exists. $\hat{w}$ is the estimate of w. Using the least error squares, we can determine w, provided that $N > M$.

Let us define an error vector $e = [e_1 \quad e_1 \quad \cdots \quad e_N]^T$ as follows:

$$e = y - Xw \tag{5.30}$$

Next, we choose $\hat{w}$ in such a way that the criterion

$$J = \sum_{i=1}^{N} e_i^2 = e^T e \tag{5.31}$$

is minimized. To proceed, we write

$$J = (y - Xw)^T (y - Xw) \\ = y^T y - w^T X^T y - yXw + w^T X^T Xw \tag{5.32}$$

Differentiating J with respect to w and equating the results to zero, we determine the conditions on the estimate $\hat{w}$ that minimizes J. Hence,

$$\left.\frac{\partial J}{\partial w}\right|_{w=\hat{w}} = -2X^T y + 2X^T X\hat{w} = \mathbf{0} \tag{5.33}$$

$$X^T X\hat{w} = X^T y \tag{5.34}$$

from which we obtain

$$\hat{w} = (X^T X)^{-1} X^T y \tag{5.35}$$

The above equation is known as the **LS estimator** (LSE) of w. Equation 5.34 is known as the **normal equation**.

5.2.3 Statistical Properties of LSEs

We rewrite (5.30) in the form (X = deterministic matrix):

$$y = Xw + e \tag{5.36}$$

and assume that e is a stationary random vector with zero mean value, $E\{e\} = 0$. Furthermore, e is assumed to be uncorrelated with y and X. Therefore, on the given statistical properties of e, we wish to know just how good or how accurate the estimates of the parameters are.

Substituting (5.36) in (5.35) and taking the ensemble average, we obtain

$$E\{\hat{w}\} = E\{w + (X^T X)^{-1} X^T e\} = E\{w\} + E\{(X^T X)^{-1} X\} E\{e\} = w \quad (E\{e\} = 0) \tag{5.37}$$

which indicates that $\hat{w}$ is **unbiased**.

The covariance matrix corresponding to the estimate error $\hat{w} - w$ is

$$(5.38)$$

$$\begin{aligned} C_w &= E\{(\hat{w} - w)(\hat{w} - w)^T\};\ C_w = E\{[(X^T X)^{-1} X^T y - w](\hat{w} - w)^T\} \\ &= E\{[(X^T X)^{-1} X^T (Xw + e) - w](\hat{w} - w)^T\} \\ &= E\{[(X^T X)^{-1}(X^T X)w + (X^T X)^{-1} e - w][\hat{w} - w]^T\} \\ &= E\{[(X^T X)^{-1} X^T e][(X^T X)^{-1} X^T e]^T\} \\ &= (X^T X)^{-1} X^T E\{ee^T\} X (X^T X)^{-1} \\ &= (X^T X)^{-1} X^T R_e X (X^T X)^{-1} \end{aligned} \tag{5.39}$$

where:

$\boldsymbol{R}_e$ is the error correlation matrix

If the noise sample $\boldsymbol{e}(i)$ for $i = 1, 2, 3,\ldots$ is normal, identically distributed with zero mean and variance $\sigma^2\,[\boldsymbol{e} = N(\boldsymbol{0}, \sigma\boldsymbol{I})]$, then

$$\boldsymbol{R}_e = E\{\boldsymbol{e}\boldsymbol{e}^T\} = \sigma^2\boldsymbol{I} \tag{5.40}$$

and hence,

$$\boldsymbol{C}_w = \sigma^2(\boldsymbol{X}^T\boldsymbol{X})^{-1} \tag{5.41}$$

Using (5.36) and taking into consideration that $\boldsymbol{e}$ is a Gaussian random vector, then the natural logarithm of its probability density is given by

$$\ln P(\boldsymbol{e};\boldsymbol{w}) = \ln\left\{\frac{1}{(2\pi)^{N/2}}\frac{1}{|\boldsymbol{C}_e|}\exp\left[-\frac{1}{2}(\boldsymbol{y}-\boldsymbol{X}\boldsymbol{w})^T\boldsymbol{C}_e^{-1}(\boldsymbol{y}-\boldsymbol{X}\boldsymbol{w})\right]\right\} \tag{5.42}$$

$$= -\ln(2\pi\sigma^2)^{N/2} - \frac{1}{2\sigma^2}(\boldsymbol{y}-\boldsymbol{X}\boldsymbol{w})^T(\boldsymbol{y}-\boldsymbol{X}\boldsymbol{w})$$

since $\boldsymbol{C}_e = \sigma^2\boldsymbol{I}$ and $|\boldsymbol{C}_e|$ implies the determinant of $\boldsymbol{C}_e$. Next, we differentiate (5.42) with respect to parameter $\boldsymbol{w}$. Hence, we find

$$\frac{\partial \ln(\boldsymbol{e};\boldsymbol{w})}{\partial \boldsymbol{w}} = -\frac{1}{2\sigma^2}\frac{\partial}{\partial \boldsymbol{w}}(\boldsymbol{y}^T\boldsymbol{y} - 2\boldsymbol{y}^T\boldsymbol{X}\boldsymbol{w} + \boldsymbol{w}^T\boldsymbol{X}^T\boldsymbol{X}\boldsymbol{w}) \tag{5.43}$$

since $\boldsymbol{y}^T\boldsymbol{X}\boldsymbol{w} = \boldsymbol{w}^T\boldsymbol{X}^T\boldsymbol{y} = \text{scalar}$. Using the identity below (see Appendix 2)

$$\frac{\partial \boldsymbol{b}^T\boldsymbol{w}}{\partial \boldsymbol{w}} = \boldsymbol{b}, \qquad \frac{\partial \boldsymbol{w}^T\boldsymbol{A}\boldsymbol{w}}{\partial \boldsymbol{w}} = 2\boldsymbol{A}\boldsymbol{w} \qquad (\boldsymbol{A}\text{ is symmetric}) \tag{5.44}$$

Equation 5.43 becomes

$$\frac{\partial \ln(\boldsymbol{e};\boldsymbol{w})}{\partial \boldsymbol{w}} = \frac{1}{\sigma^2}(\boldsymbol{X}^T\boldsymbol{y} - \boldsymbol{X}^T\boldsymbol{X}\boldsymbol{w}) \tag{5.45}$$

Assuming that $\boldsymbol{X}^T\boldsymbol{X}$ is invertible, then

$$\frac{\partial \ln(\boldsymbol{e};\boldsymbol{w})}{\partial \boldsymbol{w}} = \frac{\boldsymbol{X}^T\boldsymbol{X}}{\sigma^2}[(\boldsymbol{X}^T\boldsymbol{X})^{-1}\boldsymbol{X}^T\boldsymbol{y} - \boldsymbol{w}] = \boldsymbol{I}(\boldsymbol{w})[\boldsymbol{g}(\boldsymbol{w}) - \boldsymbol{w}] \tag{5.46}$$

From the Cramer–Rao lower bound (CRLB) theorem, $\hat{\boldsymbol{w}}$ is the minimum variance unbiased (MVU) estimator, since we have found that

$$\hat{\boldsymbol{w}} = (\boldsymbol{X}^T\boldsymbol{X})^{-1}\boldsymbol{X}^T\boldsymbol{y} = \boldsymbol{g}(\boldsymbol{w}) \tag{5.47}$$

and (5.46) becomes

$$\frac{\partial \ln(\boldsymbol{e};\boldsymbol{w})}{\partial \boldsymbol{w}} = \frac{\boldsymbol{X}^T\boldsymbol{X}}{\sigma^2}(\hat{\boldsymbol{w}} - \boldsymbol{w}) \tag{5.48}$$

The matrix

$$I(w) = \frac{X^T X}{\sigma^2} \tag{5.49}$$

is known as the **Fisher information matrix**. In the CRLB theorem, the Fisher matrix is defined by the relation:

$$(Iw)_{ij} = E\left\{\frac{\partial^2 \ln P(e;w)}{\partial w_i \partial w_j}\right\} \tag{5.50}$$

and thus, the parameters are shown explicitly. Comparing (5.41) and (5.49), the MVU estimator of w is given by (5.47) and its covariance matrix is

$$C_w = I^{-1}(w) = \sigma^2 (X^T X)^{-1} \tag{5.51}$$

The MVU estimator of the linear model (5.36) is efficient since it attains the CRLB or, in other words, the covariance matrix is equal to the inverse of the Fisher information matrix.

Let us rewrite the error covariance matrix in the form:

$$C_w = I^{-1}(w) = \sigma^2 (X^T X)^{-1} = \frac{\sigma^2}{N}\left(\frac{1}{N} X^T X\right)^{-1} \tag{5.52}$$

where:

N is the number of equations in the vector equation (5.36). Let $\lim_{N\to\infty}[(1/N)X^T X]^{-1} = A$, where A is a rectangular constant matrix. Then

$$\lim_{N\to\infty} C_w = \lim_{N\to\infty} \frac{\sigma^2 A}{N} = \mathbf{0} \tag{5.53}$$

Since the covariance becomes zero as N goes to infinity it implies that $\hat{w} = w$. The above convergence property defines $\hat{w}$ as a **consistent estimator**.

The above development shows that, if a system is modeled as linear in the presence of white Gaussian noise, the LSE approach provides estimators that are unbiased and consistent.

5.2.4 The LS Approach

Using the LS approach, we try to minimize the squared difference between the given data (or desired data) $d(n)$ and the output signal of a linear time-invariant (LTI) system. The signal $y(n)$ is generated by some system, which in turn depends on its unknown parameters w_i's. The LSE of w_i's chooses the values that make y's closest to the given data. The measure of closeness is defined by the LSE [see also (5.19)]. For the one-coefficient system model, we have

$$J(w) = \sum_{n=1}^{N} [d(n) - y(n)]^2 \tag{5.54}$$

and the dependence of J on w is via $y(n)$. The value of w that minimizes the cost function $J(w)$ is the LSE. It is apparent that the performance of LSE will depend on the statistical properties of the corrupting noise to the signal as well as any system modeling error.

Example 5.2.5

Let us assume that the signal is $y(n) = a\cos(\omega_0 n)$, where ω_0 is known and the amplitude a must be determined. Hence, the LSE minimizes the cost function:

$$J(a) = \sum_{n=1}^{N} [d(n) - a\cos(\omega_0 n)]^n \tag{5.55}$$

Therefore, we obtain

$$\frac{\partial J(a)}{\partial a} = \sum_{n=1}^{N} (-)2\cos(\omega_0 n)(d(n) - a\cos(\omega_0 n) = 0$$

$$\hat{a} = \frac{\sum_{n=1}^{N} d(n)\cos(\omega_0 n)}{\sum_{n=1}^{N} \cos(\omega_0 n)}$$

■

Let us assume that the output of a system is linear, and it is given by the relation $y(n) = x(n)w$, where $x(n)$ is a known sequence. Hence, the LSE criterion becomes

$$J(w) = \sum_{n=1}^{N} [d(n) - x(n)w]^2$$

The estimate value of w is

$$\hat{w} = \frac{\sum_{n=1}^{N} d(n)x(n)}{\sum_{n=1}^{N} x^2(n)}$$

and the minimum LS error is given by (see Problem 5.2.4)

$$J_{\min} = J(\hat{w}) = \sum_{n=1}^{N} d^2(n) - \hat{w}\sum_{n=1}^{N} d(n)x(n) = \sum_{n=1}^{N} d^2(n) - \frac{\left[\sum_{n=1}^{N} d(n)x(n)\right]^2}{\sum_{n=1}^{N} x^2(n)} \tag{5.56}$$

Example 5.2.6

Consider the experimental data shown in Figure 5.5. It is recommended that the linear model, $y(n) = a + bn$, for the data be used. Using the LSE approach, we find the cost function:

$$J(\boldsymbol{w}) = \sum_{n=1}^{N} [d(n) - a - bn]^2 = (\boldsymbol{d} - \boldsymbol{X}\boldsymbol{w})^T (\boldsymbol{d} - \boldsymbol{X}\boldsymbol{w}) \tag{5.57}$$

where:

$$\boldsymbol{w} = \begin{bmatrix} a \\ b \end{bmatrix}, \quad \boldsymbol{X} = \begin{bmatrix} 1 & 1 \\ 1 & 2 \\ \vdots & \\ 1 & N \end{bmatrix} \tag{5.58}$$

From (5.35), the estimate of $\boldsymbol{w}$ is

$$\hat{\boldsymbol{w}} = (\boldsymbol{X}^T \boldsymbol{X})^{-1} \boldsymbol{X}^T \boldsymbol{d} \tag{5.59}$$

and from the data shown in Figure 5.5

$$\hat{\boldsymbol{w}} = \begin{bmatrix} \hat{a} \\ \hat{b} \end{bmatrix} = \begin{bmatrix} 1.6147 \\ 0.0337 \end{bmatrix}$$

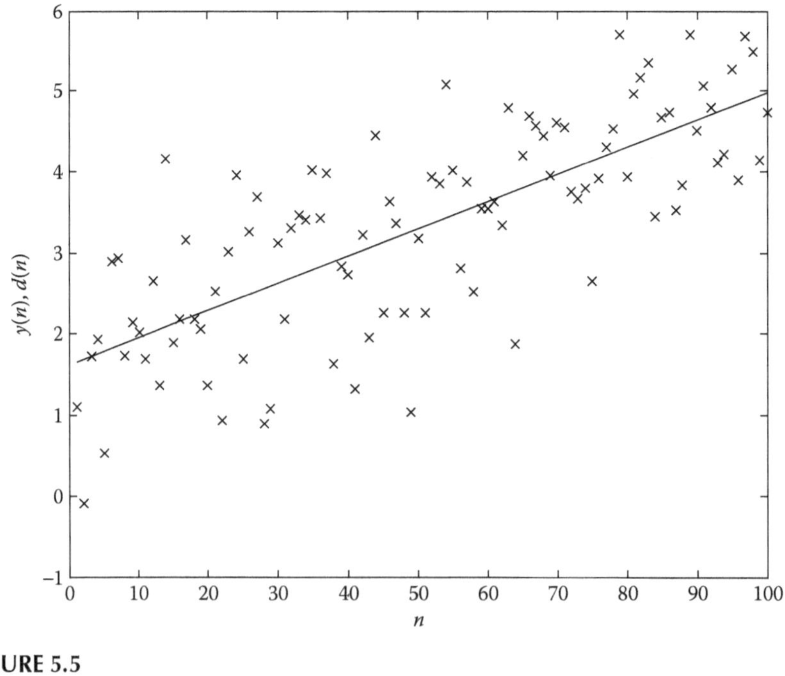

FIGURE 5.5

The straight line was also plotted to verify the procedure of LSE. The data were produced using the equation $d(n) = 1.5 + 0.035n + \mathrm{randn}(n)$ for $n = 1, 2, \ldots, 100$.

■

5.2.5 Orthogonality Principle

To obtain the orthogonality principle for the LS problem, we follow the procedure developed for the Wiener filter. Therefore, using unweighted sum of the squares of the error, we obtain

$$\frac{\partial J(w_1, w_2, \ldots, w_M)}{\partial w_k} = \frac{\partial}{\partial w_k}\left[\sum_{m=1}^{N} e(m)e(m)\right] = 2\sum_{m=1}^{N} e(m)\frac{\partial e(m)}{\partial w_k} \quad k = 1, 2, \ldots, M \quad (5.60)$$

But (5.10) and (5.11) are equal to ($\boldsymbol{w}$ has M coefficients)

$$e(m) = d(m) - \sum_{k=1}^{M} w_k x_k(m) \quad (5.61)$$

and, therefore, taking the derivative of $e(m)$ with respect of w_k and introducing the results of (5.60), we obtain

$$\frac{\partial J}{\partial w_k} = -2\sum_{m=1}^{N} e(m)x_k(m) \quad (5.62)$$

We note that when $\boldsymbol{w} = \hat{\boldsymbol{w}}$ (the optimum value), we have the relationship $\left(\partial J/\partial w_k\right) = 0$ for $k = 1, 2, \ldots, M$, and hence, (5.62) becomes

$$\sum_{m=1}^{N} \hat{e}(m)x_k(m) = \hat{\boldsymbol{e}}^T \boldsymbol{x}_k = \boldsymbol{0} \qquad k = 1, 2, \ldots, M \quad (5.63)$$

where:

$$\hat{\boldsymbol{e}} = [\hat{e}(1)\ \hat{e}(2)\ \hat{e}(3)\ \ldots\ \hat{e}(N)]^T = \boldsymbol{d} - \hat{\boldsymbol{y}} \quad (5.64)$$

$$\boldsymbol{x}_k = [x_k(1)\ \ x_k(2)\ \ \ldots\ \ x_k(N)]^T \qquad k = 1, 2, \ldots, M \quad (5.65)$$

The estimate error $\hat{e}(m)$ is optimum in the LS sense. The above result is known as the **principle of orthogonality**.

5.2.6 Corollary

Equation 5.10 may be written as the sum of the columns of $\boldsymbol{X}$ as follows:

$$\hat{\boldsymbol{y}} = \sum_{k=1}^{M} \boldsymbol{x}_k(n)\hat{w}_k \qquad n = 1, 2, \ldots, N \quad (5.66)$$

Multiplying (5.66) by $\hat{\boldsymbol{e}}^T$ and taking into consideration (5.63), we obtain

$$\hat{\boldsymbol{e}}^T \hat{\boldsymbol{y}} = 0 \tag{5.67}$$

The above corollary indicates that when the coefficients of the filter are optimum in the LS sense, then the output of the filter and the error are orthogonal.

Example 5.2.7

Using the results of Example 5.2.4, we find

$$\hat{\boldsymbol{e}} = \boldsymbol{d} - \hat{\boldsymbol{y}} = \begin{bmatrix} 1.05953819523825 \\ 0.91864528560697 \\ 0.48159639439564 \\ 1.21506254286554 \end{bmatrix}, \quad \hat{\boldsymbol{e}}^T \boldsymbol{x}_0 = 2.505 \times 10^{-15}, \quad \hat{\boldsymbol{e}}^T \boldsymbol{x}_1 = 1.457 \times 10^{-15}$$

■

5.2.7 Projection Operator

Projection operator is another form of interpretation to the solution of the LS problem. Let us, for clarity, assume that we have two vectors (N vectors in the N-dimensional case) $\boldsymbol{x}_k$ that form two-dimensional subspace (see Figure 5.6). The vectors $\boldsymbol{x}_1$ and $\boldsymbol{x}_2$ constitute the column space of the data matrix $\boldsymbol{X}$.

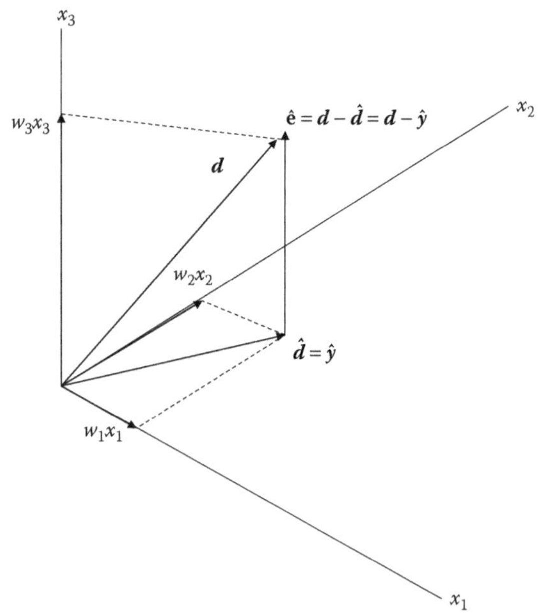

FIGURE 5.6

We note the following properties:

1. The vector $\hat{\boldsymbol{d}}$ is obtained as a linear combination of the data column space $\boldsymbol{x}_1, \boldsymbol{x}_2, \ldots, \boldsymbol{x}_M$ of $\boldsymbol{X}$ that constitutes the subspace of M.
2. From all the vectors in the subspace spanned by $\boldsymbol{x}_1, \boldsymbol{x}_2, \ldots, \boldsymbol{x}_M$, the vector $\hat{\boldsymbol{d}}$ has the minimum Euclidian distance from $\boldsymbol{d}$.
3. The distance $\hat{\boldsymbol{e}} = \boldsymbol{d} - \hat{\boldsymbol{d}}$ is a vector that is orthogonal to the subspace.

We also note that $\hat{\boldsymbol{y}}$ satisfies the above three properties. From (5.66), we observe that $\hat{\boldsymbol{y}}$ is a linear combination of the data column space, which spans the subspace. Next, minimizing $\hat{\boldsymbol{e}}^T \hat{\boldsymbol{e}}$, where $\hat{\boldsymbol{e}} = \boldsymbol{d} - \hat{\boldsymbol{d}}$, is equivalent to minimizing the Euclidian distance between $\boldsymbol{d}$ and $\hat{\boldsymbol{y}}$. The third property is satisfied by (5.67). Therefore, we can conclude that $\hat{\boldsymbol{y}}$ is the projection of $\boldsymbol{d}$ onto the subspace spanned by the vectors $\boldsymbol{x}_1, \boldsymbol{x}_2, \ldots, \boldsymbol{x}_M$.

Equation 5.66 can also be written in the matrix form:

$$\hat{\boldsymbol{y}} = \boldsymbol{X}\hat{\boldsymbol{w}} = \boldsymbol{X}(\boldsymbol{X}^T\boldsymbol{X})^{-1}\boldsymbol{X}^T\boldsymbol{d} \tag{5.68}$$

In the above equation, we set $\hat{\boldsymbol{w}} = \boldsymbol{R}^{-1}\boldsymbol{p}$ [see (5.25)], $\boldsymbol{R}^{-1} = (\boldsymbol{X}^T\boldsymbol{X})^{-1}$ [see (5.21)], and $\boldsymbol{p} = \boldsymbol{X}^T\boldsymbol{d}$ [see (5.22)]. Since the matrix

$$\boldsymbol{P} = \boldsymbol{X}(\boldsymbol{X}^T\boldsymbol{X})^{-1}\boldsymbol{X}^T \tag{5.69}$$

projects the desired vector in the N-dimensional space to $\hat{\boldsymbol{y}}$ in the M-dimensional subspace ($N > M$), it is known as the **projection matrix** or **projection operator**. The name is due to the fact that the matrix $\boldsymbol{P}$ projects the data vector $\boldsymbol{d}$ onto the column space of $\boldsymbol{X}$ to provide the LS estimate $\hat{\boldsymbol{y}}$ of $\boldsymbol{d}$.

The LS error can be expressed as

$$\hat{\boldsymbol{e}} = \boldsymbol{d} - \hat{\boldsymbol{y}} = \boldsymbol{d} - \boldsymbol{P}\boldsymbol{d} = (\boldsymbol{I} - \boldsymbol{P})\boldsymbol{d} \tag{5.70}$$

where:

$\boldsymbol{I}$ is an $N \times N$ identity matrix

The projection matrix is equal to its transpose (Hermitian for complex matrices) and **independent**, that is,

$$\boldsymbol{P} = \boldsymbol{P}^T, \quad \boldsymbol{P}^2 = \boldsymbol{P}^T\boldsymbol{P} = \boldsymbol{P} \tag{5.71}$$

The matrix $\boldsymbol{I} - \boldsymbol{P}$ is known as the orthogonal **complement projection operator**. The filter coefficients are given by

$$\hat{\boldsymbol{w}} = \boldsymbol{R}^{-1}\boldsymbol{p} = (\boldsymbol{X}^T\boldsymbol{X})^{-1}\boldsymbol{X}^T\boldsymbol{d} \tag{5.72}$$

$$\boldsymbol{X}^+ \equiv (\boldsymbol{X}^T\boldsymbol{X})^{-1}\boldsymbol{X}^T \tag{5.73}$$

is an $M \times N$ matrix known as the pseudoinverse or Moore–Penrose generalized inverse of matrix $\boldsymbol{X}$ (see Appendix 2).

Example 5.2.8

Using the data given in Example 5.2.4, we obtain

$$\boldsymbol{P} = \boldsymbol{P}\boldsymbol{X}(\boldsymbol{X}^T\boldsymbol{X})^{-1}\boldsymbol{X}^T = \begin{bmatrix} 0.7278 & -0.2156 & 0.2434 & 0.3038 \\ -0.2156 & 0.7762 & 0.0013 & 0.3566 \\ 0.2434 & 0.0013 & 0.0890 & 0.1477 \\ 0.3038 & 0.3566 & 0.1477 & 0.4068 \end{bmatrix}$$

$$\hat{y} = \boldsymbol{P}\boldsymbol{d} = [1.0595 \quad 0.9186 \quad 0.4815 \quad 1.2150]^T$$

$$\hat{e} = (\boldsymbol{I} - \boldsymbol{P})\boldsymbol{d} = [0.0595 \quad 0.0813 \quad 0.5184 \quad -0.2150]$$

■

5.2.8 LS Finite Impulse Response Filter

The error of the filter is given by

$$e(n) = d(n) - \sum_{k=1}^{M} w_k x(n-k) = d(n) - \boldsymbol{w}^T\boldsymbol{x}(n) \tag{5.74}$$

where:

$d(n)$ is the desired signal

$$\boldsymbol{x}(n) = [\,x(n) \quad x(n-1) \quad \cdots \quad x(n-M+1)]^T \tag{5.75}$$

is the input to the filter, and

$$\boldsymbol{w} = [\,w_1 \quad w_2 \quad \cdots \quad w_M\,]^T \tag{5.76}$$

is the filter coefficient vector.

It turns out that the exact form of e, d, and X depends on the range $N_i \le n \le N_f$ of the data to be used. Therefore, the range of the squared-error summation then becomes

$$J \triangleq E = \sum_{n=N_i}^{n=N_f} e^2(n) = \boldsymbol{e}^T\boldsymbol{e} \tag{5.77}$$

The LS finite impulse response (FIR) filter is found by solving the LS normal equations [(5.24) and (5.34)]:

$$(\boldsymbol{X}^T\boldsymbol{X})\,\hat{\boldsymbol{w}} = \boldsymbol{X}^T\boldsymbol{d} = \boldsymbol{p} \qquad (\text{or } \boldsymbol{R}\hat{\boldsymbol{w}} = \boldsymbol{p}) \tag{5.78}$$

with the minimum LS error

$$J_{\min} \triangleq E_{\min} = E_d - \boldsymbol{p}^T\hat{\boldsymbol{w}} \tag{5.79}$$

where:
$E_d = \boldsymbol{d}^T\boldsymbol{d}$ is the energy of the desired signal

The elements of the time-averaged correlation matrix $\boldsymbol{R}$ are given by (the real averaged correlation coefficients must be divided by $N_f - N_i$):

$$r_{ij} = \boldsymbol{x}_i^T\boldsymbol{x}_j = \sum_{n=N_i}^{n=N_f} x(n+1-i)x(n+1-j) \qquad 1 \le i,\; j \le M \tag{5.80}$$

There are two important different ways to select the summation range $N_i \le n \le N_f$ which are exploited in Problem 9.6.1: the **no-window case**, where $N_i = M-1$ and $N_f = N-1$, and the **full-window case**, where the range of the summation is from $N_i = 0$ to $N_{f=N+M-2}$. The no-window method is also known as the **autocorrelation** method and the full-window method is also known as the **covariance method**.

The covariance method, **data matrix *D***, is written as follows:

$$\boldsymbol{D}^T = [\boldsymbol{x}(M) \quad \boldsymbol{x}(M+1) \quad \cdots \quad \boldsymbol{x}(N)] = \begin{bmatrix} x(M) & x(M+1) & \cdots & x(N) \\ x(M-1) & x(M) & \cdots & x(N-1) \\ \vdots & \vdots & & \vdots \\ x(1) & x(2) & \cdots & x(N-M+1) \end{bmatrix} \tag{5.81}$$

Then the $M \times M$ time-averaged correlation matrix is given by

$$\boldsymbol{R} = \sum_{n=M}^{N} \boldsymbol{x}(n)\boldsymbol{x}^T(n) = \boldsymbol{D}^T\boldsymbol{D} \tag{5.82}$$

Book m-Function for Covariance Data Matrix

```
function[dT] = lms_data_matrix_covar_meth(x,M)
      %function[dT] = lms_data_matrix_covar_meth(x,M)
      %M = number of filter coefficients;x = data vector;
      %dT = transposed data matrix;
for m = 1:M
      for n = 1:length(x)-M+1
             dT(m,n) = x(M-m+n);
      end;
end;
```

Example 5.2.9

If the data vector is $\boldsymbol{x} = [0.7 \quad 1.4 \quad 0.4 \quad 1.3 \quad 0.1]^T$, the data filter has three coefficients ($M = 3$). Using the above Book m-function, we obtain

$$\boldsymbol{D}^T = \begin{bmatrix} 0.4 & 1.3 & 0.1 \\ 1.4 & 0.4 & 1.3 \\ 0.7 & 1.4 & 0.4 \end{bmatrix} \quad (\boldsymbol{D}^T \text{ is Toeplitz matrix})$$

$$\boldsymbol{R} = \boldsymbol{D}^T\boldsymbol{D} = \begin{bmatrix} 1.8600 & 1.2100 & 2.1400 \\ 1.2100 & 3.8100 & 2.0600 \\ 2.1400 & 2.0600 & 2.6100 \end{bmatrix}$$

■

The data matrix in (5.81) has the following properties:

Property 1: The correlation matrix $\boldsymbol{R}$ is equal to its transpose (for complex quantities $\boldsymbol{R}$ is Hermitian $\boldsymbol{R} = \boldsymbol{R}^T$). The proof is directly found from (5.82).
Property 2: The correlation matrix is nonnegative definite, $\boldsymbol{a}^T\boldsymbol{R}\boldsymbol{a} \geq 0$ for any $M \times 1$ vector $\boldsymbol{a}$.
Property 3: The eigenvalues of the correlation matrix R are all real and nonnegative (see Chapter 6).
Property 4: The correlation matrix is the product of two rectangular Toeplitz matrices that are the transpose of each other (see Example 5.2.9).

The following Book m-function produces the no-window FIR filter.

Book m-Function No-Window LS Method

```
function[R,w,Jmin] = lms_no_window_least_square_fir(x,M,d)
    %x = data of length N;M = number of filter coefficients;
    %d = desired signal = [d(M) d(M+1) ... d(N)]';
N = length(x);
for i = 1:M
    for j = 1:N-M+1
            D(i,j) = x(M-i+j);
    end;
end;
Dt = D';
R = D*Dt;
p = D*d(1,1:N-M+1)';
w = inv (R)*p;
Jmin = d'*d-p'*w;
```

5.3 THE MEAN-SQUARE ERROR

In this chapter, we develop a class of linear optimum discrete-time filters known as the **Wiener filters**. These filters are optimum in the sense of minimizing an appropriate function of the error known as the **cost function**. The cost function that is

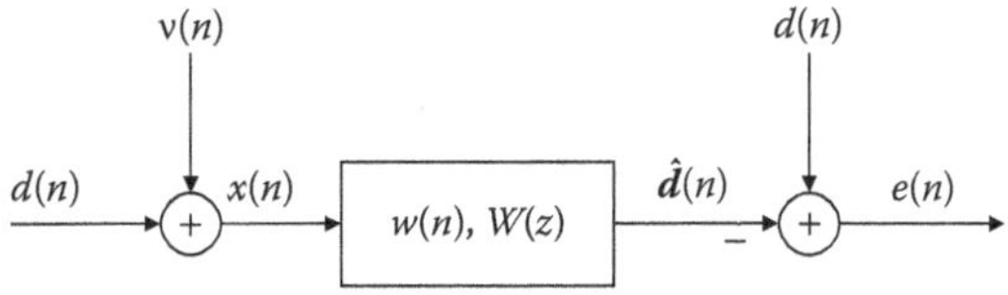

FIGURE 5.7

commonly used in filter design optimization is the **mean-square error** (MSE). Minimizing the MSE involves only second-order statistics (correlations) and leads to a theory of linear filtering that is useful in many practical applications. This approach is common to all optimum filter designs. Figure 5.7 shows the block diagram presentation of the optimum filter problem. The MSE approach is mathematically tractable and the performance function has a global minimum.

Based on the MSE approach, we try to establish a linear-invariant filter, $W(z)$, such that a signal $\{x(n)\}$ produces the minimum mean-square estimate, $\hat{d}(n)$, with a desired one, $d(n)$. The Wiener filter can consider the following problems:

1. **Filtering**: Given a noisy signal $x(n) = d(n) + v(n)$, the filter will approximate the desired one, $d(n)$, based on the present and past values of $x(n)$.
2. **Smoothing**: The Wiener filter can be designed to estimate $d(n)$ from $x(n)$ using all the available data.
3. **Prediction**: If $d(n) = x(n+1)$ and $W(z)$ are causal filters, then the Wiener filter becomes a linear predictor.
4. **Deconvolution**: When $x(n) = d(n) * g(n) + v(n)$, with $g(n)$ being the unit sample response of a linear shift-invariant filter, the Wiener filter becomes a deconvolution filter.

The basic idea is to recover a desired signal $d(n)$, given a noisy observation $x(n) = d(n) + v(n)$, where both $d(n)$ and $v(n)$ are assumed to be wide-sense (or weakly) stationary (WSS) processes. Therefore, the problem can be stated as follows:

> Design a filter that produces an estimate $\hat{d}(n)$ using a linear combination of the data $x(n)$ such that the mean square error (MSE) function (cost function)
>
> $$J = E\{(d(n) - \hat{d}(n))^2\} = E\{e^2(n)\} \tag{5.83}$$
>
> is minimized.

Depending on how the data $\{x(n)\}$ and the desired signal $d(n)$ are related, there are four basic problems that need solution: filtering, smoothing, prediction, and deconvolution.

Observe that in (5.83) the LS formula appears again but with the difference that we take the expectation (ensemble average) value of the square value of the error. To proceed with our development and crate useful formula, we will assume initially that we are able to repeat the experiment, produce the probability density function (pdf), and then find the MSE using the formula:

$$E\{e^2(n)\} = \int_{-\infty}^{\infty} e^2(n) f[e(n)] de(n) \tag{5.84}$$

where:

$f(.)$ is the pdf of the random variable (rv) $e(n)$ of the error at time n

However, when we proceed to find the final formula, we will proceed to define the signals as ergotic and then find their means, variances, autocorrelation, and cross-correlation using the approximate formulas in time domain as were developed in Chapter 4.

5.3.1 The FIR Wiener Filter

Let the sample response (filter coefficients) of the desired FIR filter, shown in Figure 5.8, be denoted by $\boldsymbol{w}$. In this chapter, we use w's for filter coefficients since it is customary to use this symbol in books involving adaptive and optimum filtering of random signals. This filter will process the real-valued stationary process $\{x(n)\}$ to produce an estimate $\hat{d}(n)$ of the desired real-valued signal $d(n)$. Without loss of generality, we will assume, unless otherwise stated, that the processes $\{x(n)\}$, $d(n)$, and so on have zero mean values. Furthermore, assuming that the filter coefficients do not change with time, the output of the filter is equal to the convolution of the input and the filter coefficients. Hence, we obtain

$$\hat{d}(n) = \sum_{m=0}^{M-1} w_m x(n-m) = \boldsymbol{w}\boldsymbol{x}^T(n) \tag{5.85}$$

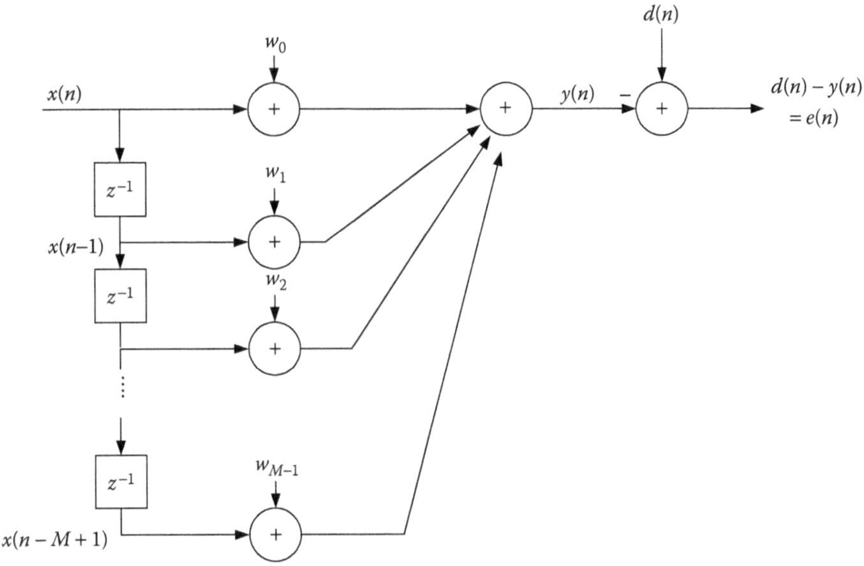

FIGURE 5.8

where:

M is the number of filter coefficients

$$\begin{aligned} \boldsymbol{w} &= [w_0 \quad w_1 \quad \cdots \quad w_{M-1}]^T \\ \boldsymbol{x}(n) &= [x(n) \quad x(n-1) \quad x(n-2) \quad \cdots \quad x(n-M+1)]^T \end{aligned} \tag{5.86}$$

The MSE [see (5.83)] is given by

$$\begin{aligned} J(\boldsymbol{w}) &= E\{e^2(n)\} = E\{[d(n)-\boldsymbol{w}^T\boldsymbol{x}(n)]^2\} \\ &= E\{[d(n)-\boldsymbol{w}^T\boldsymbol{x}(n)][d(n)-\boldsymbol{w}^T\boldsymbol{x}(n)]^T\} \\ &= E\{d^2(n)-\boldsymbol{w}^T\boldsymbol{x}(n)d(n)-d(n)\boldsymbol{x}^T(n)\boldsymbol{w}+\boldsymbol{w}^T\boldsymbol{x}(n)\boldsymbol{x}^T(n)\boldsymbol{w}\} \\ &= E\{d^2(n)\}-2\boldsymbol{w}^T E\{d(n)\boldsymbol{x}(n)\}+\boldsymbol{w}^T E\{\boldsymbol{x}(n)\boldsymbol{x}^T(n)\}\boldsymbol{w} \\ &= \sigma_d^2 - 2\boldsymbol{w}^T\boldsymbol{p}_{dx}+\boldsymbol{w}^T\boldsymbol{R}_x\boldsymbol{w} \end{aligned} \tag{5.87}$$

where:

$$\begin{aligned} &\boldsymbol{w}^T\boldsymbol{x}(n) = \boldsymbol{x}^T(n)\boldsymbol{w} = \text{number} \\ &\sigma_d^2 = \text{variance of the desired signal, } d(n) \\ &\boldsymbol{p}_{dx} = [p_{dx}(0) \quad p_{dx}(1) \quad \cdots \quad p_{dx}(M-1)]^T = \text{cross-correlation vector between } d(n) \text{ and } \boldsymbol{x}(n) \\ &p_{dx}(0) = r_{dx}(0),\ p_{dx}(1) = r_{dx}(1),\ \ldots,\ p_{dx}(M-1) = r_{dx}(M-1) \end{aligned} \tag{5.88}$$

$$\begin{aligned} \boldsymbol{R}_x &= E\left\{\begin{bmatrix} x(n) \\ x(n-1) \\ \vdots \\ x(n-M+1) \end{bmatrix} [x(n) \quad x(n-1) \quad \cdots \quad x(n-M+1)]\right\} \\ &= \begin{bmatrix} E\{x(n)x(n)\} & E\{x(n)x(n-1)\} & \cdots & E\{x(n)x(n-M+1)\} \\ E\{x(n-1)x(n)\} & E\{x(n-1)x(n-1)\} & \cdots & E\{x(n-1)x(n-M+1)\} \\ \vdots & & & \vdots \\ E\{x(n-M+1)x(n)\} & E\{x(n-M+1)x(n-1)\} & \cdots & E(x(n-M+1)x(n-M+1)\} \end{bmatrix} \\ &= \begin{bmatrix} r_x(0) & r_x(1) & \cdots & r_x(M-1) \\ r_x(-1) & r_x(0) & \cdots & r_x(M-2) \\ \vdots & & & \vdots \\ r_x(-M+1) & r_x(-M+2) & \cdots & r_x(0) \end{bmatrix} \end{aligned} \tag{5.89}$$

The above matrix is the correlation matrix of the input data, and it is symmetric because the random process is assumed to be stationary. Therefore, we have the relation $r_{xx}(k) = r_{xx}(-k)$. Since in practice we generally have one realization, we assume that the signal is ergotic and we find the correlations using the time approach. Hence, the correlation matrix based on realization will be an approximate one and not the theoretical one based on ensemble averages.

Example 5.3.1

Let us assume that we have found the sample autocorrelation coefficients $(r_{xx}(0) = 1.0, r_{xx}(1) = 0)$ from a set of data $\{x(n)\}$, which, in addition to noise, contains the desired signal. Furthermore, let the variance of the desired signal be $\sigma_d^2 = 24.50$ and the cross-correlation vector be $\boldsymbol{p}_{dx} = [2 \quad 4.5]^T$. It is desired to find the surface defined by the mean-square function $J(\boldsymbol{w})$ which is of second order with respect to the filter coefficients w_i.

Solution: Substituting the values given above in (5.87), we obtain

$$J(\boldsymbol{w}) = 24.50 - 2[w_0 \quad w_1]\begin{bmatrix} 2 \\ 4.5 \end{bmatrix} + [w_0 \quad w_1]\begin{bmatrix} 1 & 0 \\ 0 & 1 \end{bmatrix}\begin{bmatrix} w_0 \\ w_1 \end{bmatrix}$$

$$= 24.50 - 4w_0 - 9w_1 + w_0^2 + w_1^2$$

Note that the equation is quadratic with respect to filter coefficients and it is true for any number of filter coefficients. This is because we used the MSE approach for minimizing the error. Figure 5.9 shows the schematic representation of the Wiener filter. The data are the sum of the desired signal and the noise. From the data, we find the correlation matrix and the cross-correlation between the desired signal and the data. Note that to find the optimum Wiener filter coefficients the desired signal is needed. Figure 5.10 shows the **MSE surface**. This surface is found by inserting different values of w_0 and w_1 into the function $J(\boldsymbol{w})$. The values of the coefficients that correspond to the bottom of the surface are the **optimum** Wiener coefficients. The vertical distance from the $w_0 - w_1$ plane to the bottom of the surface is known as the **minimum error**, $J_{\min}$, and corresponds to the optimum Wiener coefficients. We observe that the minimum height of the surface corresponds to about $w_0 = 2$ and $w_1 = 4.5$, which are the optimum coefficients as

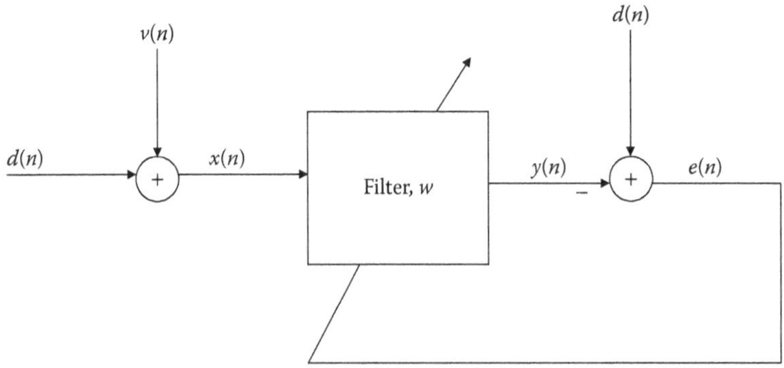

FIGURE 5.9

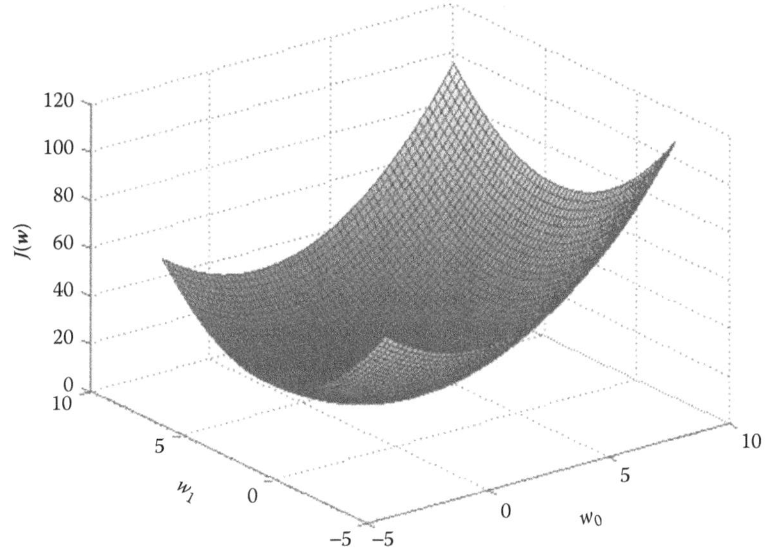

FIGURE 5.10

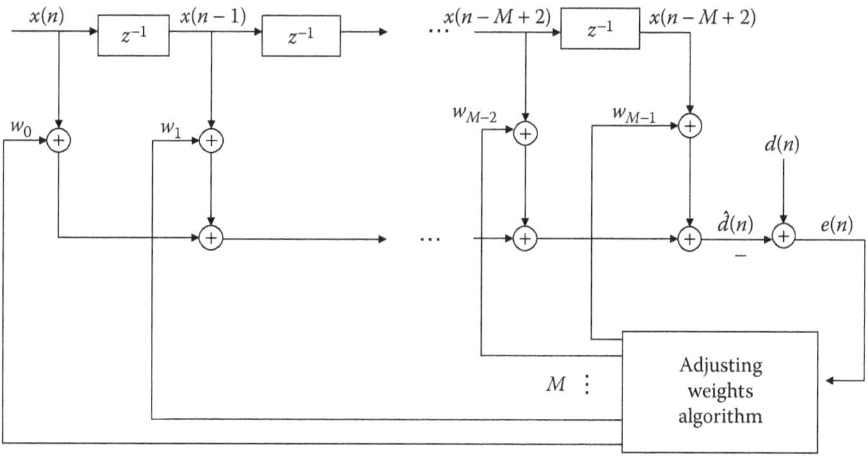

FIGURE 5.11

we will learn how to find them in Section 5.4. Figure 5.11 shows an adaptive FIR filter. The following Book MATLAB program is used to produce the MSE surface as shown in Figure 5.10.

Book MATLAB Program

```
>>x = -2:0.2:10; y = -1:0.2:10;
>>{X,Y] = meshgrid(x,y);
```

```
>>Z = 24.50-4*X-9*Y+X.^2+Y.^2;
>>colormap(gray);
>>surf(X,Y,Z);
>>hold on; xlabel(w_0');ylabel('w_1');zlabel('J(w)');
```

■

5.4 THE WIENER SOLUTION

In Figure 5.10, we observe that there exists a plane touching the parabolic surface at its minimum point, and is parallel to the w-plane. Furthermore, we observe that the surface is concave upward, and therefore, calculus tell us that the first derivative of the MSE with respect to w_0 and w_1 must be zero at the minimum point and the second derivative must be positive. Hence, we write

$$\frac{\partial J(w_0, w_1)}{\partial w_0} = 0 \qquad \frac{\partial J(w_0, w_1)}{\partial w_1} = 0 \tag{5.90a}$$

$$\frac{\partial^2 (w_0, w_1)}{\partial^2 w_0} > 0 \qquad \frac{\partial^2 J(w_0, w_1)}{\partial^2 (w_1)} > 0 \tag{5.90b}$$

For a two-coefficient filter, (5.87) becomes

$$J(w_0, w_1) = w_0^2 r_x(0) + 2w_0 w_1 r_x(1) + w_1^2 r_x(0) - 2w_0 r_{dx}(0) - 2w_1 r_{dx}(1) + \sigma_d^2 \tag{5.91}$$

Introducing next the above equation into (5.90a) produces the following set of equations:

$$2w_0^o r_x(0) + 2w_1^o r_x(1) - 2r_{dx}(0) = 0 \tag{5.92a}$$

$$2w_1^o r_x(0) + 2w_0^o r_x(1) - 2r_{dx}(1) = 0 \tag{5.92b}$$

The above system can be written in the following matrix form called the Wiener–Hopf equation:

$$\boldsymbol{R}_x \boldsymbol{w}^o = \boldsymbol{p}_{dx} \tag{5.93}$$

where:

the superscript "o" indicates the optimum Wiener solution for the filter

Note that to find the correlation matrix we must know the second-order statistics. Furthermore, we must reiterate that mostly we have only one realization at hand and, as a consequence, the data matrix is an approximate one. If, in addition, the matrix is invertible, the optimum filter is given by

$$\boldsymbol{w}^o = \boldsymbol{R}_x^{-1} \boldsymbol{p}_{dx} \tag{5.94}$$

For an M-order filter, $\boldsymbol{R}_x$ is an $M \times M$ matrix, $\boldsymbol{w}^o$ is an $M \times 1$ vector, and $\boldsymbol{p}_{dx}$ is also an $M \times 1$ vector.

If we differentiate again (5.92) with respect to w_0^o and w_1^o, which is equivalent in differentiating $J(\boldsymbol{w})$ twice, we find that it is equal to $2r_x(0)$. But $r_x(0) = E\{x(m)x(m)\} = \sigma_x^2 > 0$, and then the surface is concave upward. Therefore, the extremum is the minimum point of the surface. If we, next, introduce (5.94) in (5.87), we obtain the minimum error:

$$J_{\min} = \sigma_d^2 - \boldsymbol{p}_{dx}^T \boldsymbol{w}^o = \sigma_d^2 - \boldsymbol{p}_{dx}^T \boldsymbol{R}_x^{-1} \boldsymbol{p}_{dx} \tag{5.95}$$

which indicates that the minimum point of the error surface is at a distance $J_{\min}$ above the w-plane. The above equation shows that if no correlation exists between the desired signal and the data, the error is equal to the variance of the desired signal.

The problem we are facing is how to choose the length of the filter M. In the absence of *a priori* information, we compute the optimum coefficients, starting from a small reasonable number. As we increase the number, we check the minimum MSE (MMSE), and if its value is small enough, for example, MMSE <0.01, we accept the corresponding number of the coefficients.

Example 5.4.1

It is desired to obtain the filter coefficients of an unknown filter (system), which is shown in Figure 5.12, together with the Wiener filter. The coefficients of the unknown filter are $b_0 = 1$ and $b_1 = 0.38$.

Solution: The following Book m-file is used:

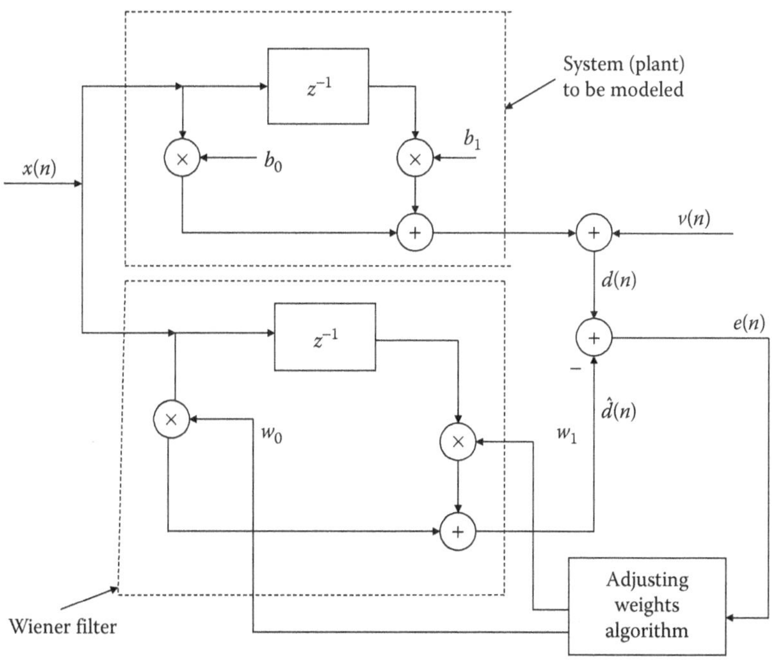

FIGURE 5.12

Book m-File: ex5_4_1

```
%Book m-file:ex5_4_1
v = 0.5*(rand(1,20)-0.5);%v = noise vector
      %(20 uniformly distributed rv's with mean zero);
x = randn(1,20);%x = data vector entering
      %the system and the Wiener filter (20 normal
      %distributed rv's with mean zero;
sysout = filter([1 0.38],1,x);%sysout = system
      %output with x as input; filter(b,a,x) is a
      %MATLAB function, where b is the
      %vector of the coefficients of the ARMA numerator,
      % a is the vector of the coefficients
      %of the ARMA denominator (see (2.4.7));
dn = sysout+v;
rx = lms_sample_biased_autoc(x,2);%book MATLAB
      %function with lag = 2;
Rx = toeplitz(rx);%toeplitz() is a MATLAB function
      %that gives the symmetric autocorrelation matrix;
pdx = xcorr(x,dn,'biased');%xcorr() is a MATLAB function
      %that gives a symmetric biased cross-correlation;
p = pdx(1,19:20);
w = inv(Rx)*p';
dnc = lms_sample_biased_autoc(dn,1);%s2dn = variance
      %of the desired signal;

jmin = dnc-p*w;
```

Some typical values obtained after running the program were $\boldsymbol{w} = \begin{bmatrix} 0.9205 & 0.3855 \end{bmatrix}$, $\boldsymbol{R}_x = [0.7447 \quad -0.0486; \quad -0.0486 \quad 0.7447]$, and $J_{\min} = 0.0114$.

■

5.4.1 Orthogonality Condition

In order for the set coefficients to minimize the cost function $J(\boldsymbol{w})$, it is necessary and sufficient that the derivatives of $J(\boldsymbol{w})$ with respect to w_k be equal to zero for $k = 0,\ 2,\ 3, \ldots,\ M-1$. Hence,

$$\frac{\partial J}{\partial w_k} = \frac{\partial}{\partial w_k} E\{e(n)e(n)\} = 2E\left\{e(n)\frac{\partial e(n)}{\partial w_k}\right\} = 0 \text{ and}$$
$$e(n) = d(n) - \sum_{m=0}^{M-1} w_m x(n-m) \tag{5.96}$$

Therefore, it follows that

$$\frac{\partial e(n)}{\partial w_k} = -x(n-k) \tag{5.97}$$

Therefore, (5.96) becomes

$$E\{e^o(n)x(n-k)\} = 0 \qquad k = 0, 1, 2, \ldots, M-1 \tag{5.98}$$

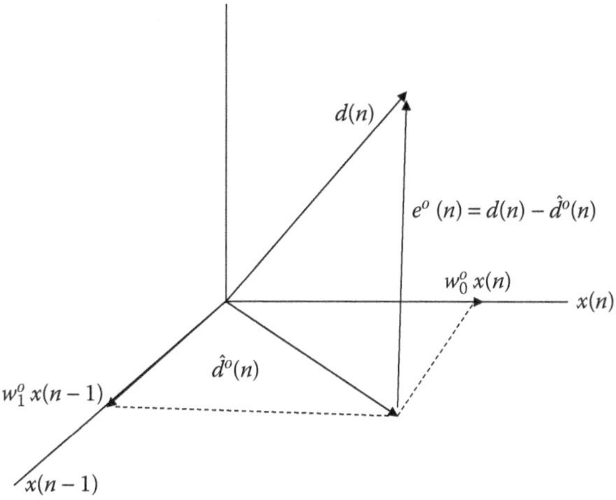

FIGURE 5.13

where:

the superscript "o" denotes that the corresponding w_k's used to find the estimation error $e^o(n)$ are the optimal ones. Figure 5.13 illustrates the orthogonality principle where the error $e^o(n)$ is orthogonal (perpendicular) to the data set $\{x(n)\}$ when the estimator employs the optimum set of filter coefficients.

Next, let us continue and write

$$E\{e^o(n)\hat{d}^{\,o}(n)\} = E\left\{e^o(n)\sum_{i=0}^{M-1} w_i^o(n)x(n-i)\right\} = \sum_{i=0}^{M-1} w_i^o(n)E\{e^o(n)x(n-i)\} \quad (5.99)$$

where:

$\hat{d}^{\,o}(n)$ is the Wiener filter output when its tap weights (coefficients) are set to their optimum values

Then, using (5.98) and (5.99), we obtain

$$E\{e^o(n)\hat{d}^{\,o}(n)\} = 0 \quad (5.100)$$

Note: The optimized Wiener filter output and the estimated error are orthogonal.

5.4.2 Normalized Performance Equation

From the error equation, we write

$$d(n) = e^o(n) + \hat{d}^{\,o}(n) \quad (5.101)$$

Squaring both sides of the above equation and taking the ensemble average (expectation), we get

$$E\{d^2(n)\} = E\{[e^o(n)]^2\} + E\{[\hat{d}^{\,o}(n)]^2\} + 2E\{e^o(n)\hat{d}^{\,o}(n)\} \quad (5.102)$$

But $E\{[e^o(n)]^2\} = J_{\min}$ (see Problem 5.4.3) and the last term is zero due to (5.100). Thus, we obtain

$$J_{\min} = E\{d^n(n)\} - E\{[\hat{d}^o(n)]\} \tag{5.103}$$

Note: The MMSE at the Wiener filter output is the difference between the MSE of the desired output and the MSE of the best estimate of that at the filter output.

The normalized performance function is defined by

$$\zeta = \frac{J}{E\{d^2(n)\}}, \qquad \zeta_{\min} = \frac{E\{d^2(n)\} - E\{\hat{d}^o(n)\}}{E\{d^2(n)\}} = 1 - \frac{E\{\hat{d}^o(n)\}}{E\{d^2(n)\}} \tag{5.104}$$

Note: The minimum performance is an indication of the ability of the filter to estimate the desired output. A value of $\zeta_{\min}$ close to zero is an indication of good performance of the filter, and a value close to one indicates poor performance.

5.4.3 Canonical Form of the Error-Performance Surface

We have developed above (5.87) the MSE in the form:

$$J(\boldsymbol{w}) = \sigma_d^2 - \boldsymbol{w}^T\boldsymbol{p} - \boldsymbol{p}^T\boldsymbol{w} + \boldsymbol{w}^T\boldsymbol{R}\boldsymbol{w} \tag{5.105}$$

The above equation can be written in the form:

$$J(\boldsymbol{w}) = \sigma_d^2 - \boldsymbol{p}^T\boldsymbol{R}^{-1}\boldsymbol{p} + (\boldsymbol{w} - \boldsymbol{R}^{-1}\boldsymbol{p})^T\boldsymbol{R}(\boldsymbol{w} - \boldsymbol{R}^{-1}\boldsymbol{p}) \tag{5.106}$$

At the optimum condition, the above equation becomes

$$J(\boldsymbol{w}) = J_{\min} + (\boldsymbol{w} - \boldsymbol{w}^o)^T\boldsymbol{R}(\boldsymbol{w} - \boldsymbol{w}^o) \tag{5.107}$$

We observe that at the optimum condition $J(\boldsymbol{w}^o) = J_{\min}$.

To simplify the error-performance surface, we proceed to change the basis on which is defined. To proceed, we use **eigendecomposition** and express the autocorrelation matrix in the form:

$$\boldsymbol{R} = \boldsymbol{Q}\Lambda\boldsymbol{Q}^T \tag{5.108}$$

The matrix Λ is a diagonal matrix consisting of the eigenvalues $\lambda_0, \lambda_1, \lambda_2, \ldots, \lambda_{M-1}$ of the autocorrelation matrix and the matrix $\boldsymbol{Q}$ has for its columns $\boldsymbol{q}_0, \boldsymbol{q}_1, \boldsymbol{q}_2, \ldots, \boldsymbol{q}_{M-1}$ associated with these eigenvalues, respectively. Substituting (5.108) in (5.107), we obtain

$$J(\boldsymbol{w}) = J_{\min} + (\boldsymbol{w} - \boldsymbol{w}^o)^T\boldsymbol{Q}\Lambda\boldsymbol{Q}^T(\boldsymbol{w} - \boldsymbol{w}^o) \tag{5.109}$$

Next, we introduce the transformation

$$\boldsymbol{v} = \boldsymbol{Q}^T(\boldsymbol{w} - \boldsymbol{w}^o) \tag{5.110}$$

into (5.109) to obtain

$$J(\boldsymbol{w}) = J_{\min} + \boldsymbol{v}^T \Lambda \boldsymbol{v} = J_{\min} + \sum_{k=0}^{M-1} \lambda_k v_k^2 \tag{5.111}$$

where:

v_k is the kth component of the vector $\boldsymbol{v}$. The components v_k constitute the principal axes of the surface.

5.5 WIENER FILTERING EXAMPLES

The examples in this section illustrate the use and utility of the Wiener filter.

Example 5.5.1 (Filtering)

Filtering of noisy signals (noise reduction) is extremely important, and the method has been used in many applications such as speech in noisy environment, reception of data across a noisy channel, and enhancement of images.

Let the received signal be $x(n) = d(n) + v(n)$, where $v(n)$ is a noise with zero mean and variance σ_v^2, and it is uncorrelated with the desired signal, $d(n)$. Hence,

$$\begin{aligned} p_{dx}(m) &= E\{d(n)x(n-m)\} = E\{d(n)d(n-m)\} + E\{d(n)\}E\{v(n-m)\} \\ &= E\{d^2(m)\} = r_{dd}(m) \end{aligned} \tag{5.112}$$

Similarly, we obtain

$$\begin{aligned} r_{xx}(m) &= E\{x(n)x(n-m)\} = E\{[d(n)+v(n)][d(n-m)+v(n-m)]\} \\ &= E\{d(n)d(n-m)\} + E\{v(n)d(n-m)\} + E\{d(n)v(n-m)\} \\ &\quad + E\{v(n)v(n-m)\} \\ &= r_{dd}(m) + r_{vv}(m) \end{aligned} \tag{5.113}$$

We have assumed that $d(n)$ and $v(n)$ are uncorrelated, $v(n)$ has zero mean value. Therefore, the Wiener equation becomes

$$(\boldsymbol{R}_d + \boldsymbol{R}_v)\boldsymbol{w}^o = \boldsymbol{p}_{dx} \tag{5.114}$$

The following Book m-file is used to produce the results shown in Figure 5.14.

Book m-File: ex5_5_1

```
%ex5_5_1 m-file
n = 0:511;
d = sin(0.1*pi*n);%desired signal
v = 0.5*randn(1,512);%white Gaussian noise
                     %with zero mean value;
x = d+v;%input signal to Wiener filter;
rd = lms_sample_biased_autoc(d,20);
     %rdx = rd = biased autoc function of the
     %desired signal (see (5.112));
rv = lms_sample_biased_autoc(v,20);
     %rv = biased autoc function of the noise;
R = toeplitz(rd(1,1:12))+toeplitz(rv(1,1:12));%see(5.114);
```

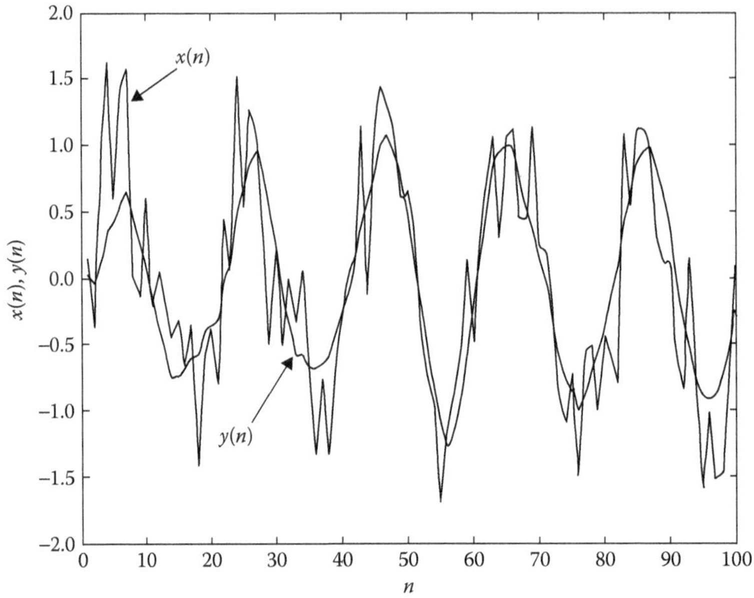

FIGURE 5.14

```
pdx = rd(1,1:12);
w = inv(R)*pdx';
y = filter(w',1,x);%output of the filter;
```

But

$$\sigma_x^2 = \sigma_d^2 + \sigma_v^2 \qquad \left[\sigma_x^2 = r_{xx}(0), \sigma_d^2 = r_{dd}(0), \sigma_v^2 = r_{vv}(0)\right] \tag{5.115}$$

and hence, from MATLAB function `var()`, we obtain for one run: `var(d) = var(x)-var(v) = 0.5094` and $J_{\min} = 0.5094 - p_{dx} * w^o = 0.0432$.

■

Example 5.5.2 (Filtering)

It is desired to find a two-coefficient Wiener filter for the communication channel that is shown in Figure 5.15. Let $v_1(n)$ and $v_2(n)$ be the white noises with zero mean, uncorrelated with each other, and with $d(n)$, and have the following variances: $\sigma_{v1}^2 = 0.31$, $\sigma_{v2}^2 = 0.12$. The desired signal produced by the first filter shown in Figure 5.15 is

$$d(n) = -0.796d(n-1) + v_1(n) \tag{5.116}$$

Therefore, the autocorrelation function of the desired signal becomes

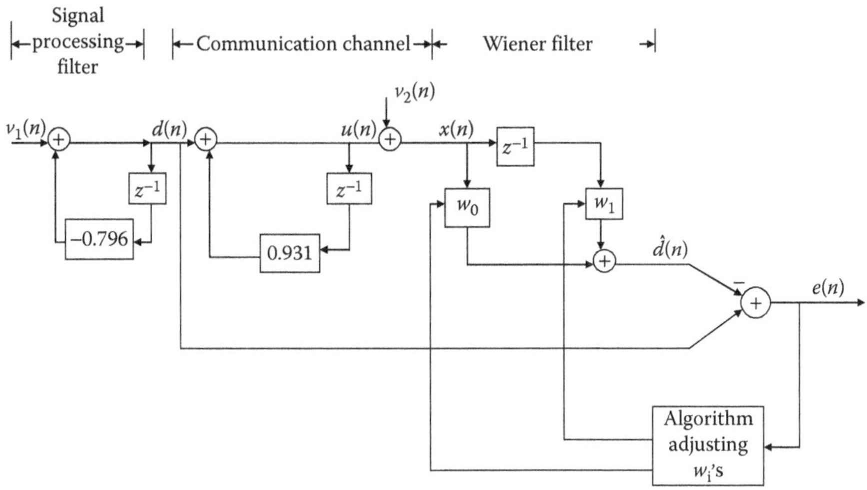

FIGURE 5.15

$$E\{d(n)d(n)\} = 0.796^2 E\{d^2(n-1)\} - 2\times 0.796 E\{d(n-1)\}E\{v_1(n)\} + E\{v_1^2(n)\}$$

$$\sigma_d^2 = 0.796^2\sigma_d^2 + \sigma_{v_1}^2 \quad \text{or} \quad \sigma_d^2 = \frac{0.31}{(1-0.796^2)} = 0.8461 \tag{5.117}$$

From the second filter, we obtain the relation:

$$d(n) = u(n) - 0.931u(n-1) \tag{5.118}$$

Introducing the above equation, for the expression of $d(n)$, into (5.116), we obtain

$$u(n) - 0.135u(n-1) - 0.7411u(n-2) = v_1(n) \tag{5.119}$$

But $x(n) = u(n) + v_2(n)$, and hence, the vector form of the set becomes $\boldsymbol{x}(n) = \boldsymbol{u}(n) + \boldsymbol{v}_2(n)$. Therefore, the autocorrelation matrix $\boldsymbol{R}_x$ becomes

$$E\left\{\boldsymbol{x}(n)\boldsymbol{x}^T(n)\right\} = \boldsymbol{R}_x = E\left\{[\boldsymbol{u}(n) + \boldsymbol{v}_2(n)][\boldsymbol{u}^T(n) + \boldsymbol{v}_2^T(n)]\right\} = \boldsymbol{R}_u + \boldsymbol{R}_{v_2} \tag{5.120}$$

We used assumption that $u(n)$ and $v_2(n)$ are uncorrelated zero-mean random sequences, which implies that $E\{\boldsymbol{v}_2(n)\boldsymbol{u}^T(n)\} = E\{\boldsymbol{u}(n)\}E\{\boldsymbol{u}^T(n)\} = 0$. Next, we multiply (5.119) by $u(n-m)$ and take the ensemble average of both sides, which results in the following expression:

$$r_{uu}(m) - 0.135r_{uu}(m-1) - 0.7411r_{uu}(m-2) = r_{v_1u}(m) = E\{v_1(n)u(n-m)\} \tag{5.121}$$

Setting $m = 1$ and then $m = 2$ in the above equation, we obtain

$$\begin{bmatrix} r_{uu}(0) & r_{uu}(-1) \\ r_{uu}(1) & r_{uu}(0) \end{bmatrix} \begin{bmatrix} -0.1350 \\ -0.7411 \end{bmatrix} = \begin{bmatrix} -r_{uu}(1) \\ -r_{uu}(2) \end{bmatrix} = \text{Yule–Walker equation} \tag{5.122}$$

because $v_1(n)$ and $u(n-m)$ are uncorrelated. If we set $m=0$ in (5.121), it becomes

$$r_{uu}(0)-0.135r_{uu}(-1)-0.7411r_{uu}(-2)=E\{v_1(n)u(n)\} \tag{5.123}$$

Next, we substitute the value of $u(n)$ from (5.119) in the above equation; taking into consideration that v and u are independent rv's, we obtain

$$r_{uu}(1)=\frac{0.135}{1-0.7411}r_{uu}(0)=\frac{0.135}{0.2589}\sigma_u^2 \tag{5.124}$$

Substitute the above equation in the second equation of the set of (5.122), we obtain

$$r_{uu}(2)=0.135\frac{0.135}{0.2589}\sigma_u^2+0.7411\sigma_u^2 \tag{5.125}$$

Therefore, the last three equations give the variance of u:

$$\sigma_u^2=\frac{\sigma_{v_1}^2}{0.3282}=\frac{0.31}{0.3282}=0.9445 \tag{5.126}$$

Using (5.126), (5.124), and the given value $\sigma_{v_2}^2=0.12$, we obtain the correlation matrix:

$$\boldsymbol{R}_x=\boldsymbol{R}_u+\boldsymbol{R}_{v_2}=\begin{bmatrix}0.9445 & 0.4925\\ 0.4925 & 0.9445\end{bmatrix}+\begin{bmatrix}0.12 & 0\\ 0 & 0.12\end{bmatrix}=\begin{bmatrix}1.0645 & 0.4925\\ 0.4925 & 1.0545\end{bmatrix} \tag{5.127}$$

We multiply (5.118) by $u(n)$ first and next by $u(n-1)$, and taking the ensemble average of the results, we obtain the vector $\boldsymbol{p}$ equal to

$$\boldsymbol{p}_{dx}=[0.4860 \quad -0.3868]^T \tag{5.128}$$

■

5.5.1 Minimum MSE

Introducing the above results in (5.87), we obtain the MSE surface (cost function) as a function of the filter coefficients, w's. Hence, we write

$$\begin{aligned}J(\boldsymbol{w})&=0.8461-2[w_0 \quad w_1]\begin{bmatrix}0.4860\\ -0.3868\end{bmatrix}+[w_0 \quad w_1]\begin{bmatrix}1.0645 & 0.4925\\ 0.4925 & 1.0645\end{bmatrix} \quad (5.129)\\ &=0.8461-0.972w_0+0.7736w_1+1.0645w_0^2+1.0645w_1^2+0.985w_0w_1\end{aligned}$$

The MSE surface and its contour plots are shown in Figure 5.16.

5.5.2 Optimum Filter (w^o)

The optimum filter was found above to be $\boldsymbol{R}_x\boldsymbol{w}^o=\boldsymbol{p}_{dx}$, and in this case, it takes the following form:

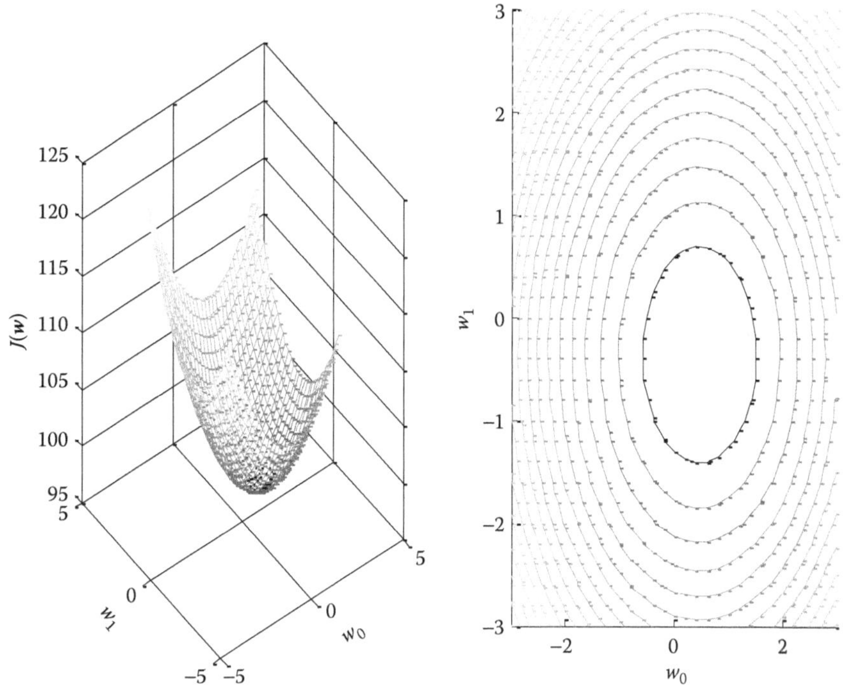

FIGURE 5.16

$$\boldsymbol{w}^o = \boldsymbol{R}_x^{-1}\boldsymbol{p}_{dx} = \begin{bmatrix} 1.1953 & -0.5531 \\ -0.5531 & 1.1953 \end{bmatrix}\begin{bmatrix} 0.4860 \\ -0.3868 \end{bmatrix} = \begin{bmatrix} 0.7948 \\ -0.7311 \end{bmatrix} = \begin{bmatrix} w_0^o \\ w_2^o \end{bmatrix} \tag{5.130}$$

The MMSE was found above, and for this case, we find

$$\begin{aligned} J_{\min} &= \sigma_d^2 - \boldsymbol{p}_{dx}^T\boldsymbol{R}_x^{-1}\boldsymbol{p}_{dx} \\ &= 0.8461 - [0.4860 \quad -0.3868]\begin{bmatrix} 1.1953 & -0.5531 \\ -0.5531 & 1.1953 \end{bmatrix}\begin{bmatrix} 0.4860 \\ -0.3868 \end{bmatrix} \\ &= 0.1770 \end{aligned} \tag{5.131}$$

Example 5.5.3 (System Identification)

It is desired, using a Wiener filter, to estimate the unknown impulse response coefficients h_i's of an unknown FIR system (see Figure 5.17). The input $\{x(n)\}$ is a set of zero-mean independent identically distributed (iid) rv's with variance σ_x^2. Let the impulse response $\boldsymbol{h}$ of the filter be $\boldsymbol{h} = [0.9 \quad 0.6 \quad 0.2]^T$. Since the input sequence $\{x(n)\}$ is a set of zero mean iid rv's, the correlation matrix $\boldsymbol{R}_x$ is a diagonal matrix with elements having values σ_x^2. The desired signal $d(n)$ is the output

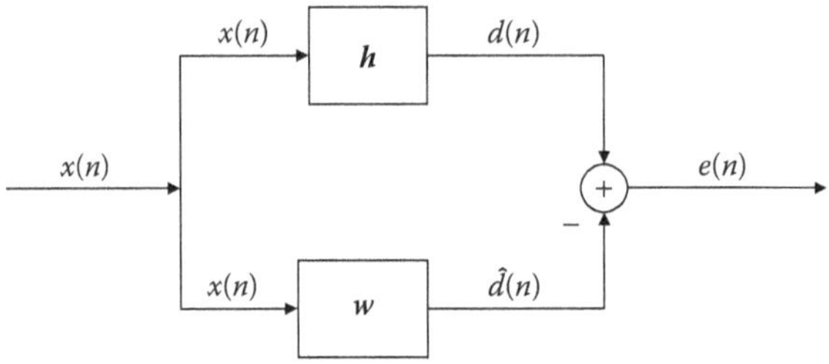

FIGURE 5.17

of the unknown filter and is given by $d(n)=0.9x(n)+0.6x(n-1)+0.2x(n-2)$ (see Figure 3.8a). Therefore, the cross-correlation output is given by

$$\begin{aligned} p_{dx}(i) &= E\{d(n)x(n-i)\} = E\{[0.9x(n)+0.6x(n-1)+0.2x(n-2)]x(n-i)\} \\ &= 0.9E\{x(n)x(n-i)\}+0.6E\{x(n-1)x(n-i)\}+0.2E\{x(n-2)x(n-i)\} \\ &= 0.9r_{xx}(i)+0.6r_{xx}(i-1)+0.2r_{xx}(i-2) \end{aligned} \quad (5.132)$$

Hence, we obtain $[r_{xx}(m)=0 \text{ for } m\neq 0]: p_{dx}(0)=0.9\sigma_x^2,\ p_{dx}(1)=0.6\sigma_x^2$. The optimum filter is

$$\boldsymbol{w}^o = \boldsymbol{R}_x^{-1}\boldsymbol{p}_{dx} = \frac{1}{\sigma_x^2}\begin{bmatrix}1 & 0\\ 0 & 1\end{bmatrix}\begin{bmatrix}0.9\\ 0.6\end{bmatrix} = \frac{1}{\sigma_x^2}\begin{bmatrix}0.9\\ 0.6\end{bmatrix}$$

and the MMSE is (assuming $\sigma_x^2 = 1$)

$$J_{\min} = \sigma_d^2 - [0.9 \quad 0.6]\begin{bmatrix}1 & 0\\ 0 & 1\end{bmatrix}\begin{bmatrix}0.9\\ 0.6\end{bmatrix}$$

But

$$\sigma_d^2 = E\{d(n)d(n)\} = E\{[0.9x(n)+0.6x(n-1)+0.2x(n-2)]^2\} = 0.81+0.36+0.04 = 1.21,$$

and hence,

$$J_{\min} = 1.21-(0.9^2+0.6^2) = 0.04.$$

Book m-Function for System Identification (Wiener filter) : [w,jm]=lms_wiener_fir_filter(x,h,M)

```
function[w,jm] = lms_wiener_fir_filter(x,h,M)
     %[w,jm] = lms_wiener_fir_filter(x,h,M);
     %x = data entering both the unknown filter(system)
     %and the Wiener filter;
     %d = the desired signal = output of the unknown
     %system; length(d) = length(x);
     %M = number of coefficients of the Wiener filter;
     %w = Wiener filter coefficients;
     %jm = minimum mean square error;
```

```
d1 = conv(x,h);%h = vector with coefficients of the
               %unknown system;
d = d1(1,1:length(x));
pdx = xcorr(d,x,'biased');
p = pdx(1,(length(pdx)+1)/2:((length(pdx)+1)/2)+M-1);
rx = lms_sample_biased_autoc(x,M);%M plays the role of
     %the lag number in correlation functions;
R = toeplitz(rx);
w = inv(R)*p';
jm = var(d)-p*w;%var() is a MATLAB function;
```

We set for the unknown system coefficients the vector $\boldsymbol{h}=[0.9 \quad 0.6 \quad 0.2]$. We also assumed $M = 6$ for the unknown filter number of coefficients. The input signal was random normally distributed with zero mean and variance of one, $N(0,1)$, or $x=\text{randn}(1,512)$. The results were as follows:

$w=\begin{bmatrix}0.8998 & 0.5998 & 0.2000 & 0.0000 & -0.0006 & -0.0016\end{bmatrix}$, $jm=-0.0052$
as we were expecting.

■

Example 5.5.4 (Noise Canceling)

In many practical applications, there exists a need to cancel the noise added to a signal. For example, we are talking on the cell phone inside the car, and the noise of the car, radio, and so on is added to the message when we are trying to transmit. Similar circumstance appears when pilots in planes and helicopters try to communicate or tank drivers try to do the same. Figure 5.18 shows pictorially the noise contamination situations. Observe that the noise added to the signal and the other component entering the Wiener filter emanate from the same source but follow different paths in the same environment. This indicates that there is somewhat difference in value. The output of the Wiener filter will approximate the noise added to the desired signal, and thus, the error will be close to the desired signal. The Wiener filter in this case is

$$\boldsymbol{R}_{v_2}\boldsymbol{w}^o = \boldsymbol{p}_{v_1v_2} \tag{5.133}$$

because the signal from the Wiener filter $\{\hat{v}_1\}$ should be equal to $\{v_1\}$ so that the desired signal is recaptured, in this case the signal $e(n)$. The individual components of the vector $\boldsymbol{p}_{v_1v_2}$ are

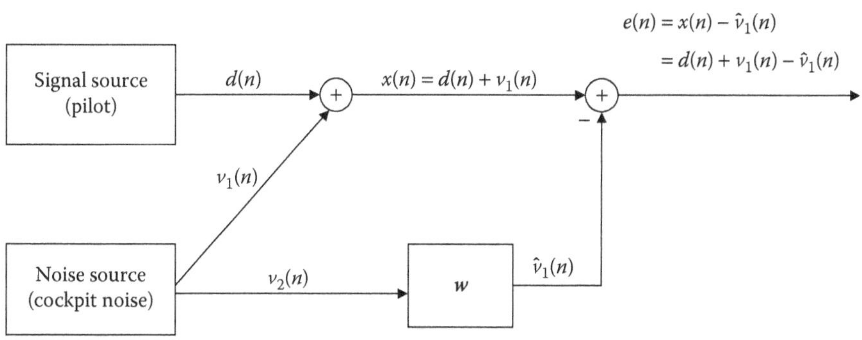

FIGURE 5.18

$$
\begin{aligned}
p_{v_1v_2}(m) &= E\{v_1(n)v_2(n-m)\} = E\{[x(n)-d(n)]v_2(n-m)\} \\
&= E\{x(n)v_2(n-m)\} - E\{d(n)v_2(n-m)\} \\
&= E\{x(n)v_2(n-m)\} - E\{d(n)\}E\{v_2(n-m)\} \\
&= E\{x(n)v_2(n-m)\} - E\{d(n)\}0 = p_{xv_2}(m)
\end{aligned} \tag{5.134}
$$

since $d(n)$ and v_2 are uncorrelated, and it was assumed that the noise had zero mean value. Therefore, (5.133) becomes

$$
\boldsymbol{R}_{v_2}\boldsymbol{w}^o = \boldsymbol{p}_{xv_2} \tag{5.135}
$$

To demonstrate the effect of the Wiener filter, let $d(n) = 2[0.99^n \sin(0.1n\pi + 0.2\pi)]$, $v_1(n) = 0.8v_1(n-1) + v(n)$, and $v_2(n) = 0.2v_1$, where $v(n)$ is white noise with zero mean value. The correlation matrix $\boldsymbol{R}_{v_2}$ and the cross-correlation vector $\boldsymbol{p}_{xv_2}$ are found using the sample biased correlation equation:

$$
\begin{aligned}
\hat{r}_{v_2v_2}(k) &= \frac{1}{N}\sum_{n=0}^{N-1} v_2(n)v_2(n-k) \qquad k = 0,1,\ldots,K-1,\ K \ll N \\
\hat{p}_{xv_2}(k) &= \frac{1}{N}\sum_{n=0}^{N-1} x(n)v_2(n-k) \qquad k = 0,1,\ldots,K-1,\ K \ll N
\end{aligned} \tag{5.136}
$$

Figure 5.19 shows simultaneous results for a sixth-order Wiener filter using the Book m-function given below. We set `a1 = 0.9`.

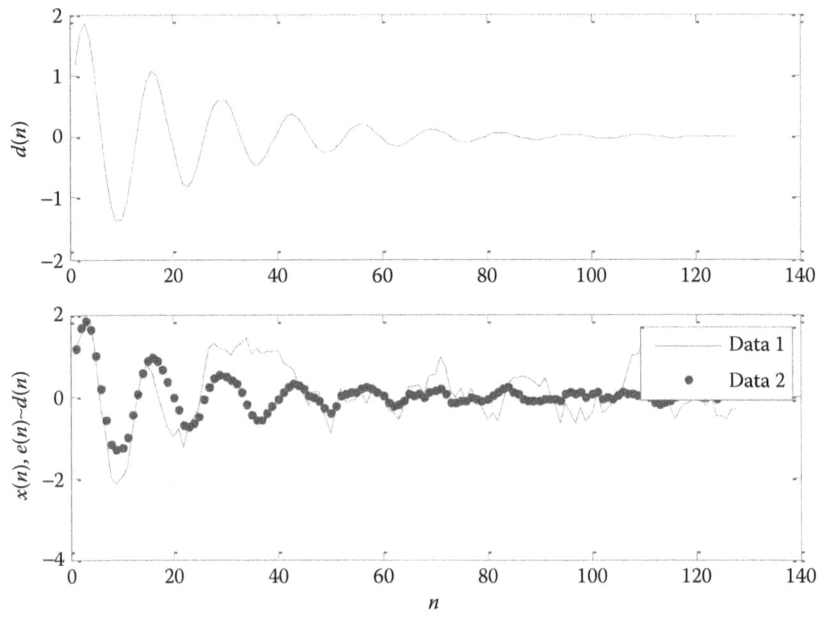

FIGURE 5.19

Book m-Function for Noise Canceling

```
function[d,w,xn] = lms_wiener_noise_canceling(dn,a1,M,N)
    %[d,w,xn] = lms_wiener_noise_canceling(dn,a1,a2,v,M,N);
    %dn = desired signal;
    %a1 = first-order IIR coefficient,a2 = first-order
    %IIR coefficient;
    %v = noise;M = number of Wiener filter coefficients;
    %N = number of sequence
    %elements of dn (desired signal) and v (noise);
    %d = output desired signal;
    %w = Wiener filter coefficients;xn = corrupted
    %signal;en = xn-v1 = d;
v1(1) = 0;v2(1) = 0;
for n = 2:N
      v1(n) = a1*v1(n-1)+(rand-0.5);
      v2(n) = 0.2*v1(n);
end;
v2autoc = lms_sample_biased_autoc(v2,M);
xn = dn+v1;
Rv2 = toeplitz(v2autoc);
p1 = xcorr(xn,v2,'biased');%MATLAB function;
if M>N
    disp(['error:M must be less than N']);
end;
R = Rv2(1:M,1:M);
p = p1(1,(length(p1)+1)/2:(length(p1)+1)/2+M-1);
w = inv(R)*p';
yw = filter(w,1,v2);%MATLAB function;
d = xn-yw(:,1:N);
```

■

Example 5.5.5 (Book Proposed Self-Correcting Wiener filter)

We can also arrange the standard single Wiener filter in a series form as shown in Figure 5.20. This book proposed configuration permits us to process the signal using filters with fewer coefficients, thus saving in computation. Figure 5.21a shows the desired signal, which is a decaying cosine wave. Figure 5.21b shows the signal with a white zero-mean additive noise. Figure 5.21c shows the output of the first stage of a self-correcting Wiener filter (SCWF), $\{x_1(n)\}$. Each stage Wiener filter has two coefficients. Figure 5.21d shows the output of the second stage of the SCWF.

■

Example 5.5.6

Use the proposed method of the extended input signal to improve the desired results.

Solution: We have used the system recognition approach (see Example 5.5.3) to verify the proposed method. The following Book MATLAB program was used.

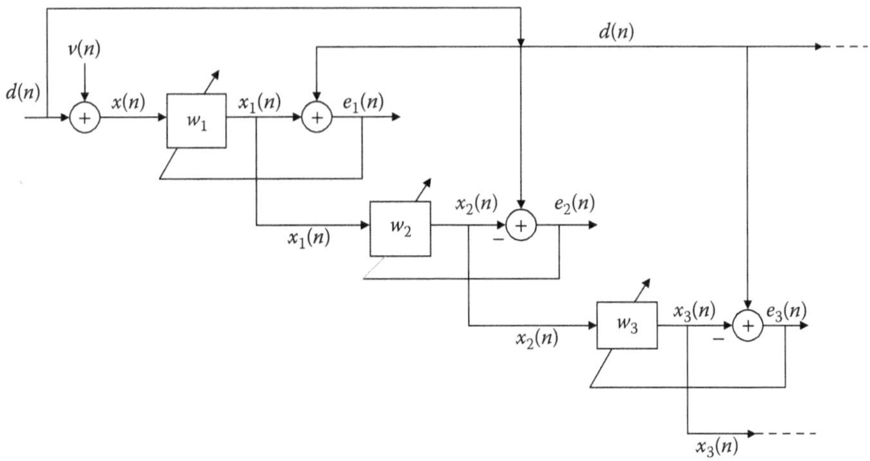

FIGURE 5.20

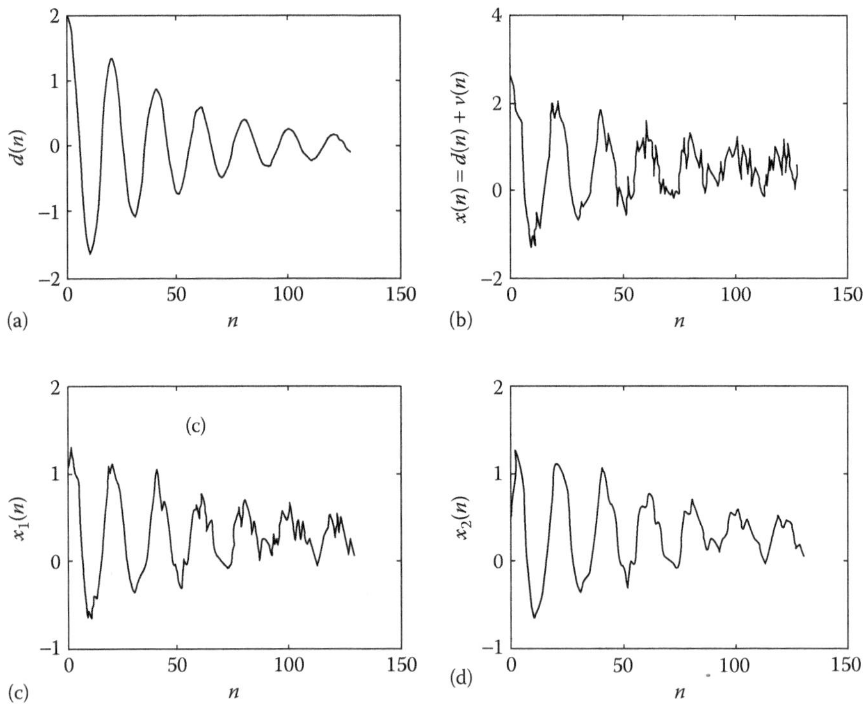

FIGURE 5.21

Book MATLAB Program

```
>>x = randn(1,8);
>>[w,jm] = lms_wiener_fir_filter(x,[0.9 0.6 0.2],5);
w = [0.7808 0.5483 -0.1444 -0.3132]T
>>x1 = [x  0.2*x  0.1*x];
```

```
>>[w1,jm1] = lms_wiener_fir_filter[x1,[0.9 0.6 0.2],5);
w1 = [0.8975 0.5990 0.1955 0.0002 -0.0040]T
```

■

5.5.3 Linear Prediction

Let us investigate the prediction problem as shown in Figure 5.22. It is desired to predict the signal in the presence of noise. The input to Wiener filter is

$$y(n) = x(n) + v(n) \tag{5.137}$$

The goal is to design a filter such that the value of the signal $x(n+1)$ is predicted based on p linear combination of previous values of the input signal $y(n)$. Hence, we write

$$\hat{x}(n+1) = \sum_{k=0}^{p-1} w(k)y(n-1) = \sum_{k=0}^{p-1} w(k)[x(n-k) - v(n-k)] \tag{5.138}$$

The Wiener equation is

$$\boldsymbol{R}_y \boldsymbol{w} = \boldsymbol{r}_{dy} \tag{5.139}$$

If the noise $v(n)$ is uncorrelated with the signal $x(n)$, then the autocorrelation matrix of the signal $y(n)$ is

$$\begin{aligned} r_{yy}(k) &= E\{y(n)y(n-k)\} = E\{(x(n)+v(n))(x(n-k)+v(n-k)\} \\ &= E\{x(n)x(n-)\} + E\{v(n)x(n-k)\} + E\{x(n)v(n-k)\} + E\{v(n)v(n-k)\} \\ &= r_{xx}(k) + 0 + 0 + r_{vv}(k) = r_{xx}(k) + r_{vv}(k) \end{aligned} \tag{5.140}$$

The vector cross-correlation between $d(n)$, in this case $x(n+1)$, and $y(n)$ is

$$\begin{aligned} r_{dy}(k) &= E\{d(n)y(n-k)\} = E\{x(n+1)y(n-k)\} \\ &= E\{x(n+1)[x(n-k)+v(n-k)]\} = E\{x(n+1)x(n-k)\} = r_{xx}(n+1) \end{aligned} \tag{5.141}$$

Equation 5.140, with $k = 0, 1, 2, \ldots, p-1$, becomes $\boldsymbol{R}_y = \boldsymbol{R}_x + \boldsymbol{R}_v$. Therefore, the MSE is

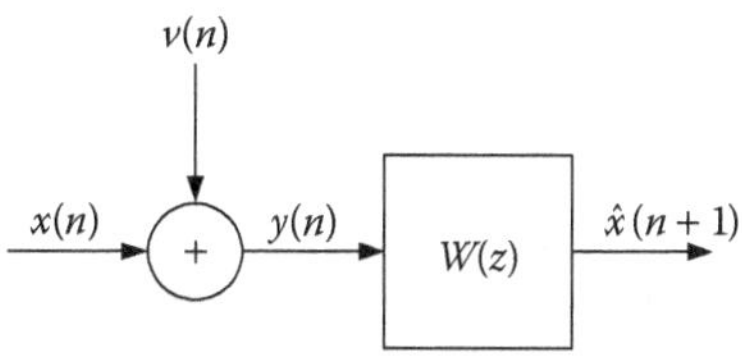

FIGURE 5.22

$$J_{\min} = \sigma_d^2 - \boldsymbol{p}_{dx}^T \boldsymbol{w}^o = \sigma_d^2 - \boldsymbol{p}_{dx}^T \boldsymbol{R}_x^{-1} \boldsymbol{p}_{dx} \tag{5.142}$$

gives the minimum error between the desired and the output of the Wiener filter.

Example 5.5.7

Let the noise be a white noise with variance σ_v^2. The noise is assumed to be uncorrelated with the signal $\{x(n)\}$. Assume that the autocorrelation function of $\{x(n)\}$ is $r_{xx}(k) = a^{|k|}$. We also assume a two-coefficient predictor filter, $[w_0 w_1]$.

Solution: The Wiener equation is (see Problem 5.5.1)

$$(\boldsymbol{R}_x + \sigma_v^2 \boldsymbol{I})\, \boldsymbol{w} = \boldsymbol{r}_{dy} \tag{5.143}$$

The above equation becomes

$$\begin{bmatrix} r_{xx}(0)+\sigma_v^2 & r_{xx}(1) \\ r_{xx}(1) & r_{xx}(0)+\sigma_v^2 \end{bmatrix} \begin{bmatrix} w_0 \\ w_1 \end{bmatrix} = \begin{bmatrix} r_{dy}(0) \\ r_{dy}(1) \end{bmatrix}, \quad r_{dy}(k) = r_{xx}(k+1)$$

or

$$\begin{bmatrix} 1+\sigma_v^2 & a \\ a & 1+\sigma_v^2 \end{bmatrix} \begin{bmatrix} w_0 \\ w_1 \end{bmatrix} = \begin{bmatrix} a \\ a^2 \end{bmatrix} \text{ or } \begin{bmatrix} w_0 \\ w_1 \end{bmatrix} = \frac{a}{(1+\sigma_v^2)^2 - a^2} \begin{bmatrix} 1+\sigma_v^2 - a^2 \\ a\sigma_v^2 \end{bmatrix}$$

Let $a = 0.8$, $\sigma_v^2 = 1.2$, $x(n) = 2\sin(0.1n\pi)$, and $y(n) = x(n) + 0.5(\text{rand} - 0.5)$. Then the Wiener filter coefficients are as follows: $w_0 = 0.2971$, $w_1 = 0.1829$. Hence, $J_{\min} = [1 \quad 0.8^2][0.2971 \quad 0.1829]^T = 0.5858$.

■

PROBLEMS

5.2.1 Find the unknown constant c, using the LS method, so that $\hat{x}(n) = ch(n)$, where $h(n)$ is known, is the best approximation to $\{x(n)\}$.

5.2.2 Verify (5.24).

5.2.3 Find the vector coefficient of the filter and the minimum error if the following are given:

$$\boldsymbol{d} = \begin{bmatrix}1 & 1 & 1 & 1 & 1\end{bmatrix}^T; \ \boldsymbol{x}_1 = \begin{bmatrix}0.7 & 1.4 & 0.4 & 1.3\end{bmatrix}^T;$$

$$\boldsymbol{x}_2 = \begin{bmatrix}1.2 & 0.6 & 0.5 & 1.1\end{bmatrix}^T; \ \boldsymbol{x}_3 = \begin{bmatrix}0.9 & 1.1 & 0.95 & 1.05\end{bmatrix}^T$$

5.2.4 The time-averaged correlation matrix $\boldsymbol{R} = \boldsymbol{X}^T \boldsymbol{X}$ (must be divided by N to be true time averaged) is invertible if the columns x_k of $\boldsymbol{X}$ are linearly independent or, equivalently, if and only if $\boldsymbol{R}$ is positive definite.

5.2.5 Let us assume that the signal from LTI system is constant, $y(n) = a$, and we observe the data $x(n)$, for $n = 1, 2, \ldots, N$. Find the estimate value of a and $\hat{a}$.

5.4.1 Find the Wiener solution using the Matrix Analysis in Appendix 2.

5.4.2 Verify that $J_{\min} = E\{[e^o(n)]\}$.

5.4.3 Find $J_{\min}$ using the orthogonality principle.

5.4.4 Find the variance $\sigma^2_{\hat{d}^o}$ of the estimate $\hat{d}^o(n)$.

5.5.1 Verify (5.143).

5.5.2 (**Multiple-step prediction**) Develop the multiple-step prediction in which $x(n+m)$ is to be predicted in terms of linear combination of the p values: $x(n), x(n+1),\ldots, x(n-p+1)$.

5.5.3 Let the data entering the Wiener filter be given by $x(n) = d(n) + v(n)$. The noise $v(n)$ has zero mean value and unit variance, and is uncorrelated with the desired signal $d(n)$. Furthermore, assume $r_{dd}(m) = 0.9^m$ and $r_{vv}(m) = 1\delta(m)$. Find $\boldsymbol{p}_{dx}, \boldsymbol{w}, \boldsymbol{J}_{\min}$, signal power, noise power, and signal-to-noise power.

5.5.4 Find the cost function and the MSE surface for the two systems shown in Figure P5.5.4. Given $E\{s^2(n)\} = 0.9$, $E[s(n)s(n-1)] = 0.4$, $E\{d^2(n)\} = 3$, $E\{d(n)s(n)\} = 0.5$, and $E\{d(n)s(n-1)\} = 0.9$.

5.5.5 Consider a two-coefficient Wiener filter with the following statistics:

$$E\{d^2(n)\} = 2, \quad \boldsymbol{R} = \begin{bmatrix} 1 & 0.6 \\ 0.6 & 1 \end{bmatrix}, \quad \boldsymbol{p} = \begin{bmatrix} 1 \\ 0.5 \end{bmatrix}$$

a. Use the above information to obtain the performance function (cost function), J, of the filter.
   By direct evaluation of the performance function, obtain the optimum values of the filter coefficients (tap weights).
b. Use the results from (a) in the performance function expression to obtain the MMSE of the filter.

5.5.6 Consider a three-coefficient Wiener filter with the following data:

$$E\{d^2(n)\} = 5, \boldsymbol{R} = \begin{bmatrix} 1 & 0.25 & 0.125 \\ 0.25 & 1 & 0.25 \\ 0.125 & 0.25 & 1 \end{bmatrix}, \boldsymbol{p} = \begin{bmatrix} 1.5 \\ 0.5 \\ 0 \end{bmatrix}$$

Repeat steps (a) and (b) of Problem 5.5.5.

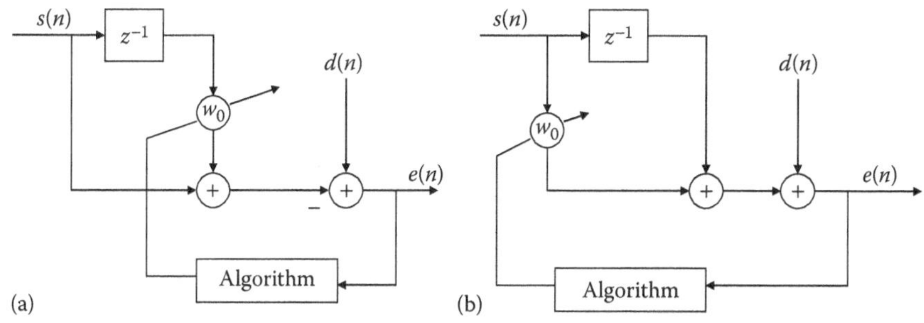

FIGURE P5.5.4

ADDITIONAL PROBLEMS

1. A Wiener filter has an input $\{x(n)\}$ with an autocorrelation matrix

$$\boldsymbol{R}_x = \begin{bmatrix} 1 & 0.5 \\ 0.5 & 1 \end{bmatrix}$$

and a cross-correlation between the input and the desired signal is

$$\boldsymbol{p} = \begin{bmatrix} 0.4 \\ 0.2 \end{bmatrix}.$$

Find (a) the coefficients, (b) the MMSE, and (c) the expansion as a function of the eigenvalues and eigenvectors.

2. Prove that the MMSE $J_{\min}$ is given by

$$J_{\min} = \sigma_d^2 - \Sigma_{k=1}^{M} \frac{\left|\boldsymbol{q}_k^H \boldsymbol{p}\right|^2}{\lambda_k}.$$

HINTS–SOLUTIONS–SUGGESTIONS

5.2.1

$$J(c) = \text{Error criterion} = \sum_{n=0}^{N-1} [x(n) - ch(n)]^2,$$

$$\frac{\partial J(c)}{\partial c} = 0 = -2\sum_{n=0}^{N-1} [x(n)h(n) - h^2(n)]$$

$$\text{or } \hat{c} = \frac{\sum_{n=0}^{N-1} x(n)h(n)}{\sum_{n=0}^{N-1} h^2(n)}$$

5.2.2 For simplicity and without loss of generality, we assume a 2×2 matrix and a 2×1 vector.

$$\frac{\partial J}{\partial \boldsymbol{w}} = \frac{\partial}{\partial \boldsymbol{w}}[E_y - 2\boldsymbol{p}\boldsymbol{w}^T + \boldsymbol{w}^T\boldsymbol{R}\boldsymbol{w}] = \frac{\partial E_y}{\partial \boldsymbol{w}} - 2\boldsymbol{p}^T\frac{\partial \boldsymbol{w}}{\partial \boldsymbol{w}} + \frac{\partial}{\partial \boldsymbol{w}}(\boldsymbol{w}^T\boldsymbol{R}\boldsymbol{w}) = \boldsymbol{0}$$

$$-2\begin{bmatrix} (\partial/\partial w_1)(p_1w_1 + p_2w_2) \\ (\partial/\partial w_2)(p_1w_1 + p_2w_2) \end{bmatrix} + \begin{bmatrix} (\partial/\partial w_1)(r_{11}w_1^2 + r_{12}w_2w_1 + r_{21}w_2w_1 + r_{22}w_2^2) \\ (\partial/\partial w_2)(r_{11}w_1^2 + r_{12}w_2w_1 + r_{21}w_2w_1 + r_{22}w_2^2) \end{bmatrix}$$

$$= -2\begin{bmatrix} p_1 \\ p_2 \end{bmatrix} + \begin{bmatrix} 2r_{11}w_1 + 2r_{12}w_2 \\ 2r_{21}w_1 + 2r_{22}w_2 \end{bmatrix} = \boldsymbol{0} \text{ or } \boldsymbol{R}\hat{\boldsymbol{w}} = \boldsymbol{p}$$

5.2.3 *Hint*: Follow Example 5.2.4.

5.2.4 If the columns of $\boldsymbol{X}$ are linearly independent, then for every $\boldsymbol{y} \neq \boldsymbol{0}$, we have $\boldsymbol{X}\boldsymbol{y} \neq \boldsymbol{0}$. Hence, $\boldsymbol{y}^T(\boldsymbol{X}^T\boldsymbol{X})\boldsymbol{y} = (\boldsymbol{X}\boldsymbol{y})^T\boldsymbol{X}\boldsymbol{y} = \|\boldsymbol{X}\boldsymbol{y}\|^2 > 0$, which indicates that $\boldsymbol{R} = \boldsymbol{X}^T\boldsymbol{X}$ is positive definite and therefore nonsingular. If the columns of $\boldsymbol{X}$ are linearly dependent, then there is a $\boldsymbol{y}_0 \neq \boldsymbol{0}$. Therefore, $\boldsymbol{X}^T\boldsymbol{X}\boldsymbol{y}_0 = 0$ and $\boldsymbol{R}$ is singular.

5.2.5

$$J(a) = \sum_{n=1}^{N}[x(n)-a]^2 \Rightarrow \frac{\partial J(a)}{\partial a} = \sum_{n=1}^{N}(-)2[x(n)-\hat{a}] = 0 \text{ or}$$

$$\sum_{n=1}^{N}x(n) - N\hat{a} = 0 \text{ or } \hat{a} = \frac{1}{N}\sum_{n=1}^{N}x(n)$$

which is the sample mean estimator.

5.4.1

$$\frac{\partial J(\boldsymbol{w})}{\partial \boldsymbol{w}} = \frac{\partial}{\partial \boldsymbol{w}}(\sigma_d^2) - 2\frac{\partial}{\partial \boldsymbol{w}}(\boldsymbol{w}^T\boldsymbol{p}_{dx}) + \frac{\partial}{\partial \boldsymbol{w}}(\boldsymbol{w}^T\boldsymbol{R}_x\boldsymbol{w}) = 0 - \underbrace{2\boldsymbol{p}_{dx}}_{\text{Eq. A2.67 App 2}} + \underbrace{2\boldsymbol{R}_x\boldsymbol{w}}_{\text{Eq. A2.73 App 2}} = 0$$

or $\boldsymbol{R}_x\boldsymbol{w}^o = \boldsymbol{p}_{dx}$

5.4.2

$$\begin{aligned} J_{\min} &= E\{[e^o(n)]^2\} \\ &= E\{e^o(n)(d(n)-\hat{d}^o(n))\} = E\{e^o(n)d(n)\} - E\{e^o(n)\,\hat{d}^o(n)\} \\ &= E\{e^o(n)d(n)\} \end{aligned}$$

where $E\{e^o(n)\hat{d}^o(n)\} = 0$ [see (5.100)]. Substituting (5.101) in the form

$$e^o(n) = d(n) - \hat{d}^o(n) = d(n) - \sum_{i=0}^{N-1} w_i^o(n)x(n-i),$$

we find

$$J_{\min} = E\left\{\left(d(n) - \sum_{i=0}^{M-1} w_i^o(n)x(n-i)\right)d(n)\right\}$$

$$= E\{d(n)\} - \sum_{i=0}^{M-1} w_i^o(n)E\{d(n)x(n-i)\} \quad \text{or}$$

$$J_{\min} = E\left\{d^2(n) - \sum_{i=0}^{M-1} w_i^o(n)p_i\right\},$$

which is the expanded form of the desired results.

5.4.3

$$J = E\left\{e(n)\left[d(n) - \sum_{m=0}^{M-1} w_m(n)x(n-m)\right]\right\} = E\{e(n)d(n)\}$$

$$-\sum_{m=0}^{M-1} w_m(n)E\{e(n)x(n-m)\}$$

If the coefficients have their optimum value, the orthogonality principle states that $E\{e(n)x(n-m)\} = 0$. Hence,

$$J_{\min} = E\{e(n)d(n)\} = E\{d(n)d(n) - \sum_{m=0}^{M-1} w_m^o x(n-m)d(n)\} = \sigma_d^2$$

$$-\sum_{m=0}^{M-1} w_m^o E\{d(n)x(n-m)\} = \sigma_d^2 - \sum_{m=0}^{M-1} w_m^o p_{dx}(m)$$

$$= r_d(0) - \boldsymbol{p}_{dx}^T \boldsymbol{w}^o = r_d(0) - \boldsymbol{p}_{dx}^T \boldsymbol{R}_x^{-1} \boldsymbol{p}_{dx}$$

5.4.4

$$\hat{d}^o(n) = (\boldsymbol{w}^o)^T \boldsymbol{x}(n)$$

$$\sigma_{\hat{d}^o}^2 = E\left\{(\boldsymbol{w}^o)^T \boldsymbol{x}(n)\boldsymbol{x}^T(n)\boldsymbol{w}^o\right\} = (\boldsymbol{w}^o)^T E\{\boldsymbol{x}(n)\boldsymbol{x}^T(n)\}\boldsymbol{w}^o = \boldsymbol{p}^T \boldsymbol{w}^o = \boldsymbol{p}^T \boldsymbol{R}^{-1} \boldsymbol{p};$$

$$J_{\min} = \sigma_d^2 - \sigma_{\hat{d}^o}^2 = \sigma_d^2 - (\boldsymbol{w}^o)^T \boldsymbol{R}\boldsymbol{w}^o = \sigma_d^2 - \boldsymbol{p}^T \boldsymbol{w}^o = \sigma_d^2 - \boldsymbol{p}^T \boldsymbol{R}^{-1} \boldsymbol{p}$$

5.5.1

$$r_{yy}(k) = r_{xx}(k) + r_{vv}(k) \Rightarrow r_{yy}(0) = r_{xx}(0) + r_{vv}(0); r_{yy}(1) = r_{xx}(1) + r_{vv}(1);$$

$$r_{yy}(-1) = r_{xx}(-1) + r_{vv}(-1) \text{ } but \text{ } r_{yy}(1) = r_{yy}(-1),\ r_{xx}(1) = r_{xx}(-1), r_{vv}(1) = r_{vv}(-1) \Rightarrow$$

$$\begin{bmatrix} r_{yy}(0) & r_{yy}(1) \\ r_{yy}(-1) & r_{yy}(0) \end{bmatrix} = \begin{bmatrix} r_{xx}(0) + r_{vv}(0) & r_{xx}(1) + r_{vv}(1) \\ r_{xx}(-1) + r_{vv}(-1) & r_{xx}(0) + r_{vv}(0) \end{bmatrix}$$

$$= \begin{bmatrix} r_{xx}(0) + \sigma_v^2 & r_{xx}(1) + 0 \\ r_{xx}(1) + 0 & r_{xx}(0) + \sigma_v^2 \end{bmatrix} = \begin{bmatrix} r_{xx}(0) & r_{xx}(1) \\ r_{xx}(1) & r_{xx}(0) \end{bmatrix} + \begin{bmatrix} \sigma_v^2 & 0 \\ 0 & \sigma_v^2 \end{bmatrix} = \boldsymbol{R}_x + \sigma_v^2 \begin{bmatrix} 1 & 0 \\ 0 & 1 \end{bmatrix}$$

ADDITIONAL PROBLEMS

1. Using MATLAB, we obtain
   a. $\text{inv}(R) = [1.3333 \quad -0.6667; -0.6667 \quad 1.3333]$,

b. $w_o = \text{inv}(R)*p = [0.4;0]; [Q, \quad L] = \text{eig}(R);$

$Q = [-0.7071 \quad 0.7071; 0.7071 \quad 0.7071], L = [0.5 \quad 0;0 \quad 1.5];$

c. $w = \{[(1/0.5)*Q(:,1)*Q(:,1)' + (1/1.5)*Q(:,2)*Q(:,2)']\}*$
$p = [0.4000;\ 0].$

2.

$$J_{\min} = \sigma_d^2 - \boldsymbol{p}^H \boldsymbol{R}^{-1}\boldsymbol{p} = \sigma_d^2 - \boldsymbol{p}^H\left(\boldsymbol{Q}\Lambda\boldsymbol{Q}^H\right)^{-1}\boldsymbol{p} = \sigma_d^2 - \boldsymbol{p}^H\left(\sum_{k=1}^{M}\lambda_k \boldsymbol{q}_k\boldsymbol{q}_k^H\right)^{-1}\boldsymbol{p}$$

$$= \sigma_d^2 - \boldsymbol{p}^H\sum_{k=1}^{M}\frac{1}{\lambda_k}\boldsymbol{q}_k\boldsymbol{q}_k^H\boldsymbol{p} = \sigma_d^2 - \sum_{k=1}^{M}\frac{1}{\lambda_k}\boldsymbol{p}^H\boldsymbol{q}_k\boldsymbol{q}_k^H\boldsymbol{p} = \sigma_d^2 - \sum_{k=1}^{M}\frac{1}{\lambda_k}\left|\boldsymbol{p}^H\boldsymbol{q}_k\right|^2$$

5.5.2 Let the noise signal $\{x(n)\}$ to enter the Wiener filter. Then the output is $\hat{x}(n+m) = \sum_{k=0}^{p-1} w(k)x(n-k)$. The desired signal is $d(n) = x(n+m)$. Therefore, $r_{dx}(k) = E\{x(n+m)x(n-k)\} = r_{xx}(m+k)$. The Wiener equation, $\boldsymbol{R}_x\boldsymbol{w} = r_{xx,m}$, becomes

$$\begin{bmatrix} r_{xx}(0) & r_{xx}(1) & \cdots & r_{xx}(p-1) \\ r_{xx}(1) & r_{xx}(0) & \cdots & r_{xx}(p-2) \\ & \vdots & & \\ r_{xx}(p-1) & r_{xx}(p-2) & \cdots & r_{xx}(0) \end{bmatrix}\begin{bmatrix} w(0) \\ w(1) \\ \vdots \\ w(p-1) \end{bmatrix} = \begin{bmatrix} r_{xx}(m) \\ r_{xx}(m+1) \\ \vdots \\ r_{xx}(m+p-1) \end{bmatrix}$$

The MMSE is $J_{\min} = r_{xx}(0) - \sum_{k=0}^{p-1} w(k)x(k+m) = r_{xx}(0) - \boldsymbol{r}_{xx,m}^T\boldsymbol{w}.$

5.5.3

$$p_{dx}(m) = E\{d(n)x(n-m)\} = E\{d(n)d(n-m)\} + E\{d(n)\}E\{v(n-m)\}$$
$$= r_{dd}(m) \qquad (1)$$

since $d(n)$ and $v(n)$ are independent and WSS. Also, $r_{xx}(m) = E\{x(n)x(n-m)\} = r_{dd}(m) + r_{vv}(m)$, where again the independence and the zero mean property of the noise were introduced. Therefore, the Wiener equation becomes $\boldsymbol{R}_d + \boldsymbol{R}_v = \boldsymbol{p}_{dx}$. From the relation $p_{dx}(m) = r_{dd}(m)$ [see (1)] and the given relations in the problem, we find that the Wiener equation and its inverse are given by

$$\begin{bmatrix} 2 & 0.9 \\ 0.9 & 2 \end{bmatrix}\begin{bmatrix} w_0 \\ w_1 \end{bmatrix} = \begin{bmatrix} 1 \\ 0.9 \end{bmatrix}, \quad \begin{bmatrix} w_0 \\ w_1 \end{bmatrix} = \frac{1}{3.19}\begin{bmatrix} 1.19 \\ 0.9 \end{bmatrix} = \begin{bmatrix} 0.373 \\ 0.282 \end{bmatrix}$$

The minimum error is $J_{\min} = r_{dd}(0) - \boldsymbol{p}_{dx}^T \boldsymbol{R}_x^{-1} \boldsymbol{w}^o = r_{dd}(0) - \boldsymbol{p}_{dx}^T \boldsymbol{w}^o = 0.373$. Since $r_{dd}(0) = \sigma_d^2 = 1$ and $\sigma_v^2 = 1$, the power of the desired signal and noise are equal, and hence, $10\log(1/1) = 0$. After filtering $x(n)$, we find that the signal power is

$$E\{\hat{d}^2(0)\} = \boldsymbol{w}^{oT} \boldsymbol{R}_d \boldsymbol{w}^o = [w_0^o \quad w_1^o] \begin{bmatrix} 1 & 0.9 \\ 0.9 & 1 \end{bmatrix} \begin{bmatrix} w_0^o \\ w_1^o \end{bmatrix} = 0.408$$

and the noise power is

$$E\{\hat{v}^2(0)\} = \boldsymbol{w}^{oT} \boldsymbol{R}_v \boldsymbol{w}^o = [w_0^o \quad w_1^o] \begin{bmatrix} 1 & 0 \\ 0 & 1 \end{bmatrix} \begin{bmatrix} w_0^o \\ w_1^o \end{bmatrix} = 0.285$$

Therefore, the signal-to-noise ratio SNR = 10log(0.408/0.285) = 1.56, which means that the Wiener filter increases the SNR by 1.56 dB.

5.5.4

a. $J = E\{(d(n) - [s(n) + w_0 s(n-1)])^2\}$

$$= E\{d^2(n) - [s(n) + w_0 s(n-1)]^2 - d(n)[s(n) + w_0 s(n-1)]\}$$

$$= E\{d^2(n)\} - E\{s^2(n)\} - w_0^2 E\{s^2(n-1)\} - 2w_0 E\{s(n)s(n-1)\}$$

$$-2E\{d(n)s(n)\} - 2w_0 E\{s(n)s(n-1)\}$$

$$= 3 - 0.9 - w_0^2 0.9 - 2w_0 0.4 + 2 \times 0.5 - 2w_0 0.4 = 3.1 - 0.9w_0^2 - 0.6w_0$$

$$\Rightarrow \frac{\partial J}{\partial w_0} = -0.9 \times 2w_0 - 1.6 = 0 \Rightarrow w_0 = 1.125$$

b. Similarly as (a).

5.5.5

a. $J = \boldsymbol{w}^T \boldsymbol{R} \boldsymbol{w} - 2\boldsymbol{w}^T \boldsymbol{p} + E\{d^2(n)\}$

$$= [w_0 \ w_1] \begin{bmatrix} 1 & 0.6 \\ 0.6 & 1 \end{bmatrix} \begin{bmatrix} w_0 \\ w_1 \end{bmatrix} - 2[w_0 \ w_1] \begin{bmatrix} 1 \\ 0.5 \end{bmatrix} + 2 \tag{1}$$

$$= w_0^2 + w_1^2 + 1.2w_0 w_1 - 2w_0 - w_1 + 2$$

$$\frac{\partial J}{\partial w_0} = 0 \Rightarrow 2w_0^o + 1.2w_1^o = 2 \tag{2}$$

$$\frac{\partial J}{\partial w_1} = 0 \Rightarrow 2w_1^o + 1.2w_0^o = 1 \tag{3}$$

Solving (2) and (3), we obtain

$$w_0^o = 1.0938; w_1^o = -0.1563 \tag{4}$$

b. Substituting (4) in (1), we get $J_{\min} = 0.9844$.

5.5.6

a. $J = \boldsymbol{w}^T \boldsymbol{R}\boldsymbol{w} - 2\boldsymbol{w}^T \boldsymbol{p} + E\{d^2(n)\}$

$$= [w_0 \quad w_1 \quad w_2]\begin{bmatrix} 1 & 0.25 & 0.125 \\ 0.25 & 1 & 0.25 \\ 0.125 & 0.25 & 1 \end{bmatrix}\begin{bmatrix} w_0 \\ w_1 \\ w_2 \end{bmatrix} - 2[w_0 \; w_1 \; w_2]\begin{bmatrix} 1.5 \\ 0.5 \\ 0 \end{bmatrix} + 5$$

$$= w_0^2 + w_1^2 + w_2^2 + 0.5w_0w_1 + 0.25w_0w_2 + 0.5w_1w_2 - 3w_0 - w_1 + 5 \tag{1}$$

$$\frac{\partial J}{\partial w_0} = 0 \Rightarrow 2w_0^o + 0.5w_1^o + 0.25w_2^o = 3 \tag{2}$$

$$\frac{\partial J}{\partial w_1} = 0 \Rightarrow 2w_1^o + 0.5w_0^o + 0.5w_2^o = 1 \tag{3}$$

$$\frac{\partial J}{\partial w_2} = 0 \Rightarrow 2w_2^o + 0.25w_0^o + 0.5w_1^o = 0 \qquad (3)$$

Solving (2) and (3), we obtain

$$w_0^o = 1.4821; \; w_1^o = 0.1875; \; w_2^o = -0.2321 \tag{4}$$

b. Substituting (4) in (1), we get $J_{\min} = 0.3661$.

6 Eigenvalues of R_x Properties of the Error Surface

6.1 THE EIGENVALUES OF THE CORRELATION MATRIX

Let $\boldsymbol{R}_x$ be an $M \times M$ correlation matrix of a wide-sense (or weakly) stationary (WSS) discrete-time process computed from a data vector $\boldsymbol{x}(n)$. This symmetric and non-negative matrix (see Problem 6.1.1) can be found as follows: We find the data vector $\boldsymbol{x}(n)$ using the MATLAB command: `x=randn(1,64)`. From this data vector, we first find its autocorrelation function $r_{xx}(m) = [1.0414 \quad 0.0136 \quad 0.0823]$ using the Book m-function: `rx=lms_sample_biased_autoc(x,3)`. We can also use the Book m-function: `rx=lms _ sample _ biased _ cross _ cor(x,x,3)`. Next, we obtain the correlation matrix $\boldsymbol{R}_x$:

$$\boldsymbol{R}_x = \begin{bmatrix} 1.0414 & 0.0136 & 0.0823 \\ 0.0136 & 1.0414 & 0.0136 \\ 0.0823 & 0.0136 & 1.0414 \end{bmatrix}$$

The above matrix was obtained using the MATLAB function: `Rx=toeplitz(rx)`. Having found the $M \times M$ correlation matrix, we desire to find an $M \times 1$ vector $\boldsymbol{q}$ such that

$$\boldsymbol{R}_x \boldsymbol{q} = \lambda \boldsymbol{q} \tag{6.1}$$

for some constant λ. The above equation indicates that the left-hand-side transformation does not change the direction of the vector $\boldsymbol{q}$ but only its length. The characteristic equation of $\boldsymbol{R}_x$ is given by

$$\det(\boldsymbol{R}_x \boldsymbol{q} - \lambda \boldsymbol{I}) = 0 \tag{6.2}$$

The roots $\lambda_1, \lambda_2, \lambda_3, \ldots, \lambda_M$ of the above equation are called **eigenvalues** of $\boldsymbol{R}_x$. Therefore, to each eigenvalue corresponds an **eigenvector** $\boldsymbol{q}_i$ such that

$$\boldsymbol{R}_x \boldsymbol{q}_i = \lambda_i \boldsymbol{q}_i \qquad i = 1, 2, \ldots, M \tag{6.3}$$

The MATLAB function

$$[\boldsymbol{Q},\boldsymbol{D}] = \text{eig}(\boldsymbol{R}) \tag{6.4}$$

gives the diagonal matrix $\boldsymbol{D}$ ($\boldsymbol{D} = \Lambda$) containing the eigenvalues of $\boldsymbol{R}(\boldsymbol{R} = \boldsymbol{R}_x)$ and gives the matrix $\boldsymbol{Q}$ with its columns equal to the eigenvectors of $\boldsymbol{R}$. Using the 3×3 matrix in (6.4), we obtain

$$\boldsymbol{Q} = \begin{bmatrix} 0.7071 & 0.1531 & 0.6903 \\ -0.0000 & -0.9763 & 0.2165 \\ -0.7071 & 0.1531 & 0.6903 \end{bmatrix} \quad \boldsymbol{D} = \begin{bmatrix} 0.9592 & 0 & 0 \\ 0 & 1.0372 & 0 \\ 0 & 0 & 1.1280 \end{bmatrix}$$

In Table 6.1, we give the eigenvalue properties of WSS processes. The reader should also refer to Section 2.6 for more details on eigenvalue properties. One of the properties is that any two different columns of the $\boldsymbol{Q}$ matrix are orthogonal. Hence, we write

```
>>Q(:,1)'*Q(:,3) = 3.6115e-016; %Q(:,1)' = row vector of the
       %first column of Q,Q(:,3) = third column of Q,
       %the result verify the orthogonality property
       %of the columns;
```

One of the properties is

$$\boldsymbol{Q} * \boldsymbol{D} * \boldsymbol{Q}^T = \begin{bmatrix} 1.0414 & 0.0136 & 0.0823 \\ 0.0136 & 1.0414 & 0.0136 \\ 0.0823 & 0.0136 & 1.0414 \end{bmatrix} = \boldsymbol{R}_x$$

as it should be.

6.1.1 Karhunen–Loeve Transformation

If the observed vector $\boldsymbol{x}(n)$ is an observation vector with the correlation $\boldsymbol{R}_x$. If $\boldsymbol{Q}$ is the matrix as defined earlier, then the elements of the vector

$$\boldsymbol{x}^{(kv)}(n) = \boldsymbol{Q}^H \boldsymbol{x}(n) \tag{6.5}$$

constitute a set of uncorrelated random variables. The above transformation is called the **Karhunen–Loeve transform**.

TABLE 6.1
Eigenvalue Properties

$\boldsymbol{x}(n) = [x(n) \quad x(n-1) \quad \cdots \quad x(n-M+1)]$	Wide-sense stationary stochastic process
$\boldsymbol{R}_x = E\{\boldsymbol{x}(n)\boldsymbol{x}^T(n)\}$	Correlation matrix
λ_i	The eigenvalues of $\boldsymbol{R}_x$ are real and positive. The eigenvalues of a Hermitian ($\boldsymbol{R}_x^H = \boldsymbol{R}_x^{*T}$) matrix are real. A Hermitian matrix is positive definite, $\boldsymbol{A} > 0$, if and only if its eigenvalues are positive, $\lambda_i > 0$.
$\boldsymbol{q}_i^T \boldsymbol{q}_j = 0$	Two eigenvectors belonging to two different eigenvalues are orthogonal. The eigenvectors of a Hermitian matrix corresponding to distinct eigenvalues are orthogonal.
$\boldsymbol{Q}^T\boldsymbol{Q} = \boldsymbol{I}$	$\boldsymbol{Q} = [\boldsymbol{q}_0 \quad \boldsymbol{q}_1 \quad \cdots \quad \boldsymbol{q}_{M-1}]$; $\boldsymbol{Q}$ is a unitary matrix.
$\boldsymbol{Q}_x^T = \boldsymbol{Q}_x^{-1}$ $(\boldsymbol{Q}_x^H = \boldsymbol{Q}_x^{-1})$	
$\boldsymbol{R}_x = \boldsymbol{Q}\Lambda\boldsymbol{Q}^T = \sum_{i=0}^{M-1} \lambda_i \boldsymbol{q}_i \boldsymbol{q}_i^T$	$\Lambda = \text{diag}[\lambda_0 \quad \lambda_1 \quad \lambda_2 \quad \cdots \quad \lambda_{M-1}]$
$\boldsymbol{R}_x = \boldsymbol{Q}\Lambda\boldsymbol{Q}^H = \sum_{i=0}^{M-1} \lambda_i \boldsymbol{q}_i \boldsymbol{q}_i^H$	$\boldsymbol{R}_x$ is a Hermitian matrix.
$\boldsymbol{R}_x^{-1} = (\boldsymbol{Q}\Lambda\boldsymbol{Q}^T)^{-1} = \sum_{i=0}^{M-1} \frac{1}{\lambda_i} \boldsymbol{q}_i \boldsymbol{q}_i^T$	
$\boldsymbol{R}_x^{-1} = (\boldsymbol{Q}\Lambda\boldsymbol{Q}^H)^{-1} = \sum_{i=0}^{M-1} \frac{1}{\lambda_i} \boldsymbol{q}_i \boldsymbol{q}_i^H$	$\boldsymbol{R}_x$ is nonsingular and Hermitian.
$\text{tr}\{\boldsymbol{R}_x\} = \sum_{i=0}^{M-1} \lambda_i$	$\text{tr}\{\boldsymbol{R}\}$ = trace of $\boldsymbol{R}$ = sum of the diagonal elements of $\boldsymbol{R}$
$\boldsymbol{A} = \boldsymbol{B} + a\boldsymbol{I}$	$\boldsymbol{A}$ and $\boldsymbol{B}$ have the same eigenvectors and the eigenvalues $\boldsymbol{A}$ are $\lambda_i + a$.
$\lambda_{\max} \le S_x^{\max} = \max_{-\pi \le \omega \le \pi} S_x(e^{j\omega})$	
$\lambda_{\min} \ge S_x^{\min} = \min_{-\pi \le \omega \le \pi} S_x(e^{j\omega})$	

Using (6.5), we obtain

$$E\{\boldsymbol{x}^{(kv)}(n)\boldsymbol{x}^{(kv)H}(n)\} = \boldsymbol{Q}^H E\{\boldsymbol{x}(n)\boldsymbol{x}^H(n)\}\boldsymbol{Q} = \boldsymbol{Q}^H \boldsymbol{R}_x \boldsymbol{Q} = \boldsymbol{Q}^H \boldsymbol{Q}\Lambda\boldsymbol{Q}^H\boldsymbol{Q} = \Lambda \quad (6.6)$$

We have assumed that the eigenvectors were normalized to unit length. Because Λ is diagonal, it is apparent that the elements of $\boldsymbol{x}^{(kv)}(n)$ are uncorrelated. Premultiply (6.5) by $\boldsymbol{Q}$ and use the relation $\boldsymbol{Q}\boldsymbol{Q}^H = \boldsymbol{I}$ to find

$$\boldsymbol{x}(n) = \sum_{i=0}^{N-1} x_i^{(kl)}(n)\boldsymbol{q}_i \quad (6.7)$$

The above expansion is known as the **Karhunen–Loeve expansion.**

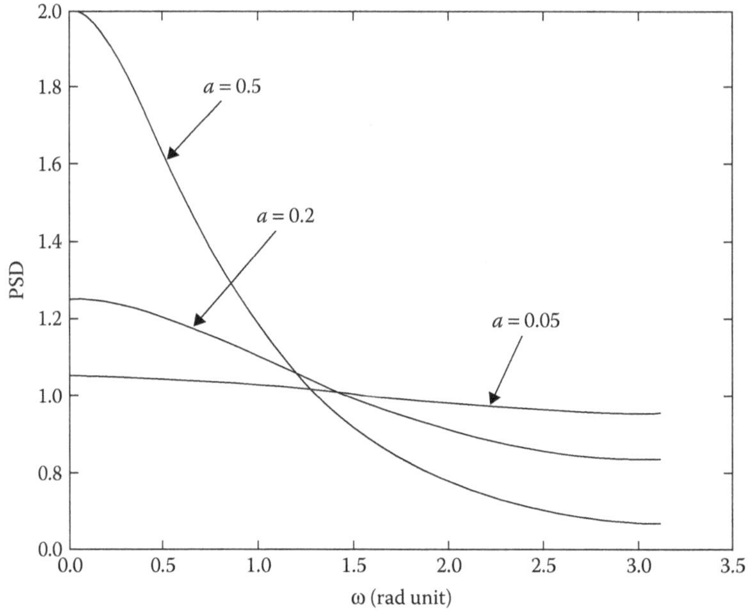

FIGURE 6.1

Suppose that a white noise with zero mean and variance one has passed a filter which for $a = 0$ has $H(z) = 1$, an all-pass filter. The output of this filter is the vector $\{\boldsymbol{x}(n)\}$ with the correlation function $r_{xx}(k) = a^k$ for $k = 0, 1, 2, \ldots, 255$. Figure 6.1 shows the power spectral density of the output signal $\{\boldsymbol{x}(n)\}$ for different values of a. It shows that as a approaches 0, the output is white noise since its spectrum becomes flat.

6.2 GEOMETRICAL PROPERTIES OF THE ERROR SURFACE

The performance function of a transversal [finite impulse response (FIR)] Wiener filter with a real-valued input sequence $\{x(n)\}$ and a desired output sequence $\{d(n)\}$ is

$$\begin{aligned}
&J = \boldsymbol{w}^T \boldsymbol{R} \boldsymbol{w} - 2\boldsymbol{p}^T \boldsymbol{w} + E\{d^2(n)\} \\
&\boldsymbol{w} = [w_0 \quad w_1 \quad \cdots \quad w_{M-1}],\ \boldsymbol{R}_x = E\{\boldsymbol{x}(n)\boldsymbol{x}^T(n)\}, \\
&\boldsymbol{x} = [x(n) \quad x(n-1) \quad \cdots \quad x(n-M+1)] \\
&\boldsymbol{p} = E\{d(n)\boldsymbol{x}(n)\}
\end{aligned} \tag{6.8}$$

The optimum Wiener filter coefficients are given by

$$\boldsymbol{R}_x \boldsymbol{w}^o = \boldsymbol{p} \tag{6.9}$$

Using the optimum solution of the Wiener filter from the above equation, and since $\boldsymbol{w}^T\boldsymbol{p} = \boldsymbol{p}^T\boldsymbol{w}$ = constant and $\boldsymbol{R}_x^T = \boldsymbol{R}_x$, the performance function in (6.8) becomes

$$
\begin{aligned}
J &= \boldsymbol{w}^T\boldsymbol{R}_x\boldsymbol{w} - \boldsymbol{w}^T\boldsymbol{R}_x\boldsymbol{w}^o - \boldsymbol{w}^{oT}\boldsymbol{R}_x^T\boldsymbol{w} + \boldsymbol{w}^{oT}\boldsymbol{R}_x\boldsymbol{w}^o + E\{d^2(n)\} - \boldsymbol{w}^{oT}\boldsymbol{R}_x\boldsymbol{w}^o \\
&= (\boldsymbol{w}-\boldsymbol{w}^o)^T\boldsymbol{R}_x(\boldsymbol{w}-\boldsymbol{w}^o) + E\{d^2(n)\} - \boldsymbol{w}^{oT}\boldsymbol{R}_x\boldsymbol{w}^o \\
&= (\boldsymbol{w}-\boldsymbol{w}^o)^T\boldsymbol{R}_x(\boldsymbol{w}-\boldsymbol{w}^o) + J_{\min} \\
&= J_{\min} + (\boldsymbol{w}-\boldsymbol{w}^o)^T\boldsymbol{R}_x(\boldsymbol{w}-\boldsymbol{w}^o)
\end{aligned}
\tag{6.10}
$$

The cost function (performance function) can be written in the form:

$$
\boldsymbol{w}^T\boldsymbol{R}_x\boldsymbol{w} - 2\boldsymbol{p}^T\boldsymbol{w} - (J - E\{d^2(n)\}) = 0 \text{ or } \boldsymbol{w}^T\boldsymbol{R}_x\boldsymbol{w} - 2\boldsymbol{p}^T\boldsymbol{w} - (J - \sigma_d^2) = 0 \tag{6.11}
$$

If we set the values of $J > J_{\min}$, the $\boldsymbol{w}$-plane will cut the second-order surface, for a two-coefficient filter, along a line whose projection on the $\boldsymbol{w}$-plane is ellipses arbitrarily oriented as shown in Figure 6.2. The contours were found using the following Book MATLAB program.

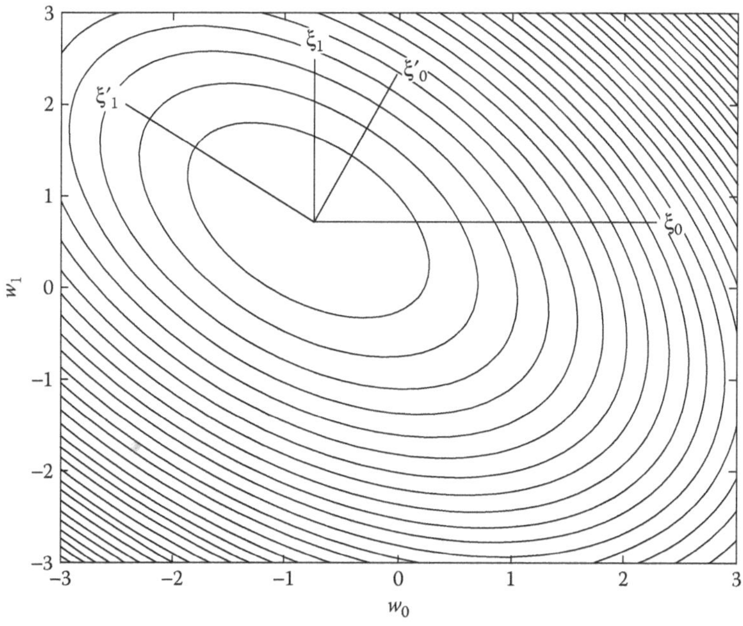

FIGURE 6.2

Book MATLAB Program

```
>>w0 = -3:0.05:3;w1 = -3:0.05:3;
>>[x,y] = meshgrid(w0,w1);%MATLAB function;
>>j=0.8461+0.972*x-0.773*y+1.0647*x.^2+1.064*y.^2+0.985*x.*y;
>>contours(x,y,j,30);%MATLAB function,30 is the number of
      %contours;
```

If we set `contour(j,[2.3 3.1 5])`, we will produce three contours at heights 2.3, 3.1, and 5. The first simplification we can do is to shift the origin of the $\boldsymbol{w}$-axis to another one whose origin is on the center of the ellipses. This is accomplished using the transformation: $\xi = \boldsymbol{w} - \boldsymbol{w}^o$. If we introduce this transformation in (6.11) and set $J = 2J_{\min}$, we obtain the relationship (see Problem 6.2.1):

$$\xi^T \boldsymbol{R}_x \xi - 2J_{\min} + \sigma_d^2 - \boldsymbol{p}^T \mathbf{w}^o = 0 \tag{6.12}$$

But $\sigma_d^2 - \boldsymbol{p}^T \boldsymbol{w}^o = J_{\min}$, and thus, the above equation becomes

$$\xi^T \boldsymbol{R}_x \xi = J_{\min} \tag{6.13}$$

The matrix is a diagonal matrix whose elements are the eigenvalues of the correlation matrix $\boldsymbol{R}_x$.

Example 6.2.1

Let $\lambda_1 = 1$, $\lambda_2 = 0.5$, and $J_{\min} = 0.67$. The ellipse in the (ξ_0', ξ_0') plane is found by solving the system:

$$\begin{aligned} &[\xi_0' \;\; \xi_0'] \begin{bmatrix} 1 & 0 \\ 0 & 0.5 \end{bmatrix} \begin{bmatrix} \xi_0' \\ \xi_1' \end{bmatrix} = 0.67 \quad \text{or} \\ &\left(\frac{\xi_0'}{\sqrt{0.67/1}} \right)^2 + \left(\frac{\xi_1'}{\sqrt{0.67/0.5}} \right)^2 = 1 \end{aligned} \tag{6.14}$$

where:

0.67/0.5 is the major axis
0.67/1 is the minor axis

Hence, for the case $J = 2J_{\min}$, the ellipse intersects the ξ_0'-axis at 0.67 and the ξ_1'-axis at 2.68. To find, for example, the intersection between the ξ_1'-axis and the ellipse, we must set ξ_0' equal to zero because the projection on the ξ_0'-axis of the intersection point is at $\xi_0' = 0$.

If we start with (6.11) and apply the shift and rotation transformations, we obtain the relationships:

$$\begin{aligned} J &= J_{\min} + (\boldsymbol{w} - \boldsymbol{w}^o)^T \boldsymbol{R}_x (\boldsymbol{w} - \boldsymbol{w}^o) = J_{\min} + \xi^T \boldsymbol{R}_x \xi \\ &= J_{\min} + \xi^T (\boldsymbol{Q} \Lambda \boldsymbol{Q}^T) \xi = J_{\min} + \xi'^T \Lambda \xi' = J_{\min} + \sum_{i=0}^{M-1} \lambda_i \xi_i'^2 \end{aligned} \tag{6.15}$$

Note:

- The contours intersect the ξ'-axes at values depending on the eigenvalues of $\boldsymbol{R}_x$ and the specific mean-square error (MSE) value chosen. The rotation and translation do not alter the shape of the MSE surface.
- If the successive contours for the values $2J_{\min}, 3J_{\min}, \ldots,$ are close to each other, the surface is steep, which in turn indicates that the mean-square estimation error is very sensitive to the choice of the filter coefficients.
- Choosing the filter values $\boldsymbol{w}$ is equivalent to choosing a point in the w-plane. The height of the MSE surface above the plane at that point is determined only by the signal correlation properties.

Let rearrange (6.15) in the form:

$$J - J_{\min} = \xi'^T \Lambda \xi' = [\xi_0' \quad \xi_1'] \begin{bmatrix} \lambda_0 & 0 \\ 0 & \lambda_1 \end{bmatrix} \begin{bmatrix} \xi_0' \\ \xi_1' \end{bmatrix} = \lambda_0 \xi_0'^2 + \lambda \xi_1'^2 \tag{6.16}$$

Next, we plot the contours of the performance surface (see Figure 6.3) for two different ratios of the eigenvalues. The contours with shapes closer to a circle were produced by the ratio 0.8/0.6 = 1.333 and the most elongated ones by the ratio 1.2/0.2 = 6. Instead of ξ_i', we renamed them to ξ_i for the plotting purpose.

■

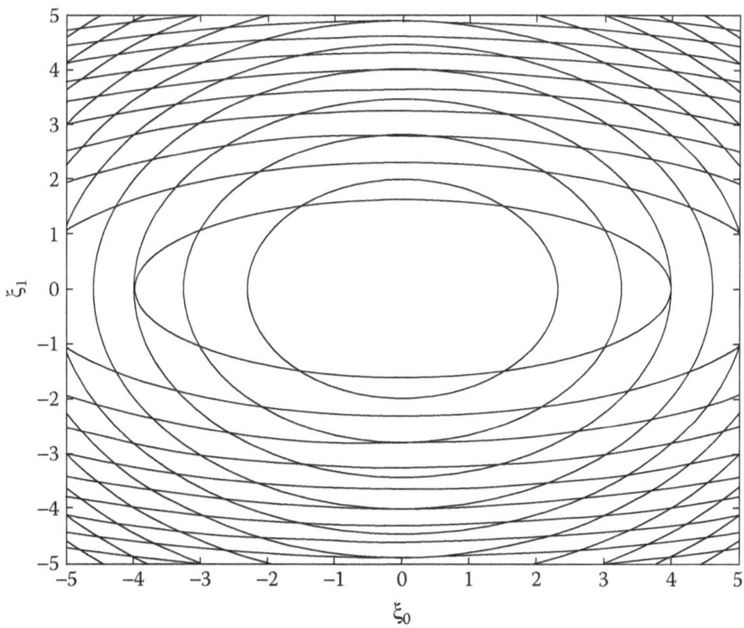

FIGURE 6.3

PROBLEMS

6.1.1 Prove that the correlation matrix of a WSS process is positive definite, that is, $\boldsymbol{a}^T\boldsymbol{R}\boldsymbol{a} \geq 0$ for any vector $\boldsymbol{a}$.

6.1.2 Verify the property $\text{tr}\{\boldsymbol{R}_x\} = \sum_{i=0}^{M-1}\lambda_i$, where $\text{tr}(\boldsymbol{A})$ stands for the trace of the matrix $\boldsymbol{A}$, which is equal to the sum of the main diagonal.

6.2.1 Verify (6.12).

6.2.2 Find the contours of the performance surface if the following data are given:

$$\boldsymbol{R}_x = \begin{bmatrix} 2 & 1 \\ 1 & 3 \end{bmatrix}, \; \boldsymbol{p}_{dx} = \begin{bmatrix} 6 \\ 7 \end{bmatrix}, \; \sigma_d^2 = 28$$

6.2.3 A Wiener filter is characterized by the following parameters:

$$\boldsymbol{R} = \begin{bmatrix} d & a \\ a & d \end{bmatrix}, \; \boldsymbol{p} = \begin{bmatrix} 1 \\ 1 \end{bmatrix}, \; \sigma_d^2 = 2$$

It is requested to explore the performance surface as the ratio λ_0/λ_1 varies. This is accomplished using different values for a and b.

6.2.4 Consider the performance function: $J_{\min} = w_0^2 + w_1^2 + w_0 w_1 - w_0 + w_1 + 2$. (a) Convert this equation to its canonical form and (b) plot the set of contour ellipses of the performance surface of J for values of $J = 1, 2, 3$, and 4.

6.2.5 If $\boldsymbol{R}$ is a correlation matrix, show that (a) $\boldsymbol{R}^n = \boldsymbol{Q}\Lambda^n\boldsymbol{Q}^T$. (b) If $\boldsymbol{R}^{1/2}\boldsymbol{R}^{1/2} = \boldsymbol{R}$, show that $\boldsymbol{R}^{1/2} = \boldsymbol{Q}\Lambda^{1/2}\boldsymbol{Q}^T$.

HINTS–SOLUTIONS–SUGGESTIONS

6.1.1

$$\boldsymbol{a}^T\boldsymbol{R}\boldsymbol{a} \geq 0 = \boldsymbol{a}^T E\{\boldsymbol{x}\boldsymbol{x}^T\}\boldsymbol{a} = E\{(\boldsymbol{a}^T\boldsymbol{x})(\boldsymbol{x}^T\boldsymbol{a})\} = E\{(\boldsymbol{a}^T\boldsymbol{x})^2\} \geq 0, \; \boldsymbol{a}^T\boldsymbol{x} = \boldsymbol{x}^T\boldsymbol{a}$$

$$= \text{Number}$$

6.1.2 $\text{tr}\{\boldsymbol{R}_x\} = \text{tr}\{\boldsymbol{Q}\Lambda\boldsymbol{Q}^T\} = \text{tr}\{\boldsymbol{Q}^T\boldsymbol{Q}\Lambda\} = \text{tr}\{\boldsymbol{I}\Lambda\} = \text{tr}\{\Lambda\}$, $\text{tr}\{\boldsymbol{A}\boldsymbol{B}\} = \text{tr}\{\boldsymbol{B}\boldsymbol{A}\}$; for Hermitian matrices, T changes to H.

6.2.1

$$(\boldsymbol{w} - \boldsymbol{w}^o + \boldsymbol{w}^o)^T\boldsymbol{R}(\boldsymbol{w} - \boldsymbol{w}^o + \boldsymbol{w}^o) - 2p(\boldsymbol{w} - \boldsymbol{w}^o + \boldsymbol{w}^o) - 2J_{\min} + \sigma_d^2$$

$$= (\xi + \boldsymbol{w}^o)^T\boldsymbol{R}(\xi + \boldsymbol{w}^o) - 2p^T(\xi + \boldsymbol{w}^o) - 2J_{\min} + \sigma_d^2$$

With $\boldsymbol{R}\boldsymbol{w}^o = \boldsymbol{p}, \boldsymbol{w}^{oT}\boldsymbol{p} = \boldsymbol{p}^T\boldsymbol{w}^o = \text{Number}$, and $\sigma_d^2 - \boldsymbol{p}^T\boldsymbol{w}^o = J_{\min}$, the equation is easily proved.

6.2.2

$J = J_{\min} + \xi'^T \Lambda \xi$ and $J_{\min} = \sigma_d^2 - \boldsymbol{p}^T \boldsymbol{w}^o$,

$$|\boldsymbol{R} - \lambda \boldsymbol{I}| = \begin{vmatrix} 2-\lambda & 1 \\ 1 & 3-\lambda \end{vmatrix} = (2-\lambda)(3-\lambda) - 1 = 0$$

$$\lambda_1 = 1.382,\ \lambda_2 = 3.618 \begin{bmatrix} 2 & 1 \\ 1 & 3 \end{bmatrix} \begin{bmatrix} w_0^o \\ w_1^o \end{bmatrix} = \begin{bmatrix} 6 \\ 7 \end{bmatrix},$$

$$\begin{bmatrix} w_0^o \\ w_1^o \end{bmatrix} = \begin{bmatrix} 11/5 \\ 8/5 \end{bmatrix},\ J_{\min} = 28 - [6\ \ 7] \begin{bmatrix} 11/5 \\ 8/5 \end{bmatrix} = 3.6$$

$$J = 3.6 + [\xi_0'\ \ \xi_1'] \begin{bmatrix} 1.328 & 0 \\ 0 & 3.618 \end{bmatrix} \begin{bmatrix} \xi_0^o \\ \xi_1^o \end{bmatrix} = 3.6 + 1.382\xi_0' + 3.618\xi_1'$$

6.2.3 The MSE surface is given by

$$J = 2 - 2[w_0\ w_1] \begin{bmatrix} 1 \\ 1 \end{bmatrix} + [w_0\ w_1] \begin{bmatrix} d & a \\ a & d \end{bmatrix} \begin{bmatrix} w_0 \\ w_1 \end{bmatrix},$$

and the optimum tap weights are

$$\begin{bmatrix} w_0^o \\ w_1^o \end{bmatrix} = \boldsymbol{R}^{-1}\boldsymbol{p} = \begin{bmatrix} d & a \\ a & d \end{bmatrix}^{-1} \begin{bmatrix} 1 \\ 1 \end{bmatrix} = \begin{bmatrix} 1/d+a \\ 1/d+a \end{bmatrix}.$$

Therefore, the minimum MSE surface is

$$J_{\min} = \sigma_d^2 - \boldsymbol{w}^{oT}\boldsymbol{p} = 2 - \begin{bmatrix} \dfrac{1}{d+a} & \dfrac{1}{d+a} \end{bmatrix} \begin{bmatrix} 1 \\ 1 \end{bmatrix} = 2\left[\frac{d+a-1}{d+a}\right].$$

We also have the relation

$$J = J_{\min} + (\boldsymbol{w} - \boldsymbol{w}^o)^T \boldsymbol{R} (\boldsymbol{w} - \boldsymbol{w}^o) = 2\left[\frac{d+a+1}{d+a}\right] + [\xi_0\ \xi_1] \begin{bmatrix} d & a \\ a & d \end{bmatrix} \begin{bmatrix} \xi_0 \\ \xi_1 \end{bmatrix}.$$

From the relation

$$\begin{vmatrix} d-\lambda & a \\ a & d-\lambda \end{vmatrix} = 0,$$

we obtain the two eigenvalues $\lambda_0 = d + a$ and $\lambda_1 = d - a$. By incorporating different values for a and d, we can create different ratios of the eigenvalues. The larger the ratio, the more elongated the ellipses are.

6.2.4

$$J = [w_0 \;\; w_1]\begin{bmatrix} 1 & 0.5 \\ 0.5 & 1 \end{bmatrix}\begin{bmatrix} w_0 \\ w_1 \end{bmatrix} - 2\begin{bmatrix} \frac{1}{2} & -\frac{1}{2} \end{bmatrix}\begin{bmatrix} w_0 \\ w_1 \end{bmatrix} + 2,\; E\{d^2(n)\} = 2,$$

$$\boldsymbol{R} = \begin{bmatrix} 1 & 0.5 \\ 0.5 & 1 \end{bmatrix},\; \boldsymbol{p} = \begin{bmatrix} 1/2 \\ -1/2 \end{bmatrix}$$

$$\boldsymbol{w}^o = \boldsymbol{R}^{-1}\boldsymbol{p} = \frac{1}{0.75}\begin{bmatrix} 1 & -0.5 \\ -0.5 & 1 \end{bmatrix}\begin{bmatrix} 1/2 \\ -1/2 \end{bmatrix} = \begin{bmatrix} 1 \\ -1 \end{bmatrix},\; J_{\min} = 2 - \boldsymbol{p}^T\boldsymbol{w}^o = 2 - 1 = 1,$$

$$J = 1 + \xi^T\begin{bmatrix} 1 & 0.5 \\ 0.5 & 1 \end{bmatrix}\xi,\; |\lambda \boldsymbol{I} - \boldsymbol{R}| = 0$$

$$\Rightarrow (\lambda - 1)^2 - 0.25 = 0 \Rightarrow \lambda_0 = 1.5,\; \lambda_1 = 0.5,$$

$$\text{Eigenvectors} \begin{bmatrix} 1 & 0.5 \\ 0.5 & 1 \end{bmatrix}\begin{bmatrix} q_{00} \\ q_{01} \end{bmatrix} = 1.5\begin{bmatrix} q_{00} \\ q_{01} \end{bmatrix}$$

$$\Rightarrow q_{00} + 0.5q_{01} = 1.5q_{00} \Rightarrow q_{00} = q_{01} \quad (\text{also } 0.5q_{00} + q_{01} = 1.5q_{01})$$

$$\Rightarrow q_{00} = q_{01} \Rightarrow \text{normalized } \boldsymbol{q}_0 = \frac{1}{\sqrt{2}}\begin{bmatrix} 1 \\ 1 \end{bmatrix};\; \begin{bmatrix} 1 & 0.5 \\ 0.5 & 1 \end{bmatrix}\begin{bmatrix} q_{10} \\ q_{11} \end{bmatrix} = 0.5\begin{bmatrix} q_{10} \\ q_{11} \end{bmatrix}$$

$$\Rightarrow q_{10} = -q_{11} \Rightarrow \text{normalized } \boldsymbol{q}_1 = \frac{1}{\sqrt{2}}\begin{bmatrix} 1 \\ -1 \end{bmatrix};$$

eigenvalue and eigenvector matrices are

$$\Lambda = \begin{bmatrix} 1.5 & 0 \\ 0 & 0.5 \end{bmatrix},\; \boldsymbol{Q} = \frac{1}{\sqrt{2}}\begin{bmatrix} 1 & 1 \\ 1 & -1 \end{bmatrix},$$

$$J = 1 + \xi^T\boldsymbol{Q}\Lambda\boldsymbol{Q}^T\xi = 1 + \xi'^T\begin{bmatrix} 1.5 & 0 \\ 0 & 0.5 \end{bmatrix}\xi'$$

$$= 1 + 1.5\xi_0'^2 + 0.5\xi_1'^2 = 1 + \left(\frac{\xi_0'}{1/\sqrt{1.5}}\right)^2 + \left(\frac{\xi_1'}{1/\sqrt{0.5}}\right)^2$$

We can use the following Book MATLAB program to produce four contours at different heights of the performance surface.

```
>>x = -5:0.1:5; y = -5:0.1:5;
>>[X,Y]=meshgrid(x,y);
>>Z=1+1.5*X.^2+0.5*Y.^2;
>>V=[1.2 2 3 4];%if we set 1 instead of 1.2, the first
     %circle will be a dot since this indicates one point
     %of the bottom of the surface, as it should be;
>>contour(X,Y,Z,V,'k');%the circles will be black;
```

6.2.5

(a) $\boldsymbol{R}^n = \boldsymbol{Q}\Lambda\boldsymbol{Q}^T\boldsymbol{Q}\Lambda\boldsymbol{Q}^T\boldsymbol{Q}\Lambda\boldsymbol{Q}^T \cdots \boldsymbol{Q}\Lambda\boldsymbol{Q}^T = \boldsymbol{Q}\Lambda^n\boldsymbol{Q}^T$, because $\boldsymbol{Q}^T\boldsymbol{Q} = \boldsymbol{I} = \boldsymbol{Q}\boldsymbol{Q}^T$

(b) Let $\boldsymbol{P} = \boldsymbol{Q}^T\boldsymbol{R}^{1/2}\boldsymbol{Q} \Rightarrow \boldsymbol{R}^{1/2} = \boldsymbol{Q}\boldsymbol{P}\boldsymbol{Q}^T \Rightarrow \boldsymbol{R}^{1/2}\boldsymbol{R}^{1/2} = \boldsymbol{R}$

$$\Rightarrow \boldsymbol{Q}\boldsymbol{P}\boldsymbol{Q}^T\boldsymbol{Q}\boldsymbol{P}\boldsymbol{Q}^T = \boldsymbol{Q}\Lambda\boldsymbol{Q}^T \Rightarrow \boldsymbol{Q}\boldsymbol{P}^2\boldsymbol{Q}^T = \boldsymbol{Q}\Lambda\boldsymbol{Q}^T$$

$$\Rightarrow \boldsymbol{Q}^T\boldsymbol{Q}\boldsymbol{Q}\boldsymbol{P}^2\boldsymbol{Q}^T\boldsymbol{Q} = \boldsymbol{Q}^T\boldsymbol{Q}\Lambda\boldsymbol{Q}^T\boldsymbol{Q} \Rightarrow \boldsymbol{P}^2 = \Lambda$$

$$\Rightarrow \boldsymbol{P} = \Lambda^{1/2} \Rightarrow \boldsymbol{R}^{1/2} = \boldsymbol{Q}\Lambda^{1/2}\boldsymbol{Q}^T$$

7 Newton's and Steepest Descent Methods

7.1 ONE-DIMENSIONAL GRADIENT SEARCH METHOD

In general, we can say that adaptive algorithms are nothing but iterative search algorithms derived from minimizing a cost function with the true statistics replaced by their estimates. To study the adaptive algorithms, it is necessary to have a thorough understanding of the iterative algorithms and their convergence properties. In this chapter, we discuss the **steepest descent** and the **Newton's method**.

The one-coefficient mean-square error (MSE) surface (line in this case) is given by [see (6.15)]

$$J(w) = J_{\min} + r_{xx}(0)(w - w^o)^2; \quad J(w^o) \le J(w) \text{ for all } w\text{'s} \tag{7.1}$$

and it is pictorially shown in Figure 7.1. The first and second derivatives are

$$\frac{\partial J(w)}{\partial w} = 2r_{xx}(0)(w - w^o) \tag{7.2a}$$

$$\frac{\partial^2 J(w)}{\partial w^2} = 2r_{xx}(0) > 0 \tag{7.2b}$$

At $w = w^o$, the first derivative is zero and the second derivative is positive and greater than zero, which indicates that the surface has a global minimum and is concave upward. To find the optimum value of w, we can use an iterative approach. We start first with an arbitrary value $w(0)$ and measure the slope of the curve $J(w)$ at $w(0)$. Next, we find $w(1)$ to be equal to $w(0)$ and the negative of an increment **proportional** to the slope of $J(w)$ at $w(0)$. Proceeding with the iteration procedure, we will eventually find the minimum value w^o. The values $w(0), w(1), \ldots$, are known as the **gradient estimates**.

7.1.1 Gradient Search Algorithm

Based on the above development in Section 7.1, the filter coefficient at iteration n, $w(n)$, is found using the relation:

$$\begin{aligned} w(n+1) &= w(n) + \mu\left[-\frac{\partial J(w)}{\partial w}\Big|_{w=w(n)}\right] \\ &= w(n) + \mu\{-\nabla J[w(n)]\} = w(n) - 2\mu r_{xx}(0)[w(n) - w^o] \end{aligned} \tag{7.3}$$

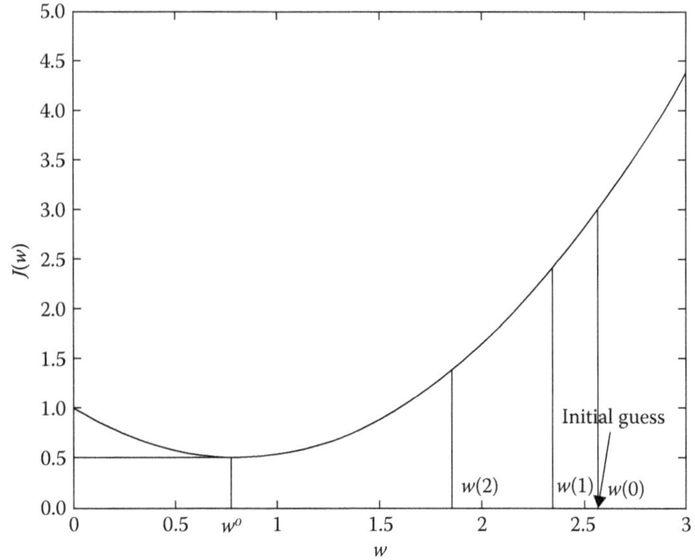

FIGURE 7.1

where:
 μ is a constant to be determined

Rearranging (7.3), we obtain

$$w(n+1) = [1 - 2\mu r_{xx}(0)]\,w(n) + 2\mu r_{xx}(0) w^o$$

$$\begin{bmatrix} \text{New} \\ \text{coefficient} \end{bmatrix} = \begin{bmatrix} \text{Old} \\ \text{coefficient} \end{bmatrix} + \begin{bmatrix} \text{Adaptation} \\ \text{gain} \end{bmatrix} \tag{7.4}$$

$$w(n+1) = aw(n) + bw^o \qquad \text{(Difference equation)}$$

The solution of the above difference equation, using the iteration approach (see Problem 7.1.1), is

$$w(n) = w^o + \left[1 - 2\mu r_{xx}(0)\right]^n [w(0) - w^o] \tag{7.5}$$

The above equation gives $w(n)$ explicitly at any iteration n in the search procedure. This is the solution to the **gradient search algorithm**. Note that if we had initially guessed $w(0) = w^o$, which is the optimum value, we would have found $w(1) = w^o$, which gives the optimum value in one step.

To have convergence of $w(n)$ in (7.5), we must impose the condition:

$$\left|1 - 2\mu r_{xx}(0)\right| < 1 \tag{7.6}$$

The above inequality defines the range of the step-size constant μ so that the algorithm will converge. Hence, we obtain

$$-1<1-2\mu r_{xx}(0)<1 \text{ or } 0<2\mu r_{xx}(0)<2 \text{ or } 0<\mu<\frac{1}{r_{xx}(0)} \tag{7.7}$$

Under the above condition, (7.5) converges to the optimum value w^o as $n\rightarrow\infty$. If $\mu>1/r_{xx}(0)$, the process is unstable and no convergence takes place.

When the filter coefficient has a value $w(n)$ (i.e., at iteration n), then the MSE surface (here a line) is [see (7.1)]

$$J(n)=J_{\min}+r_{xx}(0)[w(n)-w^o]^2 \tag{7.8}$$

Substituting the quantity $w(n)-w^o$ in (7.8) from (7.5), we obtain

$$J(n)=J_{\min}+r_{xx}(0)[w(0)-w^o]^2\left[1-2\mu r_{xx}(0)\right]^{2n} \tag{7.9}$$

The above two equations show that $w(n)-w^o$ as n increases to infinity, and the MSE undergoes a geometric progression toward $J_{\min}$. The plot of the performance surface $J(n)$ versus the iteration number n is known as the **learning curve**.

7.1.2 Newton's Method in Gradient Search

The Newton's method finds the solution (zeros) of the equation $f(w)=0$. From Figure 7.2, we observe that the slope at $w(0)$ is

$$f'[w(0)]=\left.\frac{df(w)}{dw}\right|_{w=w(0)}\cong\frac{f[w(0)]}{w(0)-w(1)} \tag{7.10}$$

where:

$w(0)$ is the initial guessing

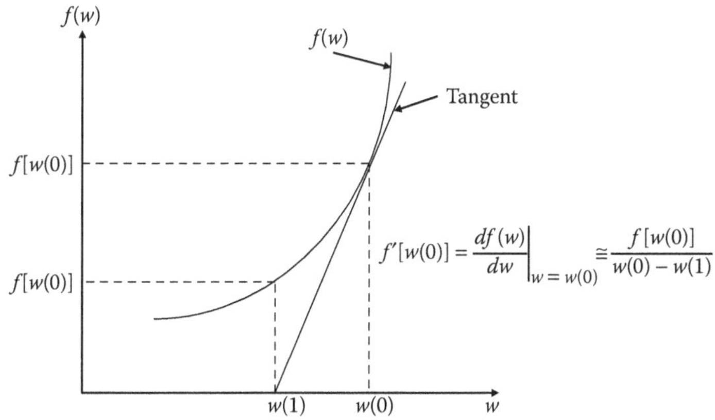

FIGURE 7.2

The above equation is the result of retaining the first two terms of Taylor's expansion and setting the rest of them equal to zero. This equation can be written in the form:

$$w(1) = w(0) - \frac{f[w(0)]}{f'[w(0)]} \tag{7.11}$$

From (7.11), it is clear that if we know the values of the function and its derivative at the same point, in this case at $n = 0$, we find $w(1)$. Hence, the nth iteration of the above equation takes the form:

$$w(n+1) = w(n) - \frac{f[w(n)]}{f'[w(n)]} \qquad n = 0, 1, 2, \ldots \tag{7.12}$$

But we have $f'[w(n)] \cong \{f[w(n)] - f[w(n-1)]\}/[w(n) - w(n-1)]$, and hence, (7.12) becomes

$$w(n+1) = w(n) - \frac{f[w(n)][w(n) - w(n-1)]}{f[w(n)] - f[w(n-1)]} \tag{7.13}$$

As we have mentioned earlier, the Newton's method finds the roots of the function $f(w)$, which solves the polynomial $f(w) = 0$. However, in our case we need to find the minimum point of the performance surface (here line). This is equivalent to setting $\partial J(w)/\partial w = 0$. Therefore, we substitute the derivative $\partial J(w)/\partial w$ for $f(w)$ in (7.12) and the second-order derivative $\partial^2 J(w)/\partial w^2$ for $f'(w)$. Hence, (7.13) becomes

$$w(n+1) = w(n) - \frac{\partial J[w(n)]/\partial w(n)}{\partial^2 [w(n)]/\partial w^2(n)} \qquad n = 0, 1, 2, \ldots \tag{7.14}$$

7.2 STEEPEST DESCENT ALGORITHM

To find the minimum value of the MSE surface, $J_{\min}$, using the steepest descent algorithm, we proceed as follows: (1) We start with the initial value $\mathbf{w}(0)$, usually using the null vector; (2) at the MSE surface point that corresponds to $\mathbf{w}(0)$, we compute the **gradient vector**, $\nabla J[\mathbf{w}(0)]$; (3) we compute the value $-\mu \nabla J[\mathbf{w}(0)]$ and add it to $\mathbf{w}(0)$ to obtain $\mathbf{w}(1)$; (4) we go back to step (2) and continue the procedure till we find the optimum value of the vector coefficient, $\mathbf{w}^o$.

If $\mathbf{w}(n)$ is the filter coefficient vector at step n (time), then its updated value $\mathbf{w}(n+1)$ is given by [see also (7.3); it is the first order of Taylor's expansion]

$$\begin{aligned} &\mathbf{w}(n+1) = \mathbf{w}(n) - \mu \nabla_{\mathbf{w}} J[\mathbf{w}(n)] \\ &\nabla J_{\mathbf{w}}[\mathbf{w}(n)] = \begin{bmatrix} \dfrac{\partial J[\mathbf{w}(n)]}{\partial w_0} \\ \dfrac{\partial J[\mathbf{w}(n)]}{\partial w_1} \\ \vdots \\ \dfrac{\partial J[\mathbf{w}(n)]}{\partial w_{M-1}} \end{bmatrix}; \quad \mathbf{w}(n) = [w_0 \;\; w_1 \;\; w_2 \ldots w_{M-1}]^T \end{aligned} \tag{7.15}$$

7.2.1 Steepest Descent Algorithm Applied to Wiener Filter

The cost function $J[\mathbf{w}(n)]$ at time n is a quadratic function of the Wiener coefficients. Hence, we write

$$J[\mathbf{w}(n)] = \sigma_d^2 - \mathbf{w}^T(n)\mathbf{p} - \mathbf{p}^T\mathbf{w}(n) + \mathbf{w}^T(n)\mathbf{R}_x\mathbf{w}(n) \tag{7.16}$$

where:
- σ_d^2 is the variance of the desired response $d(n)$
- $\mathbf{p}$ is the cross-correlation vector between the Wiener coefficients vector $\mathbf{x}(n)$ and the desired response $d(n)$
- $\mathbf{R}_x$ is the correlation matrix of the coefficient vector $\mathbf{x}(n)$

Using (7.15) and Section A2.6, we obtain

$$\nabla_w J[\mathbf{w}(n)] = -2\mathbf{p} + 2\mathbf{R}_x\mathbf{w}(n) \tag{7.17}$$

Example 7.2.1

Verify (7.17) with a two-length vector.

Solution:

$$\nabla_w\{J(\mathbf{w}(n))\} = \nabla_w\{\sigma_d^2\} - \nabla_w\{\mathbf{w}^T(n)\mathbf{p}\} - \nabla_w\{\mathbf{p}^T\mathbf{w}(n)\} + \nabla_w\{\mathbf{w}^T(n)\mathbf{R}_x\mathbf{w}(n)\}$$

Hence, we obtain

$$\nabla_w(\sigma_d^2) = 0, \quad -\nabla_w(\mathbf{w}^T\mathbf{p}) = -\begin{bmatrix} \dfrac{\partial(w_0p_0 + w_1p_1)}{\partial w_0} \\ \dfrac{\partial(w_0p_0 + w_1p_1)}{\partial w_1} \end{bmatrix} = -\begin{bmatrix} p_0 \\ p_1 \end{bmatrix} = -\mathbf{p}$$

$$-\nabla_w(\mathbf{p}^T\mathbf{w}) = -\begin{bmatrix} \dfrac{\partial(w_0p_0 + w_1p_1)}{\partial w_0} \\ \dfrac{\partial(w_0p_0 + w_1p_1)}{\partial w_1} \end{bmatrix} = -\begin{bmatrix} p_0 \\ p_1 \end{bmatrix} = -\mathbf{p}$$

$$\begin{aligned}
\nabla_w(\mathbf{w}^T\mathbf{R}_x\mathbf{w}) &= \nabla_w\left([w_0 \;\; w_1]\begin{bmatrix} r_{xx}(0) & r_{xx}(1) \\ r_{xx}(1) & r_{xx(0)} \end{bmatrix}\begin{bmatrix} w_0 \\ w_1 \end{bmatrix}\right) \\
&= \nabla_w[w_0^2r_{xx}(0) + w_0w_1r_{xx}(1) + w_0w_1r_{xx}(1) + w_1^2r_{xx}(1)] \\
&= \begin{bmatrix} 2w_0r_{xx}(0) + 2w_1r_{xx}(1) \\ 2w_0r_{xx}(1) + 2w_1r_{xx}(0) \end{bmatrix} = 2\begin{bmatrix} r_{xx}(0) & r_{xx}(1) \\ r_{xx}(1) & r_{xx}(0) \end{bmatrix}\begin{bmatrix} w_0 \\ w_1 \end{bmatrix} = 2\mathbf{R}_x\mathbf{w}
\end{aligned}$$

All vectors are taken at time n.

■

Introducing the results in Example 7.2.1 in (7.15a), we find

$$\begin{aligned} \boldsymbol{w}(n+1) &= \boldsymbol{w}(n) + 2\mu\left[\boldsymbol{p}_{dx} - \boldsymbol{R}_x\boldsymbol{w}(n)\right] = \boldsymbol{w}(n) + \mu'\left[\boldsymbol{p}_{dx} - \boldsymbol{R}_x\boldsymbol{w}(n)\right] \\ &= \left[\boldsymbol{I} - \mu'\boldsymbol{R}_x\right]\boldsymbol{w}(n) + \mu'\boldsymbol{p}_{dx}, \quad \mu' = 2\mu \quad n = 0, 1, 2, \ldots \end{aligned} \tag{7.18}$$

where:

$\boldsymbol{I}$ is the identity matrix

The value of the primed step-size parameter must be much less than 1/2 for convergence. We may view (7.18) as a **feedback model**, as illustrated by the **signal-flow graph** in Figure 7.3.

To apply the method of steepest descent, we must first find the estimate of the autocorrelation matrix $\boldsymbol{R}_x$ and the cross-correlation vector $\boldsymbol{p}_{dx}$ from the finite data. This is necessary since we do not have in practice an ensemble of data to obtain $\boldsymbol{R}_x$ and $\boldsymbol{p}_{dx}$.

7.2.2 Stability (Convergence) of the Algorithm

Let

$$\xi(n) = \boldsymbol{w}(n) - \boldsymbol{w}^o \tag{7.19}$$

be the difference between the filter coefficient vector and its optimum Wiener value $\boldsymbol{w}^o$. Next we write the first part of (7.18) in the form:

$$\begin{aligned} &\boldsymbol{w}(n+1) - \boldsymbol{w}^o = \boldsymbol{w}(n) - \boldsymbol{w}^o + \mu'\left[\boldsymbol{R}_x\boldsymbol{w}^o - \boldsymbol{R}_x\boldsymbol{w}(n)\right] \quad \text{or} \\ &\xi(n+1) = \left[\boldsymbol{I} - \mu'\boldsymbol{R}_x\right]\xi(n) \end{aligned} \tag{7.20}$$

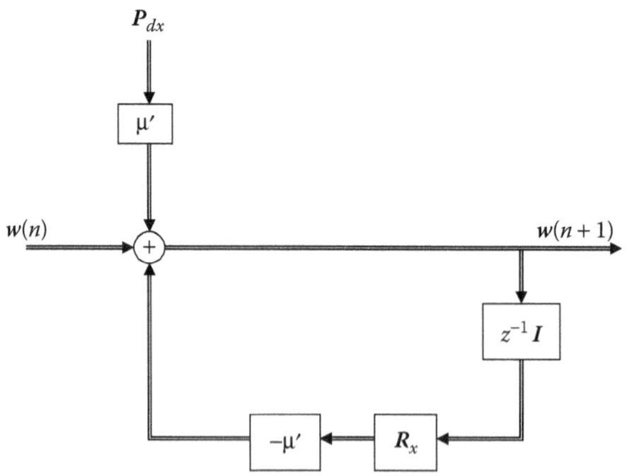

FIGURE 7.3

But since $\boldsymbol{R}_x = \boldsymbol{Q}\Lambda\boldsymbol{Q}^T$ (see Table 6.1.1) and $\boldsymbol{I} = \boldsymbol{Q}\boldsymbol{Q}^T$, (7.20) becomes

$$\xi(n+1) = [\boldsymbol{I} - \mu'\boldsymbol{Q}\Lambda\boldsymbol{Q}^T]\xi(n)$$

or

$$\boldsymbol{Q}^T\xi(n+1) = [\boldsymbol{I} - \mu'\Lambda]\boldsymbol{Q}^T\xi(n) \tag{7.21}$$

or

$$\xi'(n+1) = [\boldsymbol{I} - \mu'\Lambda]\xi'(n)$$

where the coordinate axis defined by ξ' is orthogonal to ellipsoids (see Chapter 6) and is a diagonal matrix with its diagonal elements equal to the eigenvalues of $\boldsymbol{R}_x$. The kth row of (7.21), which represents the kth **natural mode** of the steepest descent algorithm, is (see Problem 7.2.1)

$$\xi'_k(n+1) = \left[1 - \mu'\lambda_k\right]\xi'_k(n) \tag{7.22}$$

The above equation is a homogeneous difference equation, which has the following solution (see Problem 7.1.1):

$$\xi'_k(n) = (1 - \mu'\lambda_k)^n \xi'_k(0) \quad k = 1, 2, \ldots, M-1, \qquad n = 1, 2, 3, \ldots \tag{7.23}$$

For the above equation to converge (to be stable) as $n \to \infty$, we must set

$$-1 < 1 - \mu'\lambda_k < 1 \qquad k = 0, 1, 2, \ldots, M-1 \tag{7.24}$$

First, subtract -1 from all the three elements of the inequality. Next, multiply by -1 and reverse the inequalities. Hence, we obtain

$$0 < \mu' < \frac{2}{\lambda_k} \quad \text{or} \quad 0 < \mu < \frac{1}{\lambda_k} \quad k = 0, 1, 2, \ldots, M-1 \tag{7.25}$$

Under the above conditions and as $n \to \infty$, (7.23) becomes $\lim_{n\to\infty} \xi'_k(n) = 0$ or $\xi = \boldsymbol{0}$ since $\boldsymbol{Q}^T \neq \boldsymbol{0}$, and thus, $\boldsymbol{w}(\infty) = \boldsymbol{w}^o$. Because (7.23) decays exponentially to zero, there exists a time constant that depends on the value of μ' and the eigenvalues of $\boldsymbol{R}_x$. Furthermore, (7.23) implies that immaterially of the initial value $\xi(0)$, $\boldsymbol{w}(n)$ always converges to $\boldsymbol{w}^o$, provided, of course, that (7.25) is satisfied. This is an important property of the steepest descent algorithm. Since each row of (7.22) must decay as $n \to \infty$, it is necessary and sufficient that μ' obeys the following relationship:

$$0 < \mu' < \frac{2}{\lambda_{\max}} \tag{7.26}$$

Since $(1 - \mu'\lambda_k)^n$ decays exponentially, there exists an exponential function with time constant τ_k such that $[e^{-(1/\tau_k)}]^n = (1 - \mu'\lambda_k)^n$ or $(1 - \mu'\lambda_k) = e^{-(1/\tau_k)} = 1 - (1/\tau_k) + (1/2!\,\tau_k^2) - \cdots$. Therefore, for small μ' and λ_k (larger τ_k), we have

$$\tau_k \cong \frac{1}{\mu'\lambda_k} \qquad \mu'\lambda_k << 1 \tag{7.27}$$

In general, the kth time constant τ_k can be expressed in the form:

$$\tau_k = -\frac{1}{\ln(1-\mu'\lambda_k)} \tag{7.28}$$

7.2.3 Transient Behavior of MSE

Using (6.15) at time n, we obtain the relation:

$$J[\mathbf{w}(n)] = J_{\min} + \xi'^T \Lambda \xi' = J_{\min} + \sum_{k=0}^{M-1} \lambda_k \xi_k'^2(n) \tag{7.29}$$

Substituting the solution of $\xi_k'(n)$ from (7.23) in (7.29), we find the relation:

$$J[\mathbf{w}(n)] = J_{\min} + \sum_{k=0}^{M-1} \lambda_k (1-\mu'\lambda_k)^{2n} \xi'^2(0) \tag{7.30}$$

It is obvious from the above equation (the factor in parentheses is less than one) that

$$\underbrace{\lim}_{n\to\infty} J[\mathbf{w}(n)] = J_{\min} \tag{7.31}$$

From (7.29), we observe that the **learning curve** [the plot of $J[\mathbf{w}(n)]$ vs. n] consists of a sum of exponentials, each one corresponding to a natural mode of the algorithm.

Example 7.2.2

Let us have the following data:

$$\boldsymbol{R}_x = \begin{bmatrix} 1 & 0.4 \\ 0.4 & 1 \end{bmatrix}, \; \boldsymbol{p}_{dx} = \begin{bmatrix} 0 \\ 0.294 \end{bmatrix}, \; w_0(1) = -1.2, \; w_1(1) = -2.2, \; \mu = 0.2 \tag{7.32}$$

It is desired (1) to find the equation of the MSE and plot the contours, and (2) to plot the convergence path of the steepest descent algorithm in the MSE surface.

Solution: The MSE function is

$$\begin{aligned} J &= E\{d^2(k)\} - 2\boldsymbol{w}^T \boldsymbol{p}_{dx} + \boldsymbol{w}^T \boldsymbol{R}_x \boldsymbol{w} \\ &= \sigma_d^2 - 2\begin{bmatrix} w_0 & w_1 \end{bmatrix}\begin{bmatrix} 0 \\ 0.294 \end{bmatrix} + \begin{bmatrix} w_0 & w_1 \end{bmatrix}\begin{bmatrix} 1 & 0.4 \\ 0.4 & 1 \end{bmatrix}\begin{bmatrix} w_0 \\ w_1 \end{bmatrix} \\ &= 0.48 + w_0^2 + w_1^2 + 0.8 w_0 w_1 - 0.588 w_1 \end{aligned}$$

The following Book MATLAB program was used to obtain the changes in the filter coefficient:

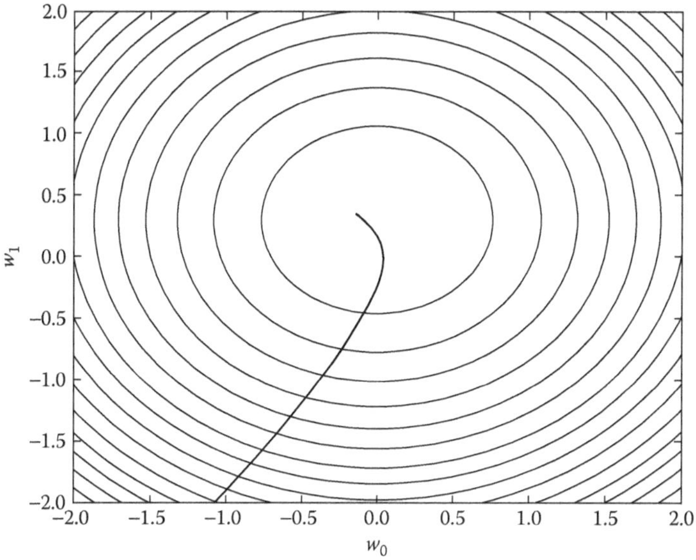

FIGURE 7.4

Book MATLAB Program

```
>>w0(1) = -1.2; w1(1) = -2.2; m = 0.2;
>>for k = 1:50
>>    w0(k+1) = w0(k)-2*m*(w0(k)+0.4*w1(k));
>>    w1(k+1) = w1(k)-2*m*(0.4*w0(k)+w1(k))+2*m*0.294;
>>end
>>x=-2:0.2:2; y = -2:0.2:2;
>>[X,Y]=meshgrid(x,y);
>>Z=0.48+X.^2+Y.^2+0.8*X*Y-0.588*Y;
>>contour(X,Y,Z,15,'k');
>>hold on; plot(w0,w1,'k');
```

Figure 7.4 shows the contours and the trajectory of the filter coefficients. The eigenvalue ratio for this case was 1.4/0.6 = 2.3333.

■

As the eigenvalue ratio increases, the contours become elongated ellipses.

7.2.4 Learning Curve

Accepting a small value of the constant μ, converging condition, the terms under the summation in (7.30) converge to zero as k increases. As a result, the minimum MSE is achieved after a sufficient number of iterations.

The curve obtained by plotting J as a function of the iteration index k is called the **learning curve**. A learning curve of the steepest descent algorithm, as can be

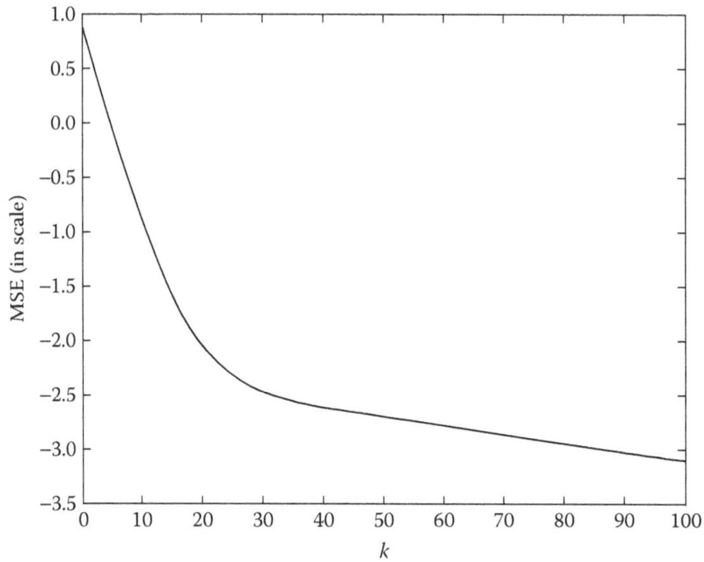

FIGURE 7.5

seen from (7.30), consists of the sum of M exponentially decaying terms, each of which corresponds to one of the modes of convergence of the algorithm. Assuming $\mu = 0.02$, $\xi'^2(0) = 1$, and the two eigenvalues, 0.1 and 2.4, we obtain the curve shown in Figure 7.5. We observe that the two straight sections in the graph indicate two different time constants (see Problem 7.2.2).

7.3 NEWTON'S METHOD

Using $\boldsymbol{p}_{dx} = \boldsymbol{R}\boldsymbol{w}^o$, (7.18) becomes

$$\boldsymbol{w}(n+1) = \boldsymbol{w}(n) - 2\mu \boldsymbol{R}_x(\boldsymbol{w} - \boldsymbol{w}^o) \tag{7.33}$$

The presence of the correlation matrix $\boldsymbol{R}_x$ in the above equation causes the eigenvalue spread problem in the steepest descent algorithm. The Newton's method overcomes this problem by replacing the scalar step-size parameter μ with the matrix step size given by $\mu \boldsymbol{R}_x^{-1}$. Using Appendix 2, we obtain from (7.16) the relation:

$$\nabla_w J[\boldsymbol{w}(n)] = -2\boldsymbol{p}_{dx} + \boldsymbol{R}_x \boldsymbol{w}(n) \tag{7.34}$$

Using the above equation in (7.33), we obtain the equation:

$$\boldsymbol{w}(n+1) = \boldsymbol{w}(n) - 2\mu \boldsymbol{R}_x[\boldsymbol{w}(n) - \boldsymbol{w}^o] = \boldsymbol{w}(n) - \mu \nabla_w J(\boldsymbol{w}) \tag{7.35}$$

Replacing μ with the Newton's relation $\mu \boldsymbol{R}_x^{-1}$ overcomes the eigenvalue spread, which may produce large values of its elements, and this will produce difficulties

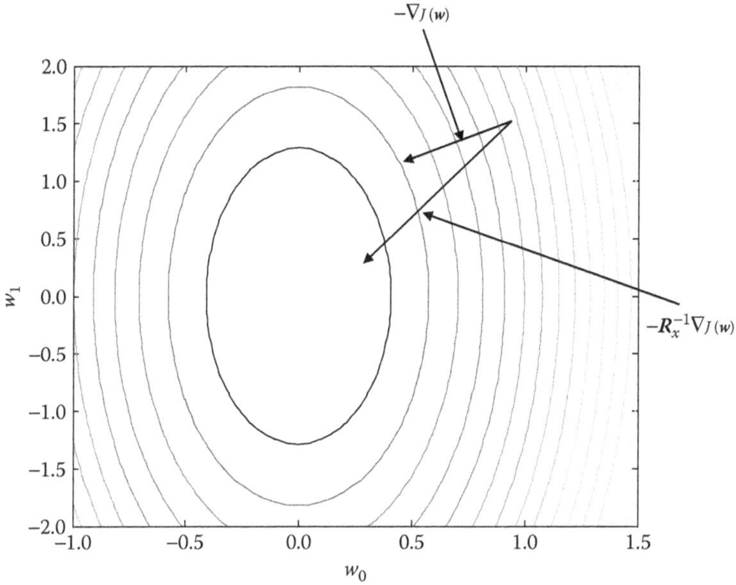

FIGURE 7.6

in solving equations that involve inverse correlation matrices. In such cases, we say that the correlation matrices are **ill-conditioned**. This substitution has the effect of rotating the gradient vector to the direction pointing toward the minimum point of the MSE surface as shown in Figure 7.6.

Substituting (7.34) in (7.35) and the Newton's relation, we obtain

$$\begin{aligned} \boldsymbol{w}(n+1) &= \boldsymbol{w}(n) - \mu \boldsymbol{R}_x^{-1}[2\boldsymbol{R}\boldsymbol{w}(n) - 2\boldsymbol{p}_{dx}] = (1-2\mu)\boldsymbol{w}(n) + 2\boldsymbol{R}_x^{-1}\boldsymbol{p}_{dx} \\ &= (1-2\mu)\boldsymbol{w}(n) + 2\mu\boldsymbol{w}^o \end{aligned} \tag{7.36}$$

Subtracting $\boldsymbol{w}^o$ from both sides of the above equation, we find

$$\boldsymbol{w}(n+1) - \boldsymbol{w}^o = (1-2\mu)[\boldsymbol{w}(n) - \boldsymbol{w}^o] \tag{7.37}$$

For $n = 0$ and $\mu = 1/2$, we obtain $\boldsymbol{w}^o$ in one step. However, in practice, $\nabla_w J(\boldsymbol{w})$ and $\boldsymbol{R}_x^{-1}$ are estimated, and therefore, the value of the step-size parameter must be less than 0.5.

Introducing $\boldsymbol{w} - \boldsymbol{w}^o = \xi$ in (7.37), we obtain the relation:

$$\xi(n+1) = (1-2\mu)\xi(n) \tag{7.38}$$

which has the solution (see Problem 7.3.1):

$$\xi(n) = (1-2\mu)^n \xi(0) \quad \text{or} \quad \boldsymbol{w}(n) - \boldsymbol{w}^o = (1-2\mu)^n[\boldsymbol{w}(0) - \boldsymbol{w}^o] \tag{7.39}$$

Using (6.15), we obtain

$$J = J_{\min} + (w - w^o)^T R_x (w - w^o) = J_{\min} + \xi^T R_x \xi$$
$$= J_{\min} + \xi^T (Q \Lambda Q^T) \xi = J_{\min} + \xi'^T \Lambda \xi'$$

In connection with (7.39), we find the equation (see Problem 7.3.2):

$$J(n) = J_{\min} + (1 - 2\mu)^{2n} [J(0) - J_{\min}] \tag{7.40}$$

where:
$J(n)$ is the value of the performance function, J, when $w = w(n)$

To obtain a decaying equivalent expression, we introduce the relation:

$$(1 - 2\mu)^{2n} = e^{-(n/\tau)} \tag{7.41}$$

where:
τ is the **time constant**

Under the condition $2\mu << 1$, we can use the approximation $\ln(1-2\mu) \cong -2\mu$. Therefore, the time constant has the value

$$\tau \cong \frac{1}{4\mu} \tag{7.42}$$

The above equation shows that the Newton's algorithm has only **one mode of convergence** (one time constant).

7.4 SOLUTION OF THE VECTOR DIFFERENCE EQUATION

If we set $n = 0$ in (7.20), we obtain

$$w(1) = \left[I - \mu' R_x \right] w(0) + \mu' p_{dx} \tag{7.43}$$

Similar, if we set $n = 1$, we find

$$w(2) = \left[I - \mu' R_x \right] w(1) + \mu' p_{dx} \tag{7.44}$$

If we, next, substitute (7.43) in (7.44), we obtain

$$\begin{aligned} w(2) &= \left[I - \mu' R_x \right]^2 w(0) + \mu' p_{dx} + [I - \mu' R_x] \mu' p_{dx} \\ &= \left[I - \mu' R_x \right]^2 w(0) + \left(\sum_{j=0}^{1} [I - \mu' R_x]^j \right) \mu' p_{dx} \end{aligned} \tag{7.45}$$

Therefore, the nth step is easily recognized to be equal to

$$w(n)=[I-\mu' R_x]^n w(0)+\left(\sum_{j=0}^{n-1}[I-\mu' R_x]^j\right)\mu' p_{dx} \tag{7.46}$$

The above equation does not provide us with a way to study the convergence of $w(n)$ to w^o as $n\to\infty$. We must decouple the equations and then describe them in a different coordinate system. To accomplish this task, we translate and then rotate the coordinate system.

Finding the eigenvalues and eigenvectors of R_x, we create a diagonal matrix Λ consisting of the eigenvalues of R_x and a matrix Q made up of the eigenvectors of R_x (see Chapter 6). Since $Q^TQ=QQ^T=I$, (7.18) takes the form:

$$w(n+1)=Q[I-\mu'\Lambda]Q^T w(n)+\mu' p_{dx} \tag{7.47}$$

To uncouple the weights (coefficients), we multiply both sides of the above equations by Q^T. Hence,

$$w'(n+1)=\left[I-\mu'\Lambda\right]w'(n)+\mu' p'_{dx} \tag{7.48}$$

where we define the following quantities:

$$w'(n+1)=Q^T w(n+1),\quad w'(n)=Q^T w(n),\quad p'_{dx}=Q^T p'_{dx},\quad w^{o\prime}=Q^T w^o \tag{7.49}$$

Next, we obtain the relation:

$$\begin{aligned} w^{o\prime} &= Q^T w^o = Q^T R_x^{-1} p_{dx} = Q^T (Q\Lambda Q^T)^{-1} p_{dx} \\ &= Q^T Q\Lambda^{-1}Q^T p_{dx} = \Lambda^{-1} p'_{dx} \end{aligned} \tag{7.50}$$

since $Q^{-1}=Q^T$ and $(Q\Lambda Q^T)^{-1}=(\Lambda Q^T)^{-1}Q^{-1}=(Q^T)^{-1}\Lambda^{-1}Q^T=Q\Lambda^{-1}Q^T$

The ith equation of the system given in (7.47) is

$$w'_i(n+1)=[1-\mu'\lambda_i]w'_i(n)+\mu' p'_{idx} \qquad 0\le i\le M-1 \tag{7.51}$$

By iteration, the above equation has the following solution (see Problem 7.4.1):

$$w'_i(n)=(1-\mu'\lambda_i)^n w'_i(0)+\mu' p'_{idx}\left[\sum_{j=0}^{n-1}(1-\mu'\lambda_i)^j\right] \tag{7.52}$$

If we set $\alpha_i=1-\mu'\lambda_i$, then (7.52) becomes

$$w'_i(n)=\alpha_i^n w'_i(0)+\mu' p'_{idx}\sum_{j=0}^{n-1}\alpha_i^j=\alpha_i^n w'_i(0)+\mu' p'_{idx}\frac{1-\alpha_i^n}{1-\alpha_i} \tag{7.53}$$

$$
\begin{aligned}
w_i'(n) &= \alpha_i^n w_i'(0) + \mu' p_{idx}' \sum_{j=0}^{n-1} \alpha_i^j = \alpha_i^n w_i'(0) + \mu' p_{idx}' \frac{1-\alpha_i^n}{1-\alpha_i} \\
&= \alpha_i^n w_i'(0) + \frac{p_{idx}'}{\lambda_i}[1-(1-\mu'\lambda_i)^n]
\end{aligned}
\tag{7.54}
$$

since the sum is a finite geometric series.

Example 7.4.1

For the development of this example, we used the MATLAB program. Hence, the data were found using `x = randn(1,10)` and the desired ones were found using `d = conv[(1  0.2),x]` or `d = filter[(1  0.2),1,x]`. The correlation matrix was found using the function `R = toeplitz [xc(1,10 :11)]`, where `xc = xcorr (x,'biased')`. Similarly, we found the *pdx* cross-correlation. Hence,

$$
\boldsymbol{R}_x = \begin{bmatrix} 0.7346 & 0.0269 \\ 0.0269 & 0.7346 \end{bmatrix}; \quad [\boldsymbol{Q},\Lambda] = \text{eig}(\boldsymbol{R}_x)
$$

$$
\boldsymbol{Q} = \begin{bmatrix} -0.7071 & 0.7071 \\ 0.7071 & 0.7071 \end{bmatrix}; \quad \Lambda = \begin{bmatrix} 0.7077 & 0 \\ 0 & 0.7071 \end{bmatrix}
$$

and $\mu' < 2/0.7071 = 2.8285$ for the solution to converge. We next choose $[w_0(0) \quad w_1(0)]' = [0 \quad 0]'$, and hence, $\boldsymbol{w}'(0) = \boldsymbol{Q}^T \boldsymbol{w}(0) = \boldsymbol{0}$. Therefore, (7.54) becomes

$$
w_i'(n) = \mu' p_{idx}' \frac{1-\alpha_i^n}{1-\alpha_i} = \frac{p_{idx}'}{\lambda_i}\left[1-(1-\mu'\lambda_i)^n\right]
$$

From (7.49), we obtain

$$
\boldsymbol{p}_{dx}' = \boldsymbol{Q}^T \boldsymbol{p}_{dx} = \begin{bmatrix} -0.7071 & 0.7071 \\ 0.7071 & 0.7071 \end{bmatrix} \begin{bmatrix} 0.7399 \\ -0.0003 \end{bmatrix} = \begin{bmatrix} -0.5234 \\ 0.5230 \end{bmatrix}
$$

Therefore, the system is (we set $\mu' = 3$ for convergence)

$$
w_0'(n) = \frac{1}{0.7077}(-0.5234)\left[1-(1-2\times 0.7071)^n\right]
$$

$$
w_1'(n) = \frac{1}{0.7614}(0.5230)\left[1-(1-2\times 0.7614)^n\right]
$$

Since $\boldsymbol{w}'(n) = \boldsymbol{Q}^T \boldsymbol{w}(n)$ and $\boldsymbol{Q}^T = \boldsymbol{Q}^{-T}$, we find $\boldsymbol{w}(n) = \boldsymbol{Q}^T \boldsymbol{w}'(n)$, and at the limit value when n approaches infinity, the filter coefficients take the values:

$$\begin{bmatrix} w_0^o \\ w_1^o \end{bmatrix} = \boldsymbol{Q}^T \boldsymbol{w}' = \begin{bmatrix} -0.7071 & 0.7071 \\ 0.7071 & 0.7071 \end{bmatrix} \begin{bmatrix} \frac{-0.5234}{0.7077} \\ \frac{0.5230}{0.7614} \end{bmatrix} = \begin{bmatrix} 1.0087 \\ -0.0373 \end{bmatrix}$$

■

PROBLEMS

7.1.1 Verify (7.5).

7.1.2 Plot (7.5) for the following positive values: (a) $0 < \mu < 1/2r_{xx}(0)$, (b) $\mu \cong 1/2r_{xx}(0)$, and (c) $1/2r_{xx}(0) < \mu < 1/r_{xx}(0)$, $r_{xx}(0) = 1$.

7.1.3 Using the one-dimensional Newton's algorithm, find the third root of 8. Start with $x(1) = 1.2$.

7.1.4 Let the MSE be given by the nonquadratic equation $J = 2 - (1/35)[(1 - w^2)(4.5 + 3.5w)]$. Find and plot the recursive Newton's algorithm for the coefficients $\{w(n)\}$ for $n = 0,\ 1,\ 2, \ldots$. Start with $w(0) = 2$.

7.2.1 Verify (7.22).

7.2.2 Find the time constant for one eigenvalue.

7.3.1 Verify (7.39).

7.3.2 Verify (7.40).

7.3.3 Find the equivalent expression of $J[w(n)] = J_{\min} + \xi'^T(n)\Lambda\xi'(n)$ as a function of $\boldsymbol{w}(n)$, $\boldsymbol{p}_{dx}$, and $\boldsymbol{R}_x$.

7.3.4 The correlation matrix of the filter input is

$$\boldsymbol{R}_x = \begin{bmatrix} 1 & 0.85 \\ 0.85 & 1 \end{bmatrix}$$

with eigenvalues $\lambda_0 = 1.85$ and $\lambda_1 = 0.15$. Plot the learning curve in a semi-log format and find the two time constants. Compare your results obtained from the graph and those given analytically [see (7.27)].

7.4.1 Verify (7.52).

7.4.2 Let

$$\boldsymbol{R}_x = \begin{bmatrix} 1 & 0.7 \\ 0.7 & 1 \end{bmatrix} \text{ and } \boldsymbol{p}_{dx} = [0.7 \quad 0.5]^T$$

are the estimates derived from data. Find the vector $\boldsymbol{w}(n)$.

EDITION PROBLEMS

1. Let $\boldsymbol{R} = [1 \quad 0.5;\ 0.25 \quad 1]$ and $\boldsymbol{p} = [0.5 \quad 0.2]^T$. (a) Find a value of the step size that ensures the convergence of the steepest descent method; (b) find the recursion for computing the elements $w_1(n)$ and $w_2(n)$ of the $\boldsymbol{w}(n)$.
2. Assuming one weight coefficient $w(n)$, find the following: (a) find the MSE $J(n)$ as a function of $w(n)$; (b) find the Wiener solution and the minimum MSE $J_{\min}(n)$; (c) sketch a plot of $J(n)$ versus $w(n)$.

HINTS–SOLUTIONS–SUGGESTIONS

7.1.1

$$w(1) = aw(0) + bw^o \ [a = 1 - 2\mu r(0),\ b = 2\mu r(0)]$$

$$w(2) = aw(1) + bw^o = a[aw(0) + bw^o] + bw^o = a^2 w(0) + abw^o + bw^o$$

$$\Rightarrow w(n) = a^n w(0) + (a^{n-1} + a^{n-2} + \cdots + 1)bw^o = a^n w(0) + \frac{1-a^n}{1-a} bw^o, \text{ but}$$

$$1 - a = b \Rightarrow w(n) = a^n w(0) + w^o - a^n w^o = w^o + [w(0) - w^o]a^n$$

7.1.2 (a) Overdamped case, (b) critically damped, and (c) underdamped case.

7.1.3 Set $f(x) = x^3 - 8$ to obtain $x(n+1) = x(n) - [x(n)^3 - 8]/[3x(n)^2]$. For $x(1) = 1.2$, MATLAB gives us the results: $x(1) = 1.2000, x(2) = 2.6519$, $x(3) = 2.1471$, $x(4) = 2.0098$, $x(5) = 2.0000$, $x(6) = 2.0000$. The Book MATLAB program is given as follows:

```
>>x(1) = 1.2;
>>for n = 1:10
>>x(n+1) = x(n)-(x(n)^3-8)/(3*x(n)^2);
>>end;
```

7.1.4

$$w(n+1) = w(n) - \{[\partial J(w)/\partial w]/[\partial J^2(w)/\partial w^2]\} = w(n)$$

$$- \frac{10.5w^2(n) + 9w(n) - 3.5}{21w(n) + 9}$$

Using MATLAB, we obtain 2.0000, 0.8922, 0.4275, 0.3014, 0.2905, and 0.2905.

7.2.1

$$\begin{bmatrix} \xi_0'(n+1) \\ \xi_1'(n+1) \\ \vdots \\ \xi_{M-1}'(n+1) \end{bmatrix} = \left(\begin{bmatrix} 1 & 0 & \cdots & 0 \\ 0 & 1 & \cdots & 0 \\ \vdots & & & \\ 0 & 0 & \cdots & 1 \end{bmatrix} - \mu \begin{bmatrix} \lambda_0 & 0 & \cdots & 0 \\ 0 & \lambda_1 & \cdots & 0 \\ \vdots & & & \\ 0 & 0 & \cdots & \lambda_{M-1} \end{bmatrix} \right) \begin{bmatrix} \xi_0'(n) \\ \xi_1'(n) \\ \vdots \\ \xi_{M-1}' \end{bmatrix}$$

$$= \begin{bmatrix} (1 - \mu\lambda_0)\xi_0'(n) \\ (1 - \mu\lambda_1)\xi_1'(n) \\ \vdots \\ (1 - \mu\lambda_{M-1})\xi_{M-1}'(n) \end{bmatrix}$$

and the results are found since the equality of two vectors is the equality of each corresponding element.

7.2.2

$$(1-2\mu\lambda_i)^{2k} = e^{-(k/\tau_i)} \Rightarrow \tau_i = \frac{-1}{2\ln(1-2\mu\lambda_i)} \quad \text{(1)for small } \mu \Rightarrow 2\mu\lambda_i << 1 \Rightarrow$$

$$\ln(1-2\mu\lambda_i) \approx -2\mu\lambda_i, \text{ substituting in (1)} \Rightarrow \tau_i \approx \frac{1}{4\mu\lambda_i}$$

7.3.1

$$w(1) = aw(0) + bw^o \; [a = 1-2\mu r(0),\, b = 2\mu r(0)]$$

$$w(2) = aw(1) + bw^o = a[aw(0) + bw^o] + bw^o = a^2 w(0) + abw^o + bw^o$$

$$\Rightarrow w(n) = a^n w(0) + (a^{n-1} + a^{n-2} + \cdots + 1)bw^o = a^n w(0) + \left(\frac{1-a^n}{1-a}\right)bw^o$$

But $1-a = b \Rightarrow w(n) = a^n w(0) + w^o - a^n w^o = w^o + [w(0) - w^o]a^n$.

7.3.2 Equations 6.5 and 7.39 were used in $J(n) = J_{\min} + \xi^T \boldsymbol{R}_x \xi = J_{\min} + (1-2\mu)^n \xi^T(0) \boldsymbol{R}_x (1-2\mu)^n \xi(0) = J_{\min} + (1-2\mu)^{2n} \xi^T(0) \boldsymbol{R}_x \xi(0)$. But $\xi^T(0)\boldsymbol{R}_x\xi(0) = J(0) - J_{\min}$, and thus Problem 7.3.2 is proved.

7.3.3

$$\begin{aligned}
J(\boldsymbol{w}(n)) &= \sigma_d^2 - \boldsymbol{p}_{dx}^T \boldsymbol{w}^o + \xi(n)\boldsymbol{Q}\boldsymbol{Q}^T \boldsymbol{R}_x \boldsymbol{Q}\boldsymbol{Q}^T \xi(n) \\
&= \sigma_d^2 - \boldsymbol{p}_{dx}^T \boldsymbol{w}^o + [\boldsymbol{w}(n) - \boldsymbol{w}^o]\boldsymbol{R}_x[\boldsymbol{w}(n) - \boldsymbol{w}^o] \\
&= \sigma_d^2 - \boldsymbol{p}_{dx}^T \boldsymbol{w}^o + \left[\boldsymbol{w}^T(n) - \boldsymbol{w}^o\right]\boldsymbol{R}_x\left[\boldsymbol{w}(n) - \boldsymbol{w}^o\right] \\
&= \sigma_d^2 - \boldsymbol{p}_{dx}^T \boldsymbol{w}^o + \boldsymbol{w}^{oT}\boldsymbol{R}_x \boldsymbol{w}(n) - \boldsymbol{w}^{oT}\boldsymbol{R}_x \boldsymbol{w}(n) - \boldsymbol{w}^T(n)\boldsymbol{R}_x \boldsymbol{w}^o + \boldsymbol{w}^{oT}\boldsymbol{R}_x \boldsymbol{w}^o
\end{aligned}$$

But $(\boldsymbol{R}_x \boldsymbol{w}^o)^T = \boldsymbol{w}^{oT}\boldsymbol{R}_x^T = \boldsymbol{w}^{oT}\boldsymbol{R}_x = \boldsymbol{p}_{dx}^T$, $\boldsymbol{R}_x\boldsymbol{w}^o = \boldsymbol{p}_{dx}$, $\boldsymbol{p}_{dx}^T\boldsymbol{w}^o = \boldsymbol{w}^{oT}\boldsymbol{p}_{dx}$, and hence, $J[\boldsymbol{w}(n)] = \sigma_d^2 - 2\boldsymbol{w}^T(n)\boldsymbol{p}_{dx} + \boldsymbol{w}^T(n)\boldsymbol{R}_x\boldsymbol{w}(n)$.

7.3.4 The time constants are found from the two slopes of the graph.

7.4.1 Start with $n = 0 \Rightarrow w_i'(1) = (1-\mu'\lambda_i)^2 w_i'(0) + \mu' p_{idx}'$, and then set

$$\begin{aligned}
n = 1 \Rightarrow w_i'(2) &= (1-\mu'\lambda_i)\left[(1-\mu'\lambda_i)w_i'(0) + \mu' p_{idx}'\right] + \mu' p_{idx}' \\
&= (1-\mu'\lambda_i)^2 w_i'(0) + \mu' p_{idx}'\left[1 + (1-\mu'\lambda_i)\right] \\
&= (1-\mu'\lambda_i)^2 w_i'(0) + \mu' p_{idx}'\left[\sum_{j=0}^{2-1}(1-\mu'\lambda_i)^j\right]
\end{aligned}$$

and therefore, at the nth iteration, we obtain (7.52).

7.4.2 Using the MATLAB function `[v,d] = eig(R),` we obtain the following eigenvalues and eigenvectors: $\lambda_1 = 0.3000$, $\lambda_2 = 1.7000$, $q_1 = [-0.7071 \quad 0.7071]^T$, $q_2 = [0.7071 \quad 0.7071]^T$. The step-size factor must be set as follows: $\mu' < 2/1.7 = 1.1765$, so that the solution converges. Choosing $w(0) = \mathbf{0}$ implies that

$$w'(0) = Q^T w(0) = \mathbf{0}$$

$$\Rightarrow p'_{dx} = Q^T p_{dx} = \begin{bmatrix} -0.7071 & 0.7071 \\ 0.7071 & 0.7071 \end{bmatrix} \begin{bmatrix} 0.7 \\ 0.5 \end{bmatrix} = \begin{bmatrix} -0.1414 \\ 0.8485 \end{bmatrix}$$

$$w'_0(n) = \frac{1}{0.3}(-0.1414)[1-(1-\mu' 0.3)^n]; \; w'_1(0) = \frac{1}{1.7} 0.8485[1-(1-\mu' 1.7)^n]$$

Since

$$w'(n) = Q^T w(n), \; Q^T = Q^{-1}, \; w(n) = Q^T w'(n)$$

$$\begin{bmatrix} w_0(n) \\ w_1(n) \end{bmatrix} = \begin{bmatrix} -0.7071 & 0.7071 \\ 0.7071 & 0.7071 \end{bmatrix} \begin{bmatrix} \frac{1}{0.3}(-0.1414)[1-(1-\mu' 0.3)^n] \\ \frac{1}{1.7} 0.8485[1-(1-\mu' 1.7)^n] \end{bmatrix}$$

$$= \begin{bmatrix} \frac{-0.7071}{0.3}(-0.1414)[1-(1-\mu' 0.3)^n] + \frac{0.7071}{1.7} 0.8485[1-(1-\mu' 1.7)^n] \\ \frac{0.7071}{0.3}(-0.1414)[1-(1-\mu' 0.3)^n] + \frac{0.7071}{1.7} 0.8485[1-(1-\mu' 1.7)^n] \end{bmatrix}$$

ADDITIONAL PROBLEMS

1.

a. $[Q, L] = \text{eig}([1 \quad 0.5; 0.25 \quad 1])$; $Q = [0.8165 \quad 0.8165; 0.5774 \quad 0.5774]$; $L = [1.3536 \quad 0.6464]$. Therefore, $0 < \mu < 1/\lambda_{max}$, and thus, $0 < \mu < 1/1.3536 = 0.7388$.

b. $w(n+1) = w(n) + \mu[p - Rw(n)]$

$$w(n+1) = w(n) + 1 * ([0.5; 0.2] - [1 \quad 0.5; 0.25 \quad 1] * w(n))$$

or

$$w(n+1) = w(n) + ([1 \quad 0; 0 \quad 1] - [1 \quad 0.5; \; 0.25 \quad 1]) * w(n) - [0.5; \quad 0.2]$$

$$= [0 \quad -0.5; -0.25 \quad 0] * w(n) - [0.5 \quad 0.2]$$

Hence,

$$\begin{bmatrix} w_1(n+1) \\ w_2(n+1) \end{bmatrix} = \begin{bmatrix} 0 & -0.5 \\ -0.25 & 0 \end{bmatrix} \begin{bmatrix} w_1(n) \\ w_2(n) \end{bmatrix} - \begin{bmatrix} 0.5 \\ 0.2 \end{bmatrix}$$

or

$$w_1(n+1) = -0.5w_2(n) - 0.5;\ w_2(n+1) = -0.25w_1(n) - 0.2$$

For $n = 0$, $w_1(1) = -0.5$ and $w_2(1) = -0.2$ with $w_1(0) = 0$ and $w_2(0) = 0$

$w_1(2) = -0.5(-0.2) - 0.5 = -0.4$, $w_2(2) = -0.25(-0.5) - 0.2 = -0.075$, etc.

2.

$$J(n) = \sigma_d^2 - 2p_{xd}w(n) + r_{xx}w^2(n);\ \frac{\partial J(n)}{\partial w(n)} = 0 - 2p_{xd} + 2r_{xx}w(n) = 0 \Rightarrow w^o = \frac{p_{dx}}{r_{xx}}$$

$$\Rightarrow J_{\min} = \sigma_d^2 - 2p_{xd}\frac{p_{dx}}{r_{xx}} + r_{xx}\left(\frac{p_{dx}}{r_{xx}}\right)^2 = \sigma_d^2 - \frac{p_{dx}^2}{r_{xx}}$$

The curve is concave upward with minimum $J_{\min}$.

8 The Least Mean-Square Algorithm

8.1 INTRODUCTION

In this chapter, we present the calibrated **least mean-square (LMS)** algorithm developed by Widrow and Hoff in 1960. This algorithm is a member of stochastic gradient algorithms, and because of its robustness and low computational complexity, it has been used in a wide spectrum of applications.

The LMS algorithm has the following most important properties:

1. It can be used to solve the Wiener–Hopf equation without finding matrix inversion. Furthermore, it does not require the availability of the autocorrelation matrix of the filter input and the cross-correlation between the filter input and its desired signal.
2. Its form is simple as well as its implementation, yet it is capable of delivering high performance during the adaptation process.
3. Its iterative procedure involves the following:
    a. Computing the output of a finite impulse response (FIR) filter produced by a set of tap inputs (filter coefficients)
    b. Generating an estimated error by computing the output of the filter to a desired response
    c. Adjusting the tap weights (filter coefficients) based on the estimation error
4. The correlation term needed to find the values of the coefficients at the $n+1$ iteration contains the stochastic product $x(n)e(n)$ without the expectation operation that is present in the steepest descent method.
5. Since the expectation operation is not present, each coefficient goes through sharp variations (noise) during the iteration process. Therefore, instead of terminating at the Wiener solution, the LMS algorithm suffers random variation around the minimum point (optimum value) of the error-performance surface.
6. It includes a step-size parameter, μ, that must be selected properly to control stability and convergence speed of the algorithm.
7. It is stable and robust for a variety of signal conditions.

8.2 THE LMS ALGORITHM

In Chapter 7, we developed the following relations using the steepest descent method:

$$\mathbf{w}(n+1) = \mathbf{w}(n) - \mu \nabla_{\mathbf{w}} J[\mathbf{w}(n)] \tag{8.1a}$$

$$\nabla_w J[\mathbf{w}(n)] = -2\mathbf{p}_{dx} + 2\mathbf{R}_x\mathbf{w}(n) \tag{8.1b}$$

The simplest choices of the estimators $\mathbf{R}_x$ and $\mathbf{p}_{dx}$ are the instantaneous estimates defined by

$$\mathbf{R}_x \cong \mathbf{x}(n)\mathbf{x}^T(n), \quad \mathbf{p}_{dx} \cong d(n)\mathbf{x}(n) \tag{8.2}$$

Substituting the above values in (8.1b) and then combining (8.1a) and (8.1b), we obtain

$$\begin{aligned} \mathbf{w}(n+1) &= \mathbf{w}(n) + 2\mu\mathbf{x}(n)[d(n) - \mathbf{x}^T(n)\mathbf{w}(n)] \\ &= \mathbf{w}(n) + 2\mu\mathbf{x}(n)[d(n) - \mathbf{w}^T(n)\mathbf{x}(n)] \\ &= \mathbf{w}(n) + 2\mu e(n)\mathbf{x}(n) \end{aligned} \tag{8.3}$$

where:

$$\begin{aligned} y(n) &= \mathbf{w}^T(n)\mathbf{x}(n) \quad \text{Filter output} \\ e(n) &= d(n) - y(n) \quad \text{Error} \end{aligned} \tag{8.4}$$

$$\begin{aligned} \mathbf{w}(n) &= [w_0 \quad w_1 \quad \cdots \quad w_{M-1}]^T \quad \text{Filter taps (coefficients) at time } n \\ \mathbf{x}(n) &= [x(n) \quad x(n-1) \quad \cdots \quad x(n-M+1)]^T \quad \text{Input data} \end{aligned} \tag{8.5}$$

The algorithms defined by (8.3) and (8.4) constitute the LMS algorithm. The algorithm at each iteration requires that $\mathbf{x}(n)$, $d(n)$, and $\mathbf{w}(n)$ are known. The LMS algorithm is a **stochastic gradient algorithm** if the input signal is a stochastic process. This results in varying the pointing direction of the coefficient vector during the iteration. An FIR adaptive filter realization is shown in Figure 8.1. Figure 8.2 presents the block diagram representation of the LMS filter. Table 8.1 presents the steps for executing the LMS algorithm.

Book LMS m-Function

```
function[w,y,e,J,w1,Js]=lms1(x,dn,mu,M)
    %function[w,y,e,J,w1]=lms1(x,dn,mu,M);
    %all quantities are real-valued;
    %x = input data to the filter; dn = desired signal;
    %M = order of the filter;
    %mu = step-size factor; x and dn must be
    %of the same length;
    %Js = smooths the learning curve;
    %w1 = a matrix of dimensions: length(x)xM,
    %each column represents the variation of
    %each filter coefficient;
N=length(x);w=zeros(1,M);w1=zeros(1,M);
for n=M:N
    x1=x(n:-1:n-M+1);%for each n the vector x1 is
        %of length M with elements from x in
        %reverse order;
```

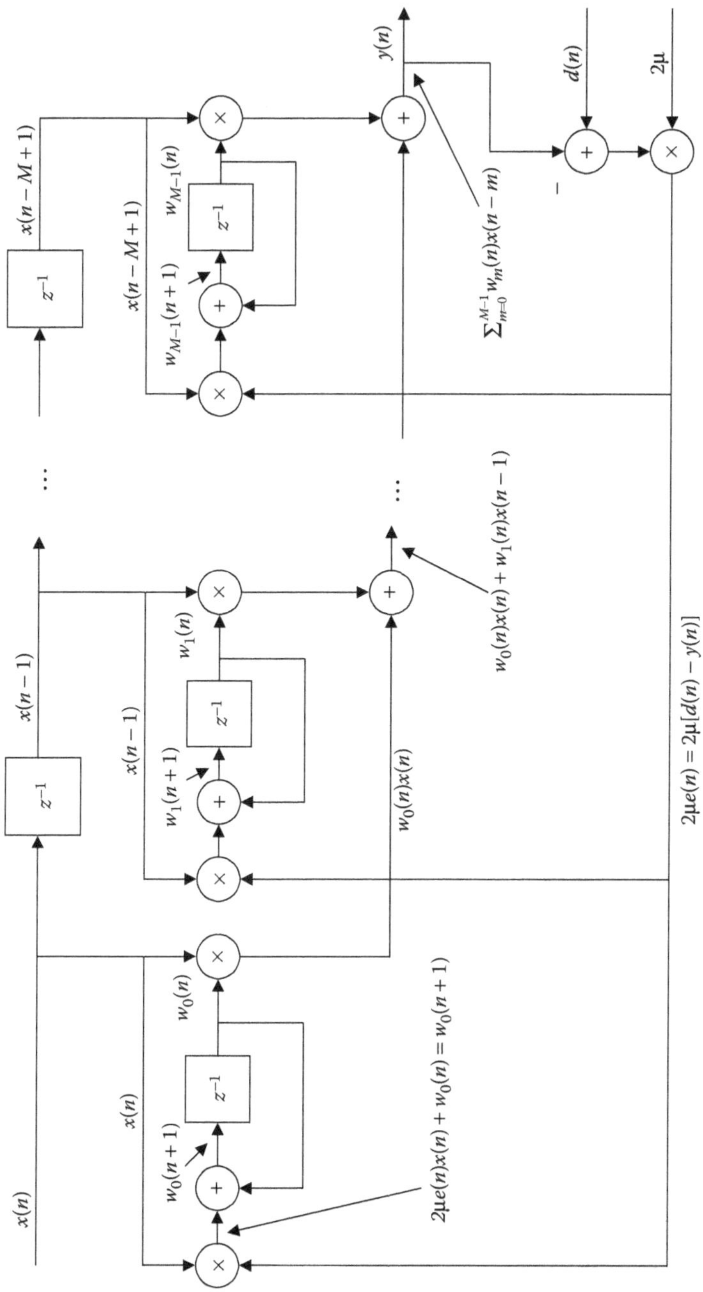
x(n)
x(n − 1)
x(n − M + 1)
z−1
x(n)
x(n − 1)
x(n − M + 1)
w0(n + 1)
w1(n + 1)
wM−1(n + 1)
w0(n)
w1(n)
wM−1(n)
w0(n)x(n)
w0(n)x(n) + w1(n)x(n − 1)
ΣM−1m=0 wm(n)x(n − m)
2μe(n)x(n) + w0(n) = w0(n + 1)
2μe(n) = 2μ[d(n) − y(n)]
y(n)
d(n)
2μ
−

FIGURE 8.1

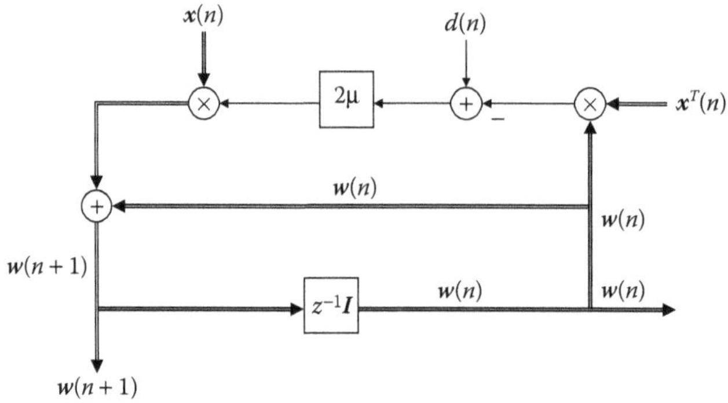

FIGURE 8.2

TABLE 8.1
The LMS Algorithm for an *M*th-Order FIR Adaptive Filter

Inputs	M = Filter length
	μ = Step-size factor
	$x(n)$ = Input data to the adaptive filter
	$w(0)$ = Initialization filter vector = 0
Outputs	$y(n)$ = Adaptive filter output = $\mathbf{w}^T(n)\mathbf{x}(n)$ approximate of the desired signal
	$e(n) = d(n) - y(n)$ = Error
	$w(n+1) = w(n) + 2\mu e(n)x(n)$

```
      y(n) = w*x1';
      e(n) = dn(n)-y(n);
      w = w+2*mu*e(n)*x1;
      w1(n-M+1,:) = w(1,:);
end;
J = e.^2;%J is the learning curve of the adaptation;
for n = 1:length(x)-5
      Js(n) = (J(n)+J(n+1)+J(n+2))/3;
end;
```

8.3 EXAMPLES USING THE LMS ALGORITHM

The following examples will elucidate the use of the LMS algorithm to different areas of engineering and will bring forth the versatility of this important algorithm.

Example 8.3.1 (Linear Prediction)

We can use an adaptive LMS filter as a predictor as shown in Figure 8.3. The data $\{x(n)\}$ were created by passing a zero-mean white noise $\{v(n)\}$

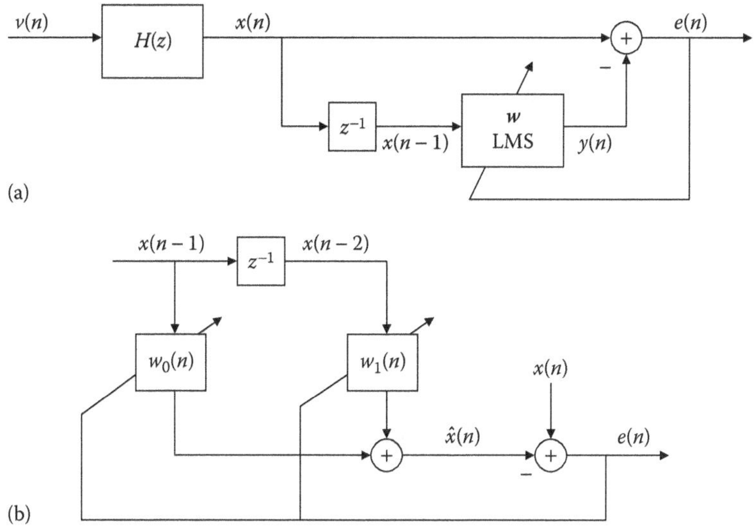

FIGURE 8.3

through an autoregressive (AR) process described by the difference equation: $x(n) = 0.6010x(n-1) - 0.7225x(n-2) + v(n)$. The LMS filter is used to predict the values of the AR filter parameters 0.6010 and −0.7225. A two-coefficient LMS filter predicts $x(n)$ by

$$\hat{x}(n) = \sum_{i=0}^{1} w_i(n)x(n-1-i) \equiv y(n) \tag{8.6}$$

Figure 8.4 shows the trajectory of w_0 and w_1 versus the number of iterations for two different values of step-size parameter ($\mu = 0.02$ and $\mu = 0.005$). The adaptive filter is a two-coefficient filter. The noise is white and Gaussian distributed. The figure shows fluctuations in the values of coefficients as they converge to a neighborhood of their optimum values 0.6010 and −0.7225. As the step-size parameter μ

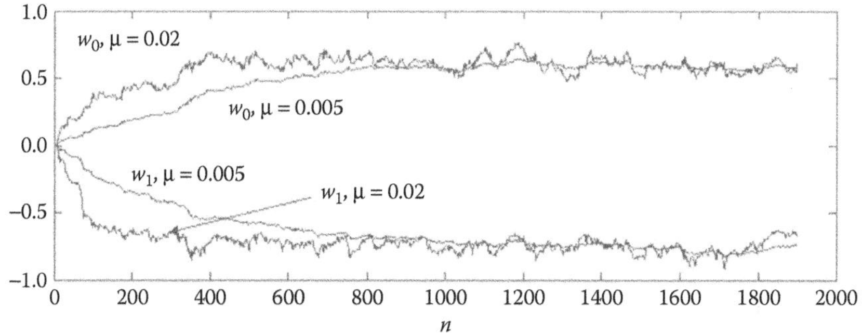

FIGURE 8.4

becomes smaller, the fluctuations are not as large, but the convergence speed to the optimal values is slower.

Book One-Step LMS Predictor m-Function

```
function[w,y,e,J,w1,Js] = lms_one_step_predictor(x,mu,M)
     %function[w,y,e,J,w1,Js] = lms_one_step_predictor(x,mu,M);
     %x = data = signal plus noise;mu = step-size factor;
     %M = number of filter
     %coefficients;w1 is a matrix and each column
     %is the history of each
     %filter coefficient versus time (iteration) n;
     %N = number of elements
     %of the vector x;Js = smoother learning curve;
     %w1 = a matrix and each column depicts the
     %variation of each coefficient;
N = length(x);
y = zeros(1,N);
w = zeros(1,M);
for n = M:N-1
     x1=x(n:-1:n-M+1);
     y(n)=w*x1';
     e(n)=x(n+1)-y(n);
     w=w+2*mu*e(n)*x1;
     w1(n-M+1,:)=w(1,:);
end;
J=e.^2;%J is the learning curve of the
     %adaptive process;
for n=1:length(x)-5
     Js(n)=(J(n)+J(n+1)+J(n+2))/3;
end;%Js=smoothed out learning curve J;
```

For example, to produce two of the four curves, we used the following Book program:

```
>>mu=0.02; x(1)=0; x(2)=0; M=2;
>>for n=1:2000
     x(n+2)=0.6010*x(n+1)-0.7225*x(n)+0.2*randn;
>>end;
>>[w,y,e,J,w1,Js]=lms_one_step_predictor(x,mu,M);
>>plot(w1(:,1),'k');
>>hold on;
>>plot(w1(:,2),'k');xlabel('n');
```

We can use a similar program for the other two curves.

■

Example 8.3.2 (Modeling)

Adaptive filtering can also be used to find the coefficients of an unknown filter as shown in Figure 8.5. The data $\boldsymbol{x}(n)$ were created similar to those in Example 8.3.1. The desired signal is given by $d(n) = x(n)-2x(n-1) + 4x(n-2)$. If the output $y(n)$

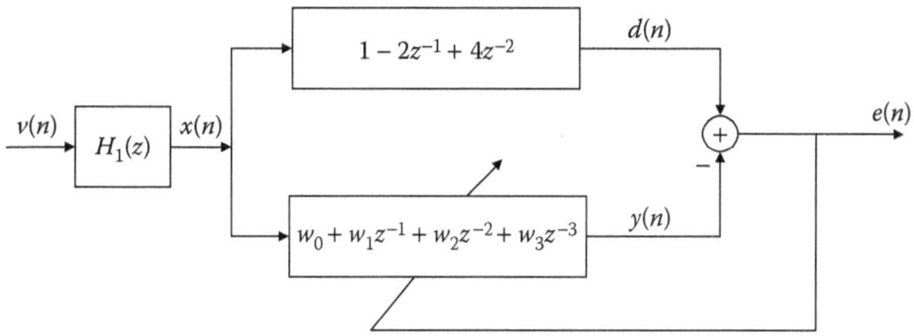

FIGURE 8.5

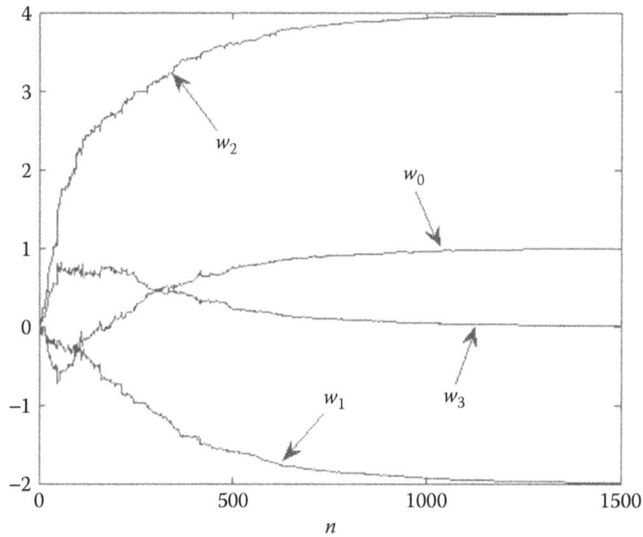

FIGURE 8.6

is approximately equal to $d(n)$, it implies that the coefficients of the LMS filter are approximately equal to those of the unknown system. Figure 8.6 shows the ability of the LMS filter to identify the unknown system. After 1000 iterations, the system is practically identified. In this example, we used $\mu = 0.15$ and $M = 4$. It is observed that the fourth coefficient is zero as it should be since the system to be identified has only three coefficients and the rest are zero. The Book program to produce Figure 8.6 is given as follows:

Book Program for System Identification

```
>>x(1)=0; x(2)=0; mu=0.15; M=4;
>>for n=1:1500
>>    x(n+2)=0.6010*x(n+1)-0.7225*x(n)+0.2*randn;
>>end;
>>for n=1:1500
>>    d(n+2)=x(n+2)-2*x(n+1)+4*x(n);
>>end;
```

```
>>[w,y,e,J,w1,Js]=lms1(x,d,mu,M);
>>plot(w1(:,1),'k'); hold on; plot(w1(:,2),'k');hold on;
>>plot(w1(:,3),'k'); hold on; plot(w1(:,4),'k'); xlabel('n');
```

■

Example 8.3.3 (Noise Cancellation)

A noise cancellation scheme is shown in Figure 8.7. We introduce in this example the following values: $H_1(z)=1$ [or $h(n)=\delta(n)$], $v1(n)=\text{white noise}=v(n)$, $L=1$, $s(n)=0.98^n\sin(0.2\pi n)$. Therefore, the input signal to the filter is $x(n)=s(n-1)+v(n-1)$ and the desired signal is $d(n)=s(n)+v(n)$. The Book LMS function algorithm **lms1** was used. Figure 8.8 shows the signal, the signal plus noise, and the outputs of the filter for two different sets of coefficients: $M=4$ and $M=14$. The following noise was used: $v=0.4[\text{rand}(1,300)-0.5]$.

■

Example 8.3.4 (Power Spectrum Approximation)

If a stochastic process is the output of an AR system when its input is a white noise with variance σ_v^2, for example,

$$x(n)=\sum_{k=1}^{M-1}a_k x(n-k)+v(n) \tag{8.7}$$

The power of the spectrum corresponding to the stochastic process is given by (see Kay 1993)

$$S_x(e^{j\omega})=\frac{\sigma_v^2}{\left|A(e^{j\omega})\right|^2} \qquad A(e^{j\omega})=1-\sum_{k=1}^{M-1}a_k e^{-j\omega k} \tag{8.8}$$

Equation 8.7 is also written in the form:

$$v(n)=x(n)-\sum_{k=1}^{M-1}a_k x(n-k) \tag{8.9}$$

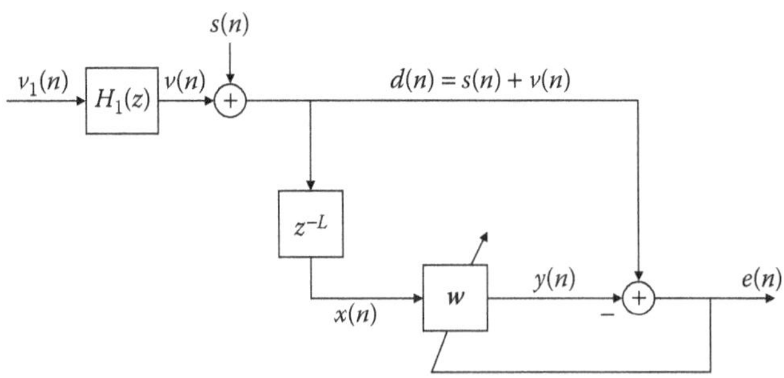

FIGURE 8.7

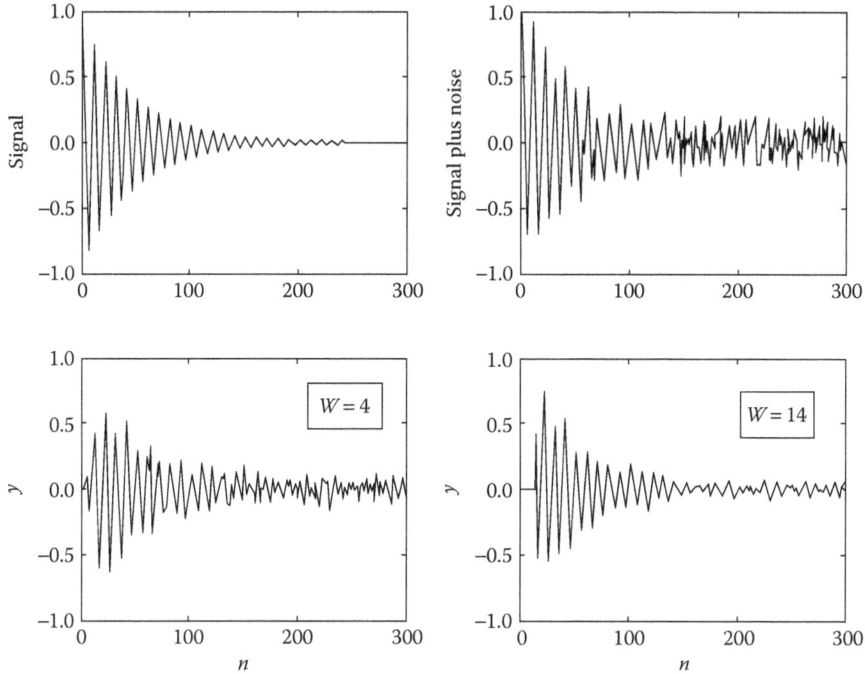

FIGURE 8.8

where:

$v(n)$ is the nonpredictable portion of the signal or the **innovations** of the AR process

Because $v(n)$ is the nonpredictable portion of $x(n)$, it suggests to use an adaptive linear predictor for spectrum estimation. If the stochastic process is the result of an AR process, the LMS filter coefficients will be very close to those of the AR system and the two spectra will also be very close to each other.

The steps that closely approximate the coefficients of (8.7) using an LMS adaptive filter are as follows:

1. Use the adaptive LMS filter in the predictive mode (see Figure 8.9).
2. Average the K most recent values of $\hat{\boldsymbol{w}}$.
3. Compute the power spectrum.

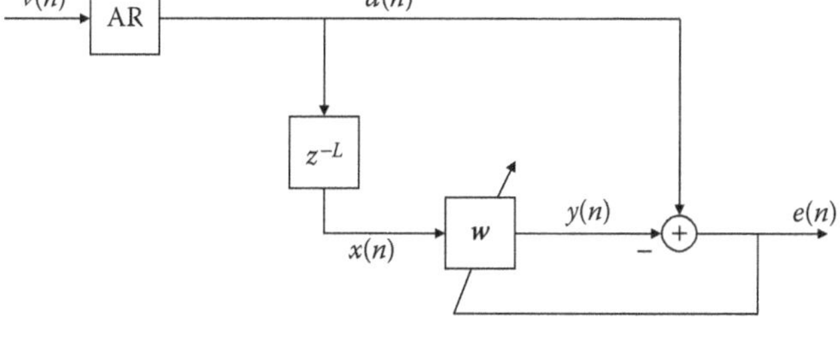

FIGURE 8.9

$$\hat{S}_x(e^{j\omega}) = 1/\left|1-\sum_{k=1}^{M-1}\hat{w}_k e^{-j\omega k}\right|^2 = 1/\left|1-\sum_{k=1}^{M-1}\hat{w}_k z^{-k}\right|^2 \tag{8.10}$$

Let an exact stochastic process be created by an AR system having poles at $(z_1, z_2) = 0.95e^{\pm j(\pi/4)}$. To find the difference equation, which characterizes the AR system, apply the definition of the system in the z-domain. Hence, we write

$$\begin{aligned} H(z) &= \frac{\text{Output}}{\text{Input}} = \frac{X(z)}{\sigma_v^2 V(z)} \\ &= \frac{1}{(1-0.95e^{j(\pi/4)}z^{-1})(1-0.95e^{-j(\pi/4)}z^{-1})} = \frac{1}{(1-1.3435z^{-1})+(0.9025z^{-2})} \end{aligned} \tag{8.11}$$

The above equation can be written as

$$X(z) - 1.3435z^{-1}X(z) + 0.9025z^{-2}X(z) = \sigma_v^2 V(z)$$

Taking the inverse z-transform of both sides of the above equation (remember that a negative exponent of z means time shifting in the time domain), we obtain the difference equation describing the AR system given by

$$x(n) = 1.3435x(n-1) - 0.9025x(n-2) + \sigma_v^2 v(n) \tag{8.12}$$

The power spectrum is given by [see (8.8)]

$$\begin{aligned} S_x(e^{j\omega}) &= \frac{\sigma_v^2}{\left|1-1.3435e^{-j\omega}+0.9025e^{-j2\omega}\right|^2} \\ &= \frac{\sigma_v^2}{\left|1-1.3435z^{-1}+0.9025z^{-2}\right|^2_{z=e^{j\omega}}} \end{aligned} \tag{8.13}$$

Figure 8.10 shows the true spectrum and the approximate one. We assumed that the desired signal was produced by the AR filter given by (8.12). The approximate spectrum was found using the following constants: $\mu = 0.02$, $M = 3$, $N = 1000$, avn = 3, $x(n) = dn(n-1)$, and $\sigma_v^2 = 0.3385$. The function `lms_power_spectra_av1` will average the output $\boldsymbol{w}$ over a number of times as desired by the reader. If we had guessed $M = 2$ and avn = 5, the two curves will be approximately equal and the filter coefficients are also approximately equal: 1.3452 and −0.8551.

Book m-Function for Obtaining Power Spectra

```
function[wa1,v]=lms_power_spectra_av1(a1,a2,a3,mu,M,N,avn,vr)
     %lms_power_spectra_av1(a1,a2,a3,mu,M,N,avn,vr);
wa = zeros(1,M);
dn = zeros(1,N);x = zeros(1,N);
for k=1:avn
     for n=4:N
     v(n)=vr*(rand-0.5);
```

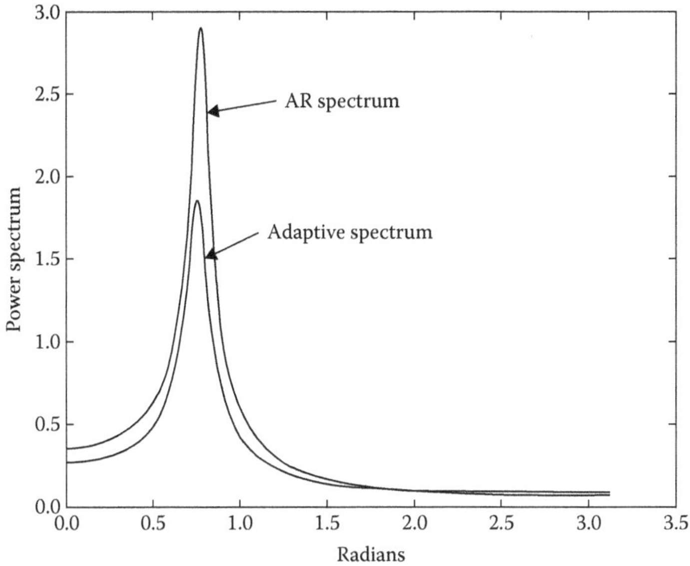

FIGURE 8.10

```
          dn(n)=-a1*dn(n-1)-a2*dn(n-2)-a3*dn(n-3)+v(n);
          x(n)=dn(n-1);
          end;
          w=lms1(x,dn,mu,M);
          wa=wa+w;
     end;
     wa1=wa/avn;
          %this function gives averaged w's to be used for finding
          %the approximate spectrum of the output of an AR filter
          %up to the third order;M = number of LMS coefficients;
          %N = length of desired signal and input signal to LMS
          %filter;
          %avn = number of times w's are averaged;the function is
          %easily
          %modified for AR filter with more coefficients; vr =
          %controls
          %the variance of the white noise, multiplies the
          %quantity (rand-0.5);the function up to the first end
          %produces
          %the output from the AR filter with coefficients a1,a2,
          %and a3;
     [sx,W]=freqz(var(v),[1 -[wa1]],512);%freqz() is a MATLAB
          %function;
     n=0:pi/512:pi-(pi/512);
     plot(n,abs(sx),'k'); xlabel('Radians');ylabel('Power
          Spectrum');
          %this function is useful up to three coefficients AR
          %model;
          %var controls the variance of the input noise;
```

The constants that were used in the above example and produced Figure 8.10 were as follows: $a1 = -1.3435$, $a2 = 0.9025$, $a3 = 0.051$, $vr = 1.5$, avn = 5, $N = 1000$, and $M = 3$. The plot of (8.13) was done using the following expressions:

```
>>om=0:pi/512:pi-(pi/512);
>>plot(om,abs(0.2./(1-1.3435*exp(-j*om)+
     0.9025*exp(-j*om*2))),'k');
```

■

Example 8.3.5 (Sinusoid Plus Noise)

Assume that the noisy signal is given as $x(n) = a\sin(\omega n) + v(n)$, where $v(n)$ is a white noise sequence. It is desired to obtain the sinusoid with the smallest amount of noise.

Solution: Figure 8.11 shows such a situation. The constants and the signals that were used were as follows: $\mu = 0.01$, $M = 30$, $x = \sin(0.1\pi n) + 2(\text{rand} - 0.5)$, and $d = \sin(0.1\pi n)$.

■

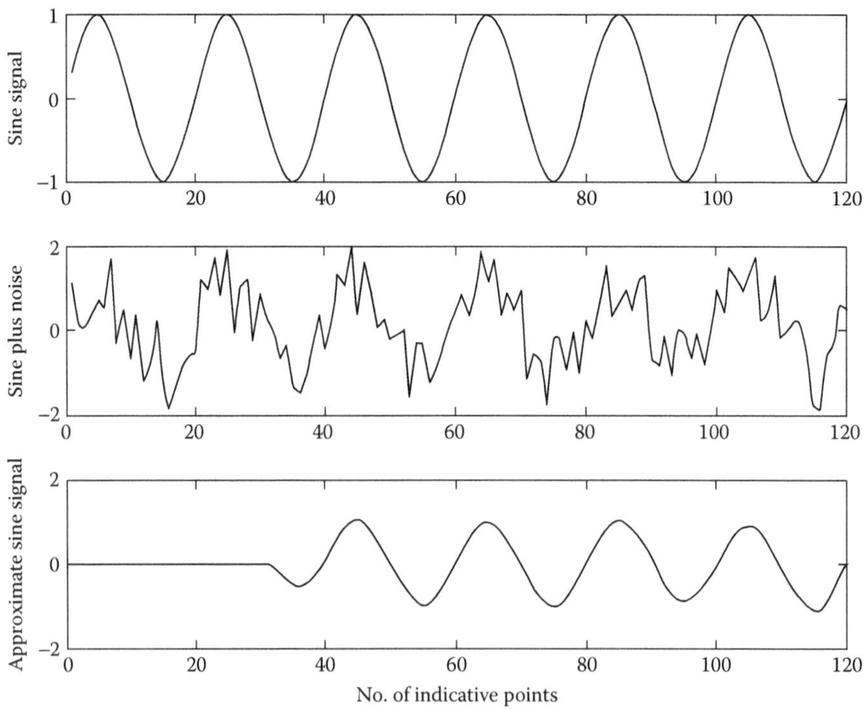

FIGURE 8.11

Example 8.3.6 (Adaptive Channel Estimation)

Figure 8.12 shows a communication noisy channel that is required to be estimated by the adaptive processing technique.

Solution: In this example, we assumed that the channel is represented by an FIR system with four unknown coefficients $c = [0.6 \quad -0.9 \quad 0.2 \quad -0.1]$. In addition, we used the following values: $M = 20$, random channel noise $0.5(\text{rand} - 0.5)$, and the signal $s(n) = \sin(0.1\pi n)$. Figure 8.13 presents the noisy channel output signal and the reproduction of the input signal to the channel. It is obvious that we recapture the input to the channel signal. In practice, we can create an adaptive filter by sending a known signal, and after the establishment of the proper adaptive filter and assuming that the channel varies slowly, we can use the adaptive filter for a while. Then, periodically, we can find the appropriate adaptive filters.

The Book m-function with which we produced Figure 8.13 is given below.

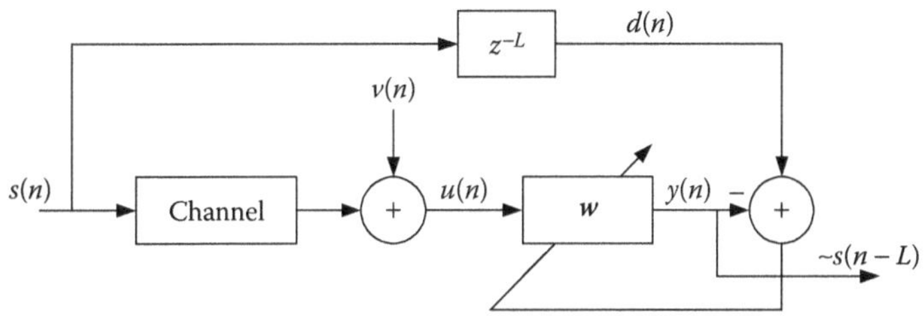

FIGURE 8.12

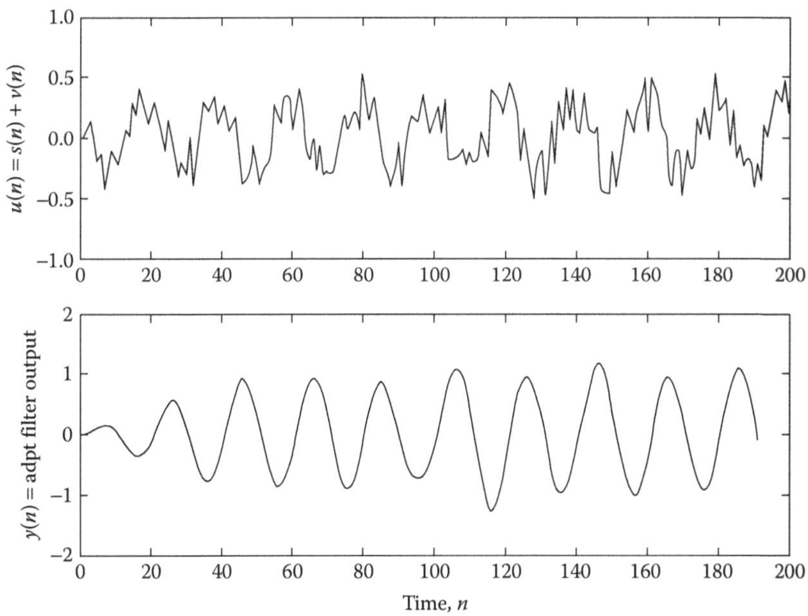

FIGURE 8.13

Book m-Function: [w,e,J,y,u]=lms_ex8_3_6(s,mu,c,M)

```
function[w,e,J,y,u]=lms_ex8_3_6(s,mu,c,M)
     %s = signal;mu = step-size coefficient;
     %c = vector with the channel coefficients;
     %M = number of adaptive filter coefficients
w=zeros(1,M);
u=conv(s,c)+0.5*(rand(1,length(conv(s,c)))-0.5);
for n=M:length(u)-length(c)
     u1=u(n:-1:n-M+1);
     y(n)=w*u1';
     e(n)=s(n+1)-y(n);
     w=w+2*mu*e(n)*u1;
end;
J=e.^2;
```

■

Example 8.3.7 (Inverse System Identification)

To find the inverse of an unknown filter, we place the adaptive filter in series with the unknown system as shown in Figure 8.14. The delay is needed so that the system is causal. Figure 8.15 shows the noisy signal, the output from the adaptive filter, and a typical learning curve. In this example, we used four FIR filter coefficients, and the input to the unknown system was a sine function with a white Gaussian noise. The following Book script m-file was used:

Book Script m-File

```
%script file ex8_3_7
n=1:1500;xi=zeros(1,1500);
xi=xi+sin(0.1*pi*n)+0.06*randn(1,1500);
x=conv(xi,[0.6 -0.9 0.2 -0.1]);
for n=1:1500
     d(n+1)=x(n);
end;
[w,y,e,J]=lms1(x(1,1:1450),d(1,1:1450),0.02,8);
```

The functions and constant used were as follows: $s(n) = \sin(0.1\pi n)$, $v(n) = 0.06\text{randn}(1{,}1500)$, unknown system $c = [0.6 \quad -0.9 \quad 0.2 \quad -0.1]$, $\mu = 0.02$, and $M = 8$.

■

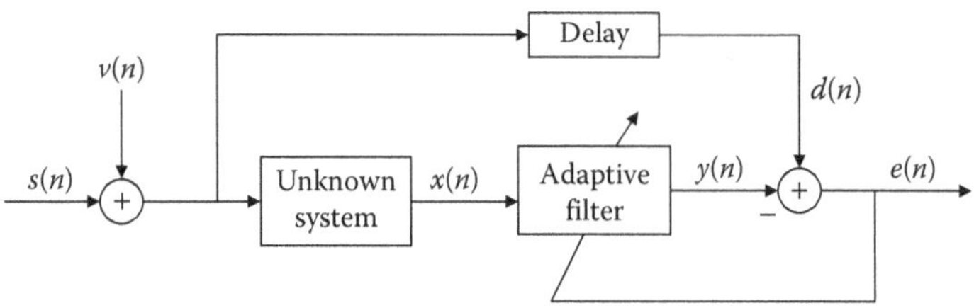

FIGURE 8.14

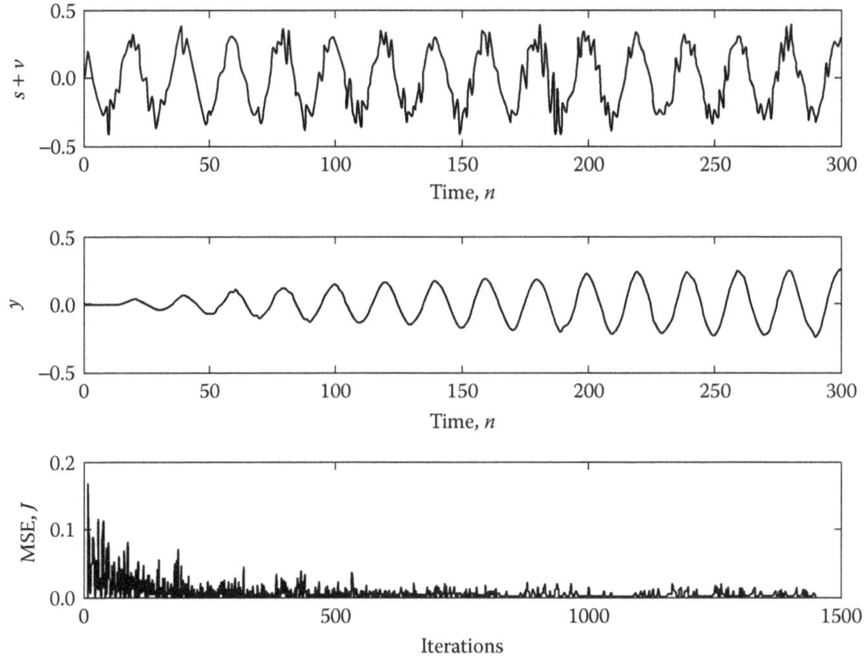

FIGURE 8.15

Example 8.3.8 (Adaptive Prediction)

Based on the Figure 8.16, which presents a one-tap AR filter, find the mean-square error (MSE) of a single and an ensemble average result.

Solution: The results in Figures 8.17 and 8.18 are found using the Book m-function given below.

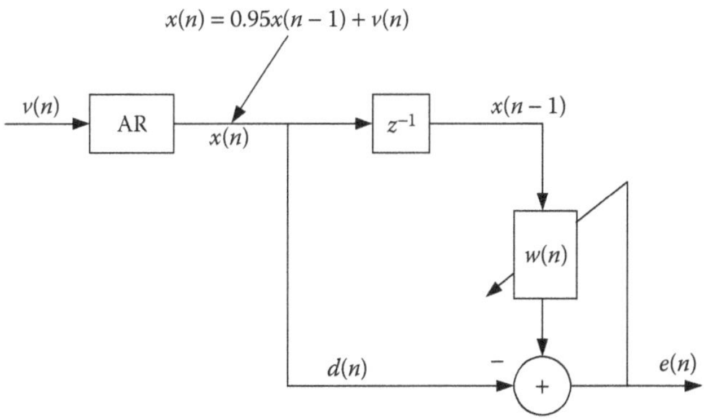

FIGURE 8.16

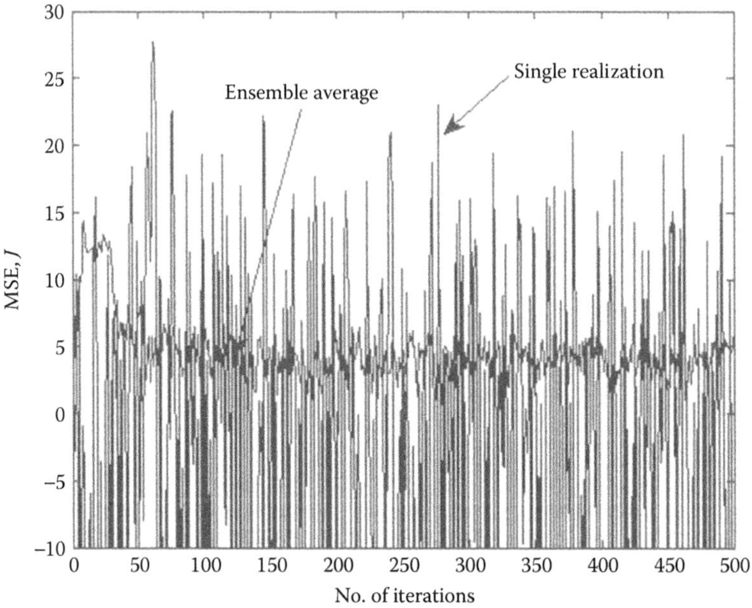

FIGURE 8.17

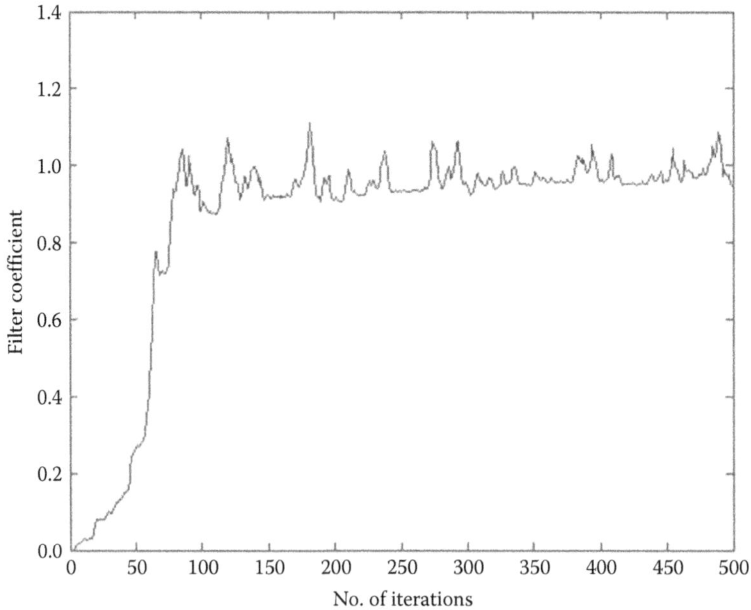

FIGURE 8.18

Book m-Function: [w,y,e,Jav]=lms_ex8_3_8(mu,av,N)

```
function[w,y,e,Jav]=lms_ex8_3_8(mu,av,N)
for m=1:av
```

```
        %all quantities are real-valued;
        %x = input data to the filter; d = desired signal;
        %M = order of the filter;
        %mu = step-size factor; x and d must be
        %of the same length;
        w(1)=0;x(1)=0;w(2)=0;
        for r=1:N+2
             x(r+1)=0.95*x(r)+3*(rand-0.5);
             d(r+1)=x(r+1);
        end;
        for n=1:N
             y(n+1)=w(n+1)*x(n);
             e(n+1,m)=x(n+1)-y(n);
             w(n+2)=w(n+1)+mu*x(n)*e(n+1,m);
        end;
        y=zeros(1,N);
    end;
    Jav=sum((e.^2),2)/av;%Jav is the learning curve;
```

Figure 8.17 was produced using $mu = 0.005$ and $N = 1000$. The number of the ensemble was 100. Figure 8.18 shows the creation of the final value of the adaptive filter coefficient versus the number of iterations.

■

8.4 PERFORMANCE ANALYSIS OF THE LMS ALGORITHM*

By subtracting the Wiener filter $\boldsymbol{w}^o$ (column vector of the optimum adaptive filter) from both sides of (8.3), we obtain the following equation:

$$\boldsymbol{w}(n+1) - \boldsymbol{w}^o = \boldsymbol{w}(n) - \boldsymbol{w}^o + 2\mu e(n)\boldsymbol{x}(n) \tag{8.14}$$

The vectors $\xi(n+1) = \boldsymbol{w}(n+1) - \boldsymbol{w}^o$ and $\xi(n) = \boldsymbol{w}(n) - \boldsymbol{w}^o$ are known as the **weight errors**, which are described on a coordinate system shifted by $\boldsymbol{w}^o$ on the $\boldsymbol{w}$ plane. Therefore, (8.14) becomes

$$\xi(n+1) = \xi(n) + 2\mu e(n)x(n) \tag{8.15}$$

$$\begin{aligned} e(n) &= d(n) - y(n) = d(n) - \boldsymbol{w}^T(n)\boldsymbol{x}(n) = d(n) - \boldsymbol{x}^T(n)\boldsymbol{w}(n) \\ &= d(n) - \boldsymbol{x}^T(n)\boldsymbol{w}(n) - \boldsymbol{x}^T(n)\boldsymbol{w}^o + \boldsymbol{x}^T(n)\boldsymbol{w}^o \\ &= d(n) - \boldsymbol{x}^T(n)\boldsymbol{w}^o - \boldsymbol{x}^T(n)[\boldsymbol{w}(n) - \boldsymbol{w}^o] \\ &= \mathrm{e}^o(n) - \boldsymbol{x}^T(n)\xi(n) = \mathrm{e}^o(n) - \xi^T(n)\boldsymbol{x}(n) \end{aligned} \tag{8.16}$$

where:

$$\mathrm{e}^o(n) = d(n) - \boldsymbol{x}^T(n)\boldsymbol{w}^o \tag{8.17}$$

* The reader can skip the section at the first reading.

is the error when the filter is optimum. Substituting (8.16) in (8.15) and rearranging, we obtain

$$\begin{aligned}\xi(n+1) &= \xi(n)+2\mu[\mathrm{e}^{o}(n)-\boldsymbol{x}^{T}(n)\xi(n)]\boldsymbol{x}(n)\\ &= \xi(n)+2\mu\boldsymbol{x}(n)[\mathrm{e}^{o}(n)-\boldsymbol{x}^{T}(n)\xi(n)]\\ &= [\boldsymbol{I}-2\mu\boldsymbol{x}(n)\boldsymbol{x}^{T}(n)]\xi(n)+2\mu\mathrm{e}^{o}(n)\boldsymbol{x}(n)\end{aligned} \tag{8.18}$$

where:

$\boldsymbol{I}$ is the identity matrix (ones in the diagonal)
$\mathrm{e}^{o}(n)-\boldsymbol{x}^{T}(n)\xi(n)$ is a scalar

Next, we take the expectation of both sides of (8.18):

$$\begin{aligned}E\{\xi(n+1)\} &= E\{[\boldsymbol{I}-2\mu\boldsymbol{x}(n)\boldsymbol{x}^{T}(n)]\}\xi(n)+2\mu E\{\mathrm{e}^{o}(n)\boldsymbol{x}(n)\}\\ &= E\{[\boldsymbol{I}-2\mu\boldsymbol{x}(n)\boldsymbol{x}^{T}(n)]\}\xi(n)\}\end{aligned} \tag{8.19}$$

Since $\mathrm{e}^{o}(n)$ is orthogonal to all data (see Section 5.5), the last expression is identically zero. The expression

$$E\{\boldsymbol{x}(n)\boldsymbol{x}^{T}(n)\xi(n)\} = E\{\boldsymbol{x}(n)\boldsymbol{x}^{T}(n)\}E\{\xi(n)\} \tag{8.20}$$

is simplified by incorporating the independence assumption, which states that the present observation of the data $\{x(n), d(n)\}$ are independent of the past observations $[x(n-1), d(n-1)], [x(n-2), d(n-2)], \ldots,$ where

$$\boldsymbol{x}(n) = [x(n)\ \ x(n-1)\ \ x(n-2)\cdots x(n-N+1)] \tag{8.21}$$

$$\boldsymbol{x}(n-1) = [x(n-1)\ \ x(n-2)\ \ x(n-3)\cdots x(n-N)] \tag{8.22}$$

Another way to justify the independence assumption is through the following observation: The LMS coefficients $\boldsymbol{w}(n)$ at any given time are affected by the whole past history of the data $[x(n-1), d(n-1)], [x(n-2), d(n-2)], \ldots,$ and therefore, for smaller step-size parameter μ, the past N observations of the data have small contribution to $\boldsymbol{w}(n)$, and thus, we can say that $\boldsymbol{w}(n)$ and $\boldsymbol{x}(n)$ are weakly dependent. This observation clearly suggests the approximation given by (8.20). Substituting (8.20) in (8.19), we obtain

$$\begin{aligned}E\{\xi(n+1)\} &= [\boldsymbol{I}-2\mu E\{\boldsymbol{x}(n)\boldsymbol{x}^{T}(n)\}]E\{\xi(n)\}\\ &= [\boldsymbol{I}-2\mu\boldsymbol{R}_{x}]E\{\xi(n)\}\end{aligned} \tag{8.23}$$

The mathematical forms of (8.23) and (7.20) of the steepest descent method are identical except that the deterministic weight-error vector $\xi(n)$ in (7.20) has been replaced by the average weight-error vector $E\{\xi(n)\}$ of the LMS algorithm. This suggests that, on average, the present LMS algorithm behaves just like the steepest descent algorithm. Like the steepest descent method, the LMS algorithm is controlled by M modes of convergence, which are dependent on the eigenvalues of the correlation matrix $\boldsymbol{R}_x$. In particular, the convergence behavior of the LMS algorithm is directly related to the eigenvalue spread of $\boldsymbol{R}_x$, and hence to the power spectrum of the input data $\boldsymbol{x}(n)$. The more flatness of the power spectrum, the higher speed of convergence of the LMS algorithm is attained.

8.4.1 Learning Curve

In the development below, we assume the following: (1) the input signal to LMS filter $\boldsymbol{x}(n)$ is a zero-mean stationary process, (2) the desired signal $d(n)$ is a zero-mean stationary process, (3) $\boldsymbol{x}(n)$ and $d(n)$ are jointly Gaussian-distributed random variables for all n, and (4) at times n, the coefficients $\boldsymbol{w}(n)$ are independent of the input vector $\boldsymbol{x}(n)$ and the desired signal $d(n)$. The validity of (4) is justified for small values of μ (independent assumption). Assumptions (1) and (2) simplify the analysis. Assumption (3) simplifies the final results so that the third-order and higher moments that appear in the derivation are expressed in terms of the second-order moments due to their Gaussian distribution.

If we take the mean-square average of the error given by (8.16), we obtain

$$\begin{aligned} J(n) = E\{\mathrm{e}^2(n)\} &= E\{[\mathrm{e}^o(n) - \xi^T(n)\boldsymbol{x}(n)][\mathrm{e}^o(n) - \boldsymbol{x}^T(n)\xi(n)]\} \\ &= E\{\mathrm{e}^{o2}(n)\} + E\{\xi^T(n)\boldsymbol{x}(n)\boldsymbol{x}^T(n)\xi(n)\} - 2E\{\mathrm{e}^o(n)\xi^T(n)\boldsymbol{x}(n)\} \end{aligned} \tag{8.24}$$

For independent random variables, we have the following relations:

$$E\{xy\} = E\{x\}E\{y\} = E\{xE\{y\}\} \tag{8.25}$$

$$E\{x^2y^2\} = E\{x^2\}E\{y^2\} = E\{x^2E\{x^2\}\} = E\{xE\{y^2\}x\} \tag{8.26}$$

Based on the above two equations, the second term of (8.24) becomes $[\boldsymbol{x}^T(n)\xi(n) = \xi^T(n)\boldsymbol{x}(n) = \text{number}]$:

$$\begin{aligned} E\{[\xi^T(n)\boldsymbol{x}(n)]^2\} &= E\{\xi^T(n)\boldsymbol{x}(n)\boldsymbol{x}^T(n)\xi(n)\} = E\{\xi^T(n)E\{\boldsymbol{x}(n)\boldsymbol{x}^T(n)\}\xi(n)\} \\ &= E\{\xi^T\boldsymbol{R}_x\xi(n)\} = \mathrm{tr}\{E\{\xi^T(n)\boldsymbol{R}_x\xi(n)\}\} \\ &= E\{\mathrm{tr}\{\xi^T(n)\boldsymbol{R}_x\xi(n)\}\} = E\{\mathrm{tr}\{\xi(n)\xi^T(n)\boldsymbol{R}_x\}\} \\ &= \mathrm{tr}\{E\{\xi(n)\xi^T(n)\}\boldsymbol{R}_x\} = \mathrm{tr}\{\boldsymbol{K}(n)\boldsymbol{R}_x\} \end{aligned} \tag{8.27}$$

where we used the following properties: (1) the trace of a scalar is the scalar itself; (2) the trace, tr, and the expectation, E, operators are linear and can be exchanged; and (3) the trace of two matrices having $N \times M$ and $M \times N$ dimensions, respectively, is given by

$$\mathrm{tr}\{\boldsymbol{AB}\} = \mathrm{tr}\{\boldsymbol{BA}\} \tag{8.28}$$

The third term in (8.24), due to the independence assumption and due to the fact that $\mathrm{e}^o(n)$ is a constant, becomes

$$E\{\mathrm{e}^o(n)\xi^T(n)\boldsymbol{x}(n)\} = E\{\xi^T(n)\boldsymbol{x}(n)\mathrm{e}^o(n)\} = E\{\xi^T(n)\}E\{\boldsymbol{x}(n)\mathrm{e}^o(n)\} = 0 \tag{8.29}$$

The second term is equal to zero due to the orthogonality property (see Section 5.5).
Substituting (8.29) and (8.28) in (8.24), we obtain

$$J(n)=E\{\mathrm{e}^2(n)\}=J_{\min}+\mathrm{tr}\{\boldsymbol{K}(n)\boldsymbol{R}_x\},\quad J_{\min}=E\{([\mathrm{e}^o(n)]^2)\} \tag{8.30}$$

However, $\boldsymbol{R}_x=\boldsymbol{Q}\Lambda\boldsymbol{Q}^T$, where $\boldsymbol{Q}$ is the eigenvector matrix and Λ is the diagonal eigenvalue one. Hence, (8.30) becomes

$$\begin{aligned}J(n)&=J_{\min}+\mathrm{tr}\{\boldsymbol{K}(n)\boldsymbol{Q}\Lambda\boldsymbol{Q}^T\}=J_{\min}+\mathrm{tr}\{\boldsymbol{Q}^T\boldsymbol{K}(n)\boldsymbol{Q}\Lambda\}\\&=J_{\min}+\mathrm{tr}\{E\{\boldsymbol{Q}^T\xi(n)\xi^T(n)\boldsymbol{Q}\}\Lambda\}\\&=J_{\min}+\mathrm{tr}\{E\{\xi'(n)\xi'^T(n)\}\Lambda\}=J_{\min}+\mathrm{tr}\{\boldsymbol{K}'(n)\Lambda\}\end{aligned} \tag{8.31}$$

where:

$$\boldsymbol{K}'(n)=E\{\xi'(n)\xi'^T(n)\} \tag{8.32}$$

Recall that $\xi'(n)$ is the weight-error vector in the coordinate system defined by the basis vectors, which are specified by the eigenvectors of $\boldsymbol{R}_x$. Since Λ is diagonal, (8.31) becomes

$$J(n)=J_{\min}+\sum_{i=0}^{M-1}\lambda_i k'_{ii}(n)=J_{\min}+\sum_{i=0}^{M-1}\lambda_i E\{\xi'^2_{ii}(n)\} \tag{8.33}$$

where:
$k'_{ij}(n)$ is the ijth element of the matrix $\boldsymbol{K}'(n)$

The learning curve can be obtained by any one of the equations: (8.30), (8.31), and (8.33). It turns out that, on average, the learning curve above is similar to the one given by the steepest descent algorithm.
The general solution of $\xi'(n)$ is given by (7.30), and hence, (8.33) becomes

$$J(n)=J_{\min}+\sum_{i=0}^{M-1}\lambda_i\left(1-2\mu\lambda_i\right)^{2n}E\{\xi'_{ii}(0)\} \tag{8.34}$$

Example 8.4.1

The filter $H_1(z)$, which produces the desired signal $d(n)=x(n)$, is represented by the difference equation $x(n)+a_1x(n-1)+a_2x(n-2)=v(n)$, where a_1 and a_2 are the system coefficients and $v(n)$ is a zero-mean white noise process of variance σ_v^2. To simulate the system coefficients a_1 and a_2, we use the adaptive predictor (see Figure 8.3a). The LMS algorithm is

$$\begin{aligned}&\boldsymbol{w}(n)=\boldsymbol{w}(n-1)+2\mu\boldsymbol{x}(n-1)e(n)\\&e(n)=d(n)-\boldsymbol{w}^T(n)\boldsymbol{x}(n-1)=\boldsymbol{x}(n)-\boldsymbol{w}^T(n)\boldsymbol{x}(n-1)\\&J(n)=\mathrm{e}^2(n)=\text{Learning curve}\end{aligned} \tag{8.35}$$

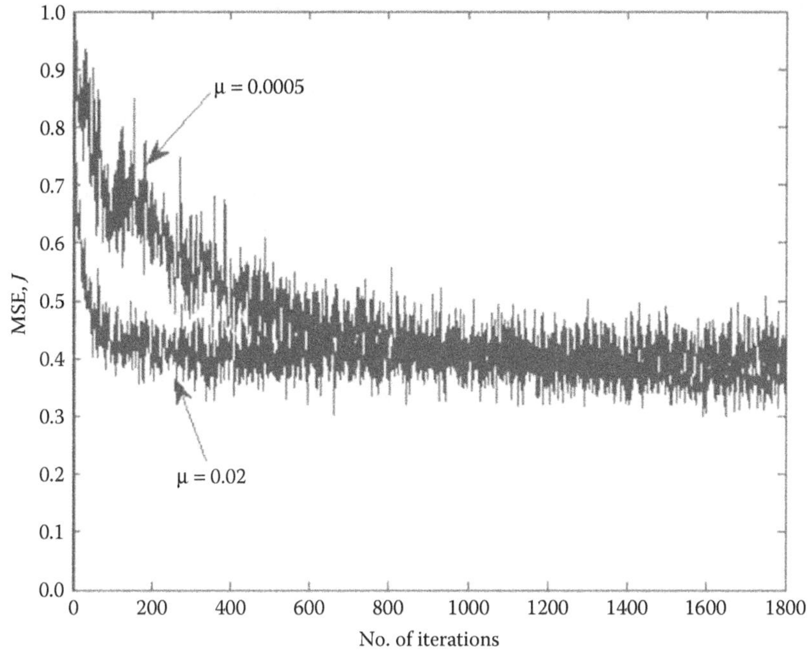

FIGURE 8.19

In this example, we used the following constants: $M = 6$, $a_1 = -0.96$, $a_2 = 0.2$, averaging number = 200, $\sigma_v^2 = 0.33$. Figure 8.19 shows the results. The following Book m-function was used.

Book m-Function

```
function[J]=lms_ex8_4_1(mu,M,avn)
     %function[J] = lms_ex8_4_1(mu,M,avn);
     %M = number of filter coefficients;
     %avn = number of times the MSE (J) to be averaged;
dn(1)=0;dn(2)=0;x(1)=0;x(2)=0;
for k=1:avn
     for n=3:2000
            dn(n)=0.96*dn(n-1)-0.2*dn(n-2)+2*(rand-0.5);
            x(n)=dn(n-1);
     end;
     [w,y,e,J1]=lms1(x,dn,mu,M);
     Jk(k,:)=J1;%this expression creates a
            %matrix avn by 2000;
end;
J=sum(Jk,1)/100;
```

■

8.4.2 The Coefficient-Error or Weighted-Error Correlation Matrix

Since the MSE is related to weight-error correlation matrix $\boldsymbol{K}(n)$, the matrix is closely related to the stability of the LMS algorithm. Therefore, $J(n)$ is bounded if the elements of $\boldsymbol{K}(n)$ are also bounded. Since $\boldsymbol{K}(n)$ and $\boldsymbol{K}'(n)$ are related, the stability can be studied by using either one. Multiply both sides of (8.18) by $\boldsymbol{Q}^T$ and use the definitions $\xi'(n) = \boldsymbol{Q}^T\xi(n)$ and $\boldsymbol{x}'(n) = \boldsymbol{Q}^T\boldsymbol{x}(n)$ to obtain (see Problem 7.4.1)

$$\xi'(n+1) = [\boldsymbol{I} - 2\mu\boldsymbol{x}'(n)\boldsymbol{x}'^T(n)]\xi'(n) + 2\mu e^o(n)\boldsymbol{x}'(n) \tag{8.36}$$

Next, multiply (8.36) by its transpose and take the ensemble average of both sides to obtain (see Problem 8.4.2)

$$\begin{aligned}\boldsymbol{K}'(n+1) = \boldsymbol{K}'(n) &- 2\mu E\{\boldsymbol{x}'(n)\boldsymbol{x}'^T(n)\xi'(n)\xi'^T(n)\}\\ &-2\mu E\{\xi'(n)\xi'^T(n)\boldsymbol{x}'(n)\boldsymbol{x}'^T(n)\}\\ &+4\mu^2 E\{\boldsymbol{x}'(n)\boldsymbol{x}'^T(n)\xi'(n)\xi'^T(n)\boldsymbol{x}'(n)\boldsymbol{x}'^T(n)\}\\ &+2\mu E\{e^o(n)\boldsymbol{x}'(n)\xi'^T(n)\}\\ &+2\mu E\{e^o(n)\xi'(n)\boldsymbol{x}'^T(n)\}\\ &-4\mu^2 E\{e^o(n)\boldsymbol{x}'(n)\xi'^T(n)\boldsymbol{x}'(n)\boldsymbol{x}'^T(n)\}\\ &-4\mu^2 E\{e^o(n)\boldsymbol{x}'(n)\boldsymbol{x}'^T(n)\xi'(n)\boldsymbol{x}'^T(n)\}\\ &+4\mu^2 E\{e^{o2}(n)\boldsymbol{x}'(n)\boldsymbol{x}'^T(n)\}\end{aligned} \tag{8.37}$$

Based on the previous independent assumption, we note the following: (1) $\xi(n)$ is independent of the data $\boldsymbol{x}(n)$ and the desired signal $d(n)$, which is also true for the transformed variables $\boldsymbol{x}'(n)$; (2) $\xi'(n)$ is independent of $\boldsymbol{x}'(n)$ and $d(n)$; (3) $d(n)$ and $\boldsymbol{x}'(n)$ are zero mean and are jointly Gaussian since $d(n)$ and $\boldsymbol{x}(n)$ have the same properties; (4) by applying the orthogonality relationship, we obtain $E\{e^o(n)\boldsymbol{x}'(n)\} = E\{e^o(n)\boldsymbol{Q}^T\boldsymbol{x}(n)\} = \boldsymbol{Q}^T E\{e^o(n)\boldsymbol{x}(n)\} = \boldsymbol{Q}^T\boldsymbol{0} = \boldsymbol{0}$; (5) $e^o(n)$ depends only on $d(n)$ and $\boldsymbol{x}(n)$; (6) from (4) $e^o(n)$ and $\boldsymbol{x}'(n)$ are uncorrelated (Gaussian variables are also independent); (7) $e^o(n)$ has zero mean. With the above assumptions in mind, the factors of (8.37) become (see Problem 8.4.3)

$$\begin{aligned}E\{\boldsymbol{x}'(n)\boldsymbol{x}'^T(n)\xi'(n)\xi'^T(n)\} &= E\{\boldsymbol{x}'(n)\boldsymbol{x}'^T(n)\}E\{\xi'(n)\xi'^T(n)\}\\ &= E\{\boldsymbol{Q}^T\boldsymbol{x}(n)\boldsymbol{x}^T(n)\boldsymbol{Q}\}\boldsymbol{K}'(n) = \boldsymbol{Q}^T\boldsymbol{R}_{xx}\boldsymbol{Q}\boldsymbol{K}'(n) = \Lambda\boldsymbol{K}'(n)\end{aligned} \tag{8.38}$$

$$E\{\xi'(n)\xi'^T(n)\boldsymbol{x}'(n)\boldsymbol{x}'^T(n)\} = \boldsymbol{K}'(n)\Lambda \tag{8.39}$$

$$E\{\boldsymbol{x}'(n)\boldsymbol{x}'^T(n)\xi'(n)\xi'^T(n)\boldsymbol{x}'(n)\boldsymbol{x}'(n)\} = 2\Lambda\boldsymbol{K}'(n)\Lambda + \text{tr}\{\Lambda\boldsymbol{K}'(n)\Lambda\} \tag{8.40}$$

Because $e^o(n)$, $\boldsymbol{x}'(n)$, and $\xi'(n)$ are mutually independent, $E\{e^o(n)\} = 0$ are then true:

$$E\{e^o(n)\boldsymbol{x}'(n)\xi'^T(n)\} = E\{e^o(n)\}E\{\boldsymbol{x}'(n)\xi'^T(n)\} = \boldsymbol{0} \quad (M \times M \text{ matrix}) \tag{8.41}$$

$$E\{e^o(n)\xi'(n)\boldsymbol{x}'^T(n)\} = 0 \tag{8.42}$$

$$E\{e^o(n)\boldsymbol{x}'(n)\xi'^T(n)\boldsymbol{x}'(n)\boldsymbol{x}'^T(n)\} = 0 \tag{8.43}$$

$$E\{e^o(n)\boldsymbol{x}'(n)\boldsymbol{x}'^T(n)\xi'(n)\boldsymbol{x}'^T(n)\} = 0 \tag{8.44}$$

$$E\{e^{o2}(n)\boldsymbol{x}'(n)\boldsymbol{x}'^T(n)\} = E\{e^{o2}(n)\}E\{\boldsymbol{x}'(n)\boldsymbol{x}'^T(n)\} = J_{\min}\Lambda \tag{8.45}$$

Substituting (8.38) through (8.45) in (8.37), we obtain

$$\begin{aligned} \boldsymbol{K}'(n+1) &= \boldsymbol{K}'(n) - 2\mu(\Lambda\boldsymbol{K}'(n) + \boldsymbol{K}'(n)\Lambda) + 8\mu^2\Lambda\boldsymbol{K}'(n)\Lambda \\ &\quad + 4\mu^2\mathrm{tr}\{\Lambda\boldsymbol{K}'(n)\}\boldsymbol{\Lambda} + 4\mu^2 J_{\min}\Lambda \end{aligned} \tag{8.46}$$

Concentrating on the iith component of both sides of (8.46), we obtain (see Problem 8.4.4)

$$k'_{ii}(n+1) = (1 - 4\mu\lambda_i + 8\mu^2\lambda_i^2)k'_{ii}(n) + 4\mu^2\lambda_i\sum_{j=0}^{M-1}\lambda_j k'_{jj}(n) + 4\mu^2 J_{\min}\lambda_i \tag{8.47}$$

Since $\boldsymbol{K}'(n)$ is a correlation matrix, $k'^2_{ij} \le k'_{ii}(n)k'_{jj}(n)$ for all values of i and j, and since the update of $k'_{ii}(n)$ is independent of the off-diagonal elements of $\boldsymbol{K}'(n)$, the convergence of the diagonal elements is sufficient to secure the convergence of all the elements, and thus guarantees the stability of the LMS algorithm. Therefore, we concentrate on (8.47) with $i = 0, 1, 2, \ldots, M-1$.

Equation 8.47 can be written in the following matrix form (see Problem 8.4.4):

$$\boldsymbol{k}'(n+1) = \boldsymbol{F}\boldsymbol{k}'(n) + 4\mu^2 J_{\min}\boldsymbol{\lambda} \tag{8.48}$$

where:

$$\boldsymbol{k}'(n) = \begin{bmatrix} k'_{00}(n) & k'_{11} & \cdots & k'_{M-1,M-1}(n) \end{bmatrix}^T \tag{8.49}$$

$$\boldsymbol{\lambda} = \begin{bmatrix} \lambda_0 & \lambda_1 & \cdots & \lambda_{M-1} \end{bmatrix}^T \tag{8.50}$$

$$\begin{aligned} \boldsymbol{F} &= \mathrm{diag}[f_0 \quad f_1 \cdots f_{M-1}]^T \\ f_i &= 1 - 4\mu\lambda_i + 8\mu^2\lambda_i^2 \end{aligned} \tag{8.51}$$

It has been found that if the eigenvalues of $\boldsymbol{F}$ are less than one, the LMS algorithm is stable or equivalently the elements $\boldsymbol{k}'(n)$ remain bounded. An indirect approach to obtain stability is given in Section 8.4.3.

8.4.3 Excess MSE and Misadjustment

We may write the expression in (8.30) as the difference between the MSE and the minimum MSE as follows:

$$J_{ex}(n) = J(n) - J_{\min} = \mathrm{tr}[\boldsymbol{K}(n)\boldsymbol{R}_x] \tag{8.52}$$

The steady-state form of (8.52) is

$$J_{ex}(\infty) = J(\infty) - J_{min} = \text{tr}[\boldsymbol{K}(\infty)\boldsymbol{R}_x] \tag{8.53}$$

and it is known as **excess** MSE. Equation 8.33 gives another equivalent form, which is

$$J_{ex}(\infty) = \sum_{i=0}^{M-1} \lambda_i k'_{ii}(\infty) = \xi^T \boldsymbol{k}'(\infty) \tag{8.54}$$

As $n \to \infty$, we set $\boldsymbol{k}'(n+1) = \boldsymbol{k}'(n)$, and hence, (8.48) gives another equivalent form, which is

$$\boldsymbol{k}'(\infty) = 4\mu^2 J_{min}(\boldsymbol{I} - \boldsymbol{F})^{-1}\lambda \tag{8.55}$$

As a result, (8.54) becomes

$$J_{ex}(\infty) = 4\mu^2 J_{min} \boldsymbol{\lambda}^T (\boldsymbol{I} - \boldsymbol{F})^{-1}\lambda \tag{8.56}$$

which indicates that J_{ex} is proportional to J_{min}. The normalized $J_{ex}(\infty)$ is equal to

$$\mathcal{M} = \frac{J_{ex}(\infty)}{J_{min}} = 4\mu^2 \lambda^T (\boldsymbol{I} - \boldsymbol{F})^{-1}\lambda \tag{8.57}$$

which is known as the **misadjustment** factor.

If $\boldsymbol{A}(N \times N)$, $\boldsymbol{B}(M \times M)$, and $\boldsymbol{C}(N \times M)$ are matrices that have an inverse, then

$$(\boldsymbol{A} + \boldsymbol{C}\boldsymbol{B}\boldsymbol{C}^T)^{-1} = \boldsymbol{A}^{-1} - \boldsymbol{A}^{-1}\boldsymbol{C}(\boldsymbol{B}^{-1} + \boldsymbol{C}^T\boldsymbol{A}^{-1}\boldsymbol{C})^{-1}\boldsymbol{C}^T\boldsymbol{A}^{-1} \tag{8.58}$$

But

$$\boldsymbol{I} - \boldsymbol{F} = \begin{bmatrix} 1-f_0 & 0 & \cdots & 0 \\ 0 & 1-f_1 & \cdots & 0 \\ \vdots & & & \vdots \\ 0 & 0 & \cdots & 1-f_{M-1} \end{bmatrix} = 4\mu^2\lambda\lambda^T = \boldsymbol{F}_1 + a\lambda\lambda^T \tag{8.59}$$

where:

$a = -4\mu^2$

Therefore, (8.57) takes the form (see Problem 8.4.5)

$$\begin{aligned} \mathcal{M} &= -a\lambda^T(\boldsymbol{F}_1 + a\lambda\lambda^T)^{-1}\lambda = -a\lambda^T\left[\boldsymbol{F}_1^{-1} - \frac{a\boldsymbol{F}_1^{-1}\lambda\lambda^{\mathbf{T}}\boldsymbol{F}_1^{-1}}{1 + a\lambda^T\boldsymbol{F}_1^{-1}\lambda}\right]\lambda \\ &= a'\lambda^T\left[\boldsymbol{F}_1^{-1} - \frac{a'\boldsymbol{F}_1^{-1}\lambda\lambda^{\mathbf{T}}\boldsymbol{F}_1^{-1}}{1 + a'\lambda^T\boldsymbol{F}_1^{-1}\lambda}\right]\lambda = \frac{\sum_{i=0}^{M-1}(\mu\lambda_i/1-2\mu\lambda_i)}{1 - \sum_{i=0}^{M-1}(\mu\lambda_i/1-2\mu\lambda_i)} \end{aligned} \tag{8.60}$$

where in (8.58) we set $\boldsymbol{C} = \lambda$, $\boldsymbol{B} = \boldsymbol{I}$, $a' = -a$, and $\boldsymbol{A} = \boldsymbol{F}_1$. Small $\mathcal{M}$ implies that the summation on the numerator is small. If, in addition, $2\mu\lambda_i << 1$, we obtain the following results:

$$\sum_{i=0}^{M-1} \frac{\mu\lambda_i}{1-\mu\lambda_i} \cong \mu \sum_{i=0}^{M-1} \mu\,\mathrm{tr}\{\boldsymbol{R}_x\} \tag{8.61}$$

Hence, (8.60) becomes

$$\mathcal{M} = \frac{\mu\,\mathrm{tr}\{\boldsymbol{R}_x\}}{1-\mu\,\mathrm{tr}\{\boldsymbol{R}_x\}} \tag{8.62}$$

In addition, for $\mathcal{M} << 0.1$, the quantity $\mu\,\mathrm{tr}\{\boldsymbol{R}_x\}$ is small and (8.62) becomes

$$\mathcal{M} = \mu\,\mathrm{tr}\{\boldsymbol{R}_x\} \tag{8.63}$$

Since $r_{xx}(0)$ is the mean-square value of the input signal to an M-coefficient filter, we write

$$\mathcal{M} = \mu M r_{xx}(0) = \mu M E\{x^2(0)\} = \mu M\,(\text{power input}) \tag{8.64}$$

The above equation indicates that to keep the misadjustment factor small, and at a specific desired value as the signal power changes, we must adjust the value of μ.

8.4.4 Stability

If we set

$$L = \sum_{i=0}^{M-1} \frac{\mu\lambda_i}{1-\mu\lambda_i} \tag{8.65}$$

then

$$\mathcal{M} = \frac{L}{1-L} \tag{8.66}$$

We observe that L and $\mathcal{M}$ are increasing functions of μ and L, respectively (see Problem 8.4.6). Since L reaches 1 as $\mathcal{M}$ goes to infinity, this indicates that there is a value $\mu_{\max}$ that μ cannot surpass. To find the upper value of μ, we must concentrate on the expression that can easily be measured in practice. Therefore, from (8.66) we must have

$$\sum_{i=0}^{M-1} \frac{\mu\lambda_i}{1-\mu\lambda_i} \le 1 \tag{8.67}$$

The above equation indicates that the maximum value of μ must make (8.67) an equality. It can be shown that the following inequality holds:

$$\sum_{i=0}^{M-1} \frac{\mu\lambda_i}{1-\mu\lambda_i} \le \frac{\mu \sum_{i=0}^{M-1} \lambda_i}{1-2\mu \sum_{i=0}^{M-1} \lambda_i} \tag{8.68}$$

Hence, if we solve the equality

$$\frac{\mu\sum_{i=0}^{M-1}\lambda_i}{1-2\mu\sum_{i=0}^{M-1}\lambda_i} \tag{8.69}$$

for μ, then (8.67) is satisfied. The solution of the above equality is

$$\mu_{\max} = \frac{1}{3\sum_{i=0}^{M-1}\lambda_i} = \frac{1}{3\,\mathrm{tr}\{\boldsymbol{R}_x\}} = \frac{1}{3(\text{Input power})} \tag{8.70}$$

and hence,

$$0 < \mu < \frac{1}{3\sum_{i=0}^{M-1}\lambda_i} = \frac{1}{3\,\mathrm{tr}\{\boldsymbol{R}_x\}} \tag{8.71}$$

8.4.5 The LMS and Steepest Descent Methods

The following similarities and differences exist between the two methods:

1. The steepest descent method reaches the minimum MSE $J_{\min}$ as $n \to \infty$ and $\boldsymbol{w}(n) \to \boldsymbol{w}^o$.
2. The LMS method produces an error $J(\infty)$ that approaches $J_{\min}$ as $n \to \infty$ and remains larger than $J_{\min}$.
3. The LMS method produces a $w(n)$, as the iterations $n \to \infty$, which is close to the optimum w^o.
4. The steepest descent method has a well-defined learning curve consisting of a sum of decaying exponentials.
5. The LMS learning curve is a sum of noisy decaying exponentials and the noise, in general, decreases the smaller values the step-size parameter μ takes.
6. In the steepest descent method, the correlation matrix $\boldsymbol{R}_x$ of the data $\boldsymbol{x}(n)$ and the cross-correlation vector $\boldsymbol{p}_{dx}(n)$ are found using ensemble averaging operations from the realizations of the data $\boldsymbol{x}(n)$ and the desired signal $d(n)$.
7. In the LMS filter, an ensemble of learning curves is found under identical filter parameters and then averaged point by point.

8.5 COMPLEX REPRESENTATION OF THE LMS ALGORITHM

In some practical applications, it is mathematically attractable to have complex representation of the underlying signals. For example, baseband signals in a quadrature amplitude modulation (QAM) format are written as a summation of two components: real **in-phase** component and imaginary **quadrature** component. Furthermore, signals detected by a set of antennas are also written in their complex form for easy mathematical manipulation. For this reason, we shall present in this section the most rudimentary derivation of the LMS filter, assuming that the signals are complex.

In the case where complex-type signals must be processed, we write the output of the adaptive filter in the form:

$$y(n)=\sum_{k=0}^{M-1} w_k^*(n)x(n-k) \tag{8.72}$$

and the error is given by

$$e(n)=d(n)-y(n) \tag{8.73}$$

Therefore, the MSE is

$$J=E\{e(n)\mathrm{e}^*(n)\}=E\{|e(n)|^2\} \tag{8.74}$$

Let us define the complex filter coefficient as follows:

$$w_k=a_k(n)+jb_k(n) \qquad k=0,1,\ldots,M-1 \tag{8.75}$$

Then the gradient operator ∇ has the following kth element:

$$\nabla_{w_k}\triangleq\nabla_k=\frac{\partial}{\partial a_k(n)}+j\frac{\partial}{\partial b_k(n)} \qquad k=0,1,\ldots,M-1 \tag{8.76}$$

which will produce the following kth element of the multielement gradient vector $\nabla_k J$:

$$\nabla_k J=\frac{\partial J}{\partial a_k(n)}+j\frac{\partial J}{\partial b_k(n)} \qquad k=0,1,\ldots,M-1 \tag{8.77}$$

It is noted that the gradient operator is always used to find the minimum points of a function. The above equation indicates that a complex constraint must be converted to a pair of real constraints. Hence, we set

$$\frac{\partial J}{\partial a_k(n)}=0 \qquad j\frac{\partial J}{\partial b_k(n)}=0 \qquad k=0,1,\ldots,M-1 \tag{8.78}$$

The kth element of the gradient vector, using (8.74), is

$$\nabla_k J=E\left\{\frac{\partial e(n)}{\partial a_k(n)}\mathrm{e}^*(n)+\frac{\partial \mathrm{e}^*(n)}{\partial a_k(n)}e(n)+j\frac{\partial e(n)}{\partial b_k(n)}\mathrm{e}^*(n)+j\frac{\partial \mathrm{e}^*(n)}{\partial b_k(n)}e(n)\right\} \tag{8.79}$$

Taking into consideration (8.72) and (8.73), we obtain

$$\begin{aligned}\frac{\partial e(n)}{\partial a_k(n)}&=\frac{\partial d(n)}{\partial a_k(n)}-\sum_{k=0}^{M-1}\left[\frac{\partial w_k^*(n)}{\partial a_k(n)}\right]x(n-k)\\ &=0-\sum_{k=0}^{M-1}\left\{\frac{\partial[a_k(n)-jb_k(n)]}{\partial a_k(n)}\right\}x(n-k)=-x(n-k)\end{aligned} \tag{8.80}$$

TABLE 8.2
Complex Form of the LMS Algorithm

Parameters	M = Number of filter coefficients
	μ = Step-size factor
	$\boldsymbol{x}(n) = [x(n)x(n-1)\cdots x(n-M+1)]^T$
Initialization	$\boldsymbol{w}(0) = \boldsymbol{0}$
Computation	For $n = 0, 1, 2, \ldots$
	1. $y(n) = \boldsymbol{w}^H(n)\boldsymbol{x}(n)$, H = Hermitian = conjugate transpose
	2. $e(n) = d(n) - y(n)$
	3. $\boldsymbol{w}(n+1) = \boldsymbol{w}(n) + 2\mu e^*(n)\boldsymbol{x}(n)$

Similarly, we obtain

$$\frac{\partial e^*(n)}{\partial b_k(n)} = jx(n-k), \frac{\partial e^*(n)}{\partial b_k(n)} = -jx^*(n-k), \frac{\partial e(n)}{\partial b_k(n)} = -jx(n-k) \tag{8.81}$$

Introducing the last three equations into (8.79), we obtain the relationship:

$$\nabla_k J \equiv \nabla_{w_k} J[\boldsymbol{w}(n)] = -2E\{x(n-k)e^*(n)\} \tag{8.82}$$

and thus, the gradient vector becomes

$$\nabla_w J[\boldsymbol{w}(n)] = -2E\{e^*(n)[x(n) \quad x(n-1) \cdots x(n-M+1)]^T\} \tag{8.83}$$

If $\boldsymbol{w}(n)$ is the filter coefficient vector at step n (time), then its update value $\boldsymbol{w}(n+1)$ is given by [see (7.3)]

$$\boldsymbol{w}(n+1) = \boldsymbol{w}(n) - \mu\nabla_w J[\boldsymbol{w}(n)] \tag{8.84}$$

Next, we replace the ensemble average in (8.83) by the instantaneous estimate $e^*(n)\boldsymbol{x}(n)$ to obtain

$$\boldsymbol{w}(n+1) = \boldsymbol{w}(n) + 2\mu e^*(n)\boldsymbol{x}(n) \tag{8.85}$$

which is the LMS recursion formula when we are involved with complex-valued processes. The LMS algorithm for complex signals is given in Table 8.2.

Example 8.5.1

With the input signal $x(n) = \sin(0.1\pi n) + j1.5(\text{rand} - 0.5)$, the desired signal $d(n) = \sin(0.1\pi n)$, $\mu = 0.01$, and the number of coefficients $M = 16$, we obtain the results shown in Figure 8.20. The Book m-function that produced the results is given below.

Book m-Function for Complex LMS Algorithm

```
function[w,y,e,J,w1]=lms_complex_norm_lms(x,dn,mubar,M,c)
    %function[w,y,e,J,w1]=lms_complex_norm_lms(x,dn,mubar,M,c)
    %x = input data to the filter;dn = desired signal;
```

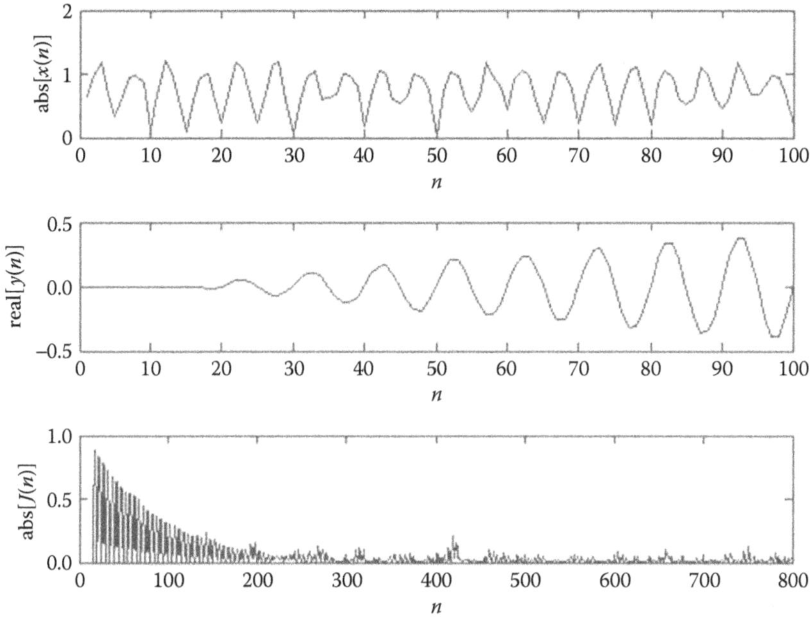

FIGURE 8.20

```
     %M = filter order;c = small constant;
     %mubar = step-size equivalent parameter;
     %x and dn must be of the same length;J = learning curve;
N=length(x);
y=zeros(1,N);
w=zeros(1,M)+j*zeros(1,M);%initialized filter coefficient
     vector;
for n=M:N
     x1=x(n:-1:n-M+1);%for each n vector x1 is of length
           %M with elements from x in reverse order;
     y(n)=conj(w)*x1';
     e(n)=dn(n)-y(n);
     w=w+(mubar/(c+conj(x1)*x1'))*conj(e(n))*x1;
     w1(n-M+1,:)=w(1,:);
end;
J=e.^2;
     %the columns of the matrix w1 depict the history of the
     %filter coefficients;
```

■

PROBLEMS

8.2.1 Develop the LMS algorithm for complex-valued functions.

8.3.1 If an AR system has poles $0.85e^{\pm j(\pi/4)}$ with input white noise $v(n)$, find its discrete-time representation.

8.4.1 Verify (8.36).

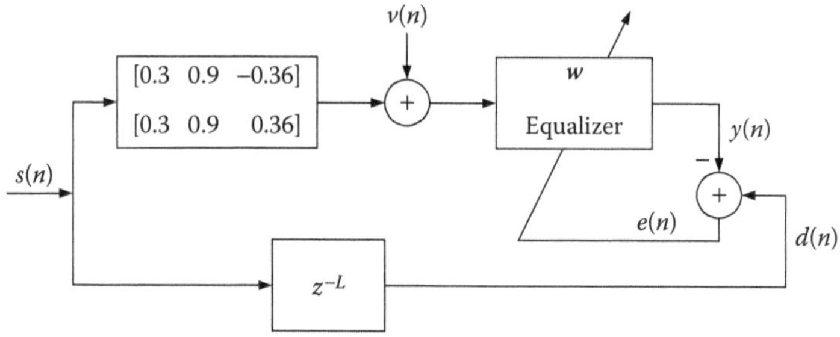

FIGURE P8.4.11

8.4.2 Verify (8.37).
8.4.3 Verify (8.40).
8.4.4 Verify (8.47).
8.4.5 Verify (8.60).
8.4.6 Verify that (8.65) and (8.66) are increasing functions of μ and L.
8.4.7 Use the average eigenvalue to obtain the average constant for the LMS algorithm and state your observations for $\mathcal{M}$.
8.4.8 Discuss the effect of the initial value $\boldsymbol{w}(0)$ on the transient behavior of the LMS algorithm.
8.4.9 Develop the LMS algorithm using the steepest descent method and setting $J(n) = \mathrm{e}^2(n)$ instead of $E\{\mathrm{e}^2(n)\}$.
8.4.10 Based on the formula $L = \sum_{i=0}^{M-1}(\mu\lambda_i/1-\mu\lambda_i)$ developed above [Misadjustment $= L/(1-L)$], plot L versus μ if the input signal to an LMS filter has the following three sets of eigenvalues given below. For stability, we must have $L < 1$. We have also developed above the relation $0 < \mu \le (1/3\sum_{i=0}^{M-1}\lambda_i) = \mu_{\max}$.

λ_0	λ_1	λ_2	λ_3	λ_4	λ_5	λ_6	λ_7	λ_8	λ_9	
1	1	1	1	1	1	1	1	1	1	(1)
1	0.1	0.1	0.1	0.1	0.1	0.1	0.1	0.1	0.1	(2)
1	1	1	1	1	1	1	1	1	0.1	(3)

8.4.11 For the communication channel shown in Figure P8.4.11. Vary the value of the delay, L, in the range from 4 to 15 for the two channels shown in the figure. For each case, find the optimal delay value L that gives the minimum MSE. The number of the coefficients is set equal to $M = 16$.

HINTS–SOLUTIONS–SUGGESTIONS

8.2.1

$\boldsymbol{R}_{ix} = \boldsymbol{x}(n)\boldsymbol{x}^H(n)$, $\boldsymbol{p}_{idx} = \boldsymbol{x}(n)d^*(n)$, $\nabla J = -2\boldsymbol{x}(n)d^*(n) + 2\boldsymbol{x}(n)\boldsymbol{x}^H(n)\boldsymbol{w}(n)$
$\boldsymbol{w}(n+1) = \boldsymbol{w}(n) + 2\mu\boldsymbol{x}(n)[d^*(n) - \boldsymbol{x}^H(n)\boldsymbol{w}(n)]$
$y(n) = \boldsymbol{w}^H(n)\boldsymbol{x}(n) = \boldsymbol{x}^H(n)\boldsymbol{w}(n)$, $e(n) = d(n) - y(n)$
$\boldsymbol{w}(n+1) = \boldsymbol{w}(n) + 2\mu\boldsymbol{x}(n)\mathrm{e}^*(n)$, H = conjugate transpose(or Hermitian).

8.3.1

$$H(z)=\frac{X(z)}{V(z)}=\frac{1}{\left\{\left[1-0.85e^{j(\pi/4)}z^{-1}\right]\left[1-0.85e^{-j(\pi/4)}z^{-1}\right]\right\}} \text{ or}$$

$$V(z)=X(z)\left[1-1.7\left(\frac{\sqrt{2}}{2}\right)z^{-1}+(0.85)^2z^{-2}\right]$$

Hence, the inverse z-transform gives $v(n) = x(n) - 1.202x(n-1) + 0.7225x(n-2)$.

8.4.1

$$\begin{aligned}\boldsymbol{Q}^T\xi(n+1) &= \boldsymbol{Q}^T[\xi(n)-2\mu\boldsymbol{x}(n)\boldsymbol{x}^T(n)\xi(n)]+2\mu e^o(n)\boldsymbol{Q}^T\boldsymbol{x}(n)\\ &= \boldsymbol{Q}^T\xi(n)-2\mu\boldsymbol{Q}^T\boldsymbol{x}(n)\boldsymbol{x}^T(n)\boldsymbol{Q}^{-T}\boldsymbol{Q}^T\xi(n)+2\mu e^o(n)\boldsymbol{x}'(n)\\ &= \xi'(n)-2\mu\boldsymbol{Q}^T\boldsymbol{x}(n)(\boldsymbol{Q}^T\boldsymbol{x}(n))^T\xi'(n)+2\mu e^o(n)\boldsymbol{x}'(n)\\ &= [\boldsymbol{I}-2\mu\boldsymbol{x}'(n)\boldsymbol{x}'^T(n)]\xi'(n)+2\mu e^o(n)\boldsymbol{x}'(n);\\ \boldsymbol{Q}^{-T} &= (\boldsymbol{Q}^{-1})^T=(\boldsymbol{Q}^T)^T=\boldsymbol{Q}\end{aligned}$$

8.4.2 Hint: $\xi'^T(n+1)=\xi'^T(n)-2\mu\xi'^T(n)\boldsymbol{x}'(n)\boldsymbol{x}'^T(n)+2\mu e^o(n)\boldsymbol{x}'^T(n)$.

8.4.3 Because $\boldsymbol{x}'(n)$ and $\xi'(n)$ are independent.

$$\begin{aligned}E\{\boldsymbol{x}'(n)\boldsymbol{x}'^T(n)\xi'(n)\xi'^T(n)\boldsymbol{x}'(n)\boldsymbol{x}'^T(n)\} &= E\{\boldsymbol{x}'(n)\boldsymbol{x}'^T(n)E\{\xi'(n)\xi'^T(n)\}\boldsymbol{x}'(n)\boldsymbol{x}'^T(n)\}\\ &= E\{\boldsymbol{x}'(n)\boldsymbol{x}'^T(n)\boldsymbol{K}'(n)\boldsymbol{x}'(n)\boldsymbol{x}'^T(n)\}\end{aligned} \quad (1)$$

$$\boldsymbol{x}'^T(n)\boldsymbol{K}'(n)\boldsymbol{x}'(n)=\sum_{i=0}^{M-1}\sum_{j=0}^{M-1}x_i'(n)x_j'(n)k_{ij}'(n) \quad (2)$$

$$\boldsymbol{C}(n)=M\times M \text{ matrix}=\boldsymbol{x}'(n)\boldsymbol{x}'^T(n)\boldsymbol{K}'(n)\boldsymbol{x}'(n)\boldsymbol{x}'^T(n) \quad (3)$$

$$c_{lm}(n)=x_l'(n)x_m'(m)\sum_{i=0}^{M-1}\sum_{j=0}^{M-1}x_i'(n)x_j'(n)k_{ij}'(n) \quad (4)$$

$$E\{c_{lm}(n)\}=\sum_{i=0}^{M-1}\sum_{j=0}^{M-1}E\{x_l'(n)x_m'(n)x_i'(n)x_j'(n)k_{ij}'(n)\} \quad (5)$$

For Gaussian random variables, we have

$$E\{x_1x_2x_3x_4\} = E\{x_1x_2\}E\{x_3x_4\} + E\{x_1x_3\}E\{x_2x_4\} + E\{x_1x_4\}E\{x_2x_3\} \tag{6}$$

$$E\{x_i'(n)x_j'(n)\} = \lambda_i\delta(i-j) \tag{7}$$

The above equation is due to the relation

$$E\{\boldsymbol{Q}^T\boldsymbol{x}(n)\boldsymbol{x}(n)^T\boldsymbol{Q}\} = \boldsymbol{Q}^T\boldsymbol{R}\boldsymbol{Q} = \Lambda, \boldsymbol{R}\boldsymbol{q}_i = \lambda_i\boldsymbol{q}_i$$

$$\begin{aligned} E\{x_l'(n)x_m'(n)x_i'(n)x_j'(n)\} &= \lambda_l\lambda_i\delta(l-m)\delta(i-j) + \lambda_l\lambda_m\delta(l-i)\delta(m-j) \\ &\quad + \lambda_l\lambda_m\delta(l-j)\delta(m-i) \end{aligned} \tag{8}$$

Substitute (8) in (5) to find

$$\begin{aligned} E\{c_{lm}(n)\} &= \sum_{i=0}^{M-1}\sum_{j=0}^{M-1}\lambda_l\lambda_i\delta(l-m)\delta(i-j)k_{ij}'(n) \\ &\quad + \sum_{i=0}^{M-1}\sum_{j=0}^{M-1}\lambda_l\lambda_m\delta(l-i)\delta(m-j)k_{ij}'(n) \\ &\quad + \sum_{i=0}^{M-1}\sum_{j=0}^{M-1}\lambda_l\lambda_m\delta(l-j)\delta(m-i)k_{ij}'(n) \\ &= \lambda_l\delta(l-m)\sum_{i=0}^{M-1}\lambda_i k_{ii}'(n) + \lambda_l\lambda_m k_{lm}'(n) + \lambda_l\lambda_m k_{ml}'(n) \end{aligned}$$

for $l = 0, 1, \ldots, M-1$ and $m = 0, 1, \ldots, M-1$. But $k_{lm}'(n) = k_{ml}'(n)$ and $\sum_{i=0}^{M-1}\lambda_i k_{ii}'(n) = \mathrm{tr}\{\Lambda\mathbf{K}'(n)\}$ and $\lambda_l\lambda_m k_{lm}'(n) + \lambda_l\lambda_m k_{ml}'(n) = 2\lambda_l\lambda_m k_{lm}'(n)$ Based on these results, (8.40) is apparent.

8.4.4

$$\begin{bmatrix} k_{00}'(n+1) & k_{01}'(n+1) & \cdots & \\ k_{10}'(n+1) & k_{11}'(n+1) & \cdots & \\ \vdots & \vdots & & \\ & & & k_{M-1,M-1}'(n+1) \end{bmatrix} = \begin{bmatrix} f_0 + 4\mu^2\lambda_0^2 & 4\mu^2\lambda_1\lambda_0 & \cdots & 4\mu^2\lambda_{M-1}\lambda_0 \\ \lambda_0\lambda_1 & f_1 + 4\mu^2\lambda_1^2 & \cdots & 4\mu^2\lambda_{M-1}\lambda_1 \\ \vdots & \vdots & & \vdots \\ \lambda_0\lambda_{M-1} & \lambda_1\lambda_{M-1} & \cdots & f_{M-1} + 4\mu^2\lambda_{M-1}^2 \end{bmatrix}$$

$$\bullet \begin{bmatrix} k_{00}'(n+1) & k_{01}'(n+1) & \cdots & \\ k_{10}'(n+1) & k_{11}'(n+1) & \cdots & \\ \vdots & & & k_{M-1,M-1}'(n+1) \end{bmatrix}$$

$$+ 4\mu^2 J_{\min}\begin{bmatrix} \lambda_0 & & & \\ & \lambda_1 & 0 & \\ & 0 & \ddots & \\ & & & \lambda_{M-1} \end{bmatrix}$$

where the iith component of both sides is $k'_{ii}(n+1) = f_i k'_{ii}(n) + 4\mu^2\lambda_i[\lambda_0 k'_{00}(n) + 4\mu^2\lambda_1 k'_{11}(n) + \cdots + 4\mu^2\lambda_{M-1}k'_{M-1}(n)] + 4\mu^2 J_{\min}\lambda_i$ which is identical to (8.47).

8.4.5

$$a[\lambda_0\,\lambda_1]\begin{bmatrix}\dfrac{1}{1-f_0} & 0\\ 0 & \dfrac{1}{1-f_1}\end{bmatrix}\begin{bmatrix}\lambda_0\\ \lambda_1\end{bmatrix}$$

$$+\frac{a^2\begin{bmatrix}\lambda_0 & \lambda_1\end{bmatrix}\begin{bmatrix}\dfrac{1}{1-f_0} & 0\\ 0 & \dfrac{1}{1-f_1}\end{bmatrix}\begin{bmatrix}\lambda_0\lambda_0 & \lambda_0\lambda_1\\ \lambda_0\lambda_1 & \lambda_1\lambda_1\end{bmatrix}\begin{bmatrix}\dfrac{1}{1-f_0} & 0\\ 0 & \dfrac{1}{1-f_1}\end{bmatrix}\begin{bmatrix}\lambda_0\\ \lambda_1\end{bmatrix}}{1-a\begin{bmatrix}\lambda_0 & \lambda_1\end{bmatrix}\begin{bmatrix}\dfrac{1}{1-f_0} & 0\\ 0 & \dfrac{1}{1-f_1}\end{bmatrix}\begin{bmatrix}\lambda_0\\ \lambda_1\end{bmatrix}}$$

$$=\frac{a\left[\dfrac{\lambda_0^2}{1-f_0}+\dfrac{\lambda_1^2}{1-f_1}\right]\left[1-a\left(\dfrac{\lambda_0^2}{1-f_0}+\dfrac{\lambda_1^2}{1-f_1}\right)\right]+()}{1-a\left(\dfrac{\lambda_0^2}{1-f_0}+\dfrac{\lambda_1^2}{1-f_1}\right)}$$

$$=\frac{a\left[(\lambda_0^2/(1-f_0))+(\lambda_1^2/(1-f_1))\right]}{1-4\mu^2\left[\lambda_0^2/(1-(1-4\mu\lambda_0+8\mu^2\lambda_0^2))\right]-4\mu^2\left[\lambda_1^2/(1-(1-4\mu\lambda_1+8\mu^2\lambda_1^2))\right]}$$

$$=\frac{\sum_{i=0}^{2-1}(\mu\lambda_i/(1-2\mu\lambda_i))}{1-\sum_{i=0}^{2-1}(\mu\lambda_i/(1-2\mu\lambda_i))}$$

which confirms (8.60) for $M = 2$.

8.4.6 *Hint*: Take derivatives with respect to μ and L, and note the positive values $(\partial L/\partial\mu)$ and $(\partial\mathcal{M}/\partial L)$.

8.4.7

$$\lambda_{av}=\frac{1}{M}\sum_{i=0}^{M-1}\lambda_i=\frac{1}{M}\mathrm{tr}\{\boldsymbol{R}_x\},\ \tau_{mse,av}\cong\frac{1}{2\mu\lambda_{av}}\quad\text{[see (7.27)]}$$

and, therefore, $\mathcal{M} \cong \mu M\lambda_{av} \cong M/(2\tau_{mse,av})$. Observations are as follows: (1) $\mathcal{M}$ increases linearly with filter length for a fixed λ_{av}; (2) the **setting time** of the LMS algorithm is proportional to $\tau_{mse,av}$, and hence, $\mathcal{M}$ is inversely proportional to the settling time; and (3) $\mathcal{M}$ is proportional to μ and inversely proportional to $\tau_{mse,av}$, but μ and $\tau_{mse,av}$ are inversely proportional to each other. Careful consideration must be given when we are considering the values for $\mathcal{M}$, μ, and $\tau_{mse,av}$.

8.4.8 The averaged learning curve of the LMS algorithm is close to the steepest descent method. Hence, we write (7.30) in the form:

$$J(n) \cong J_{\min} + \sum_{k=0}^{M-1} \lambda_k \left(1 - 2\mu\lambda_k\right)^{2n} \xi_k'^2(0) \tag{1}$$

If we set $\xi(0) = \boldsymbol{w}(n) - \boldsymbol{w}^o = \boldsymbol{0} - \boldsymbol{w}^o$, then

$$\xi'(n) = \boldsymbol{Q}^T \xi(0) = -\boldsymbol{Q}^T \boldsymbol{w}^o = -\boldsymbol{w}^o \tag{2}$$

Hence, (1) becomes

$$J(n) \cong J_{\min} + \sum_{k=0}^{M-1} \lambda_k \left(1 - \mu\lambda_k\right)^{2n} w_k^{o\prime 2} \tag{3}$$

Equation 3 indicates that if $w_k^{o\prime}$'s corresponding to the smaller eigenvalues of $\boldsymbol{R}_x$ are all close to zero, then the transient behavior of the LMS algorithm is determined by the larger eigenvalues whose associated time constants are small, and thus, a fast convergence takes place. If, however, $w_k^{o\prime}$'s corresponding to the smaller eigenvalues of the correlation matrix are significantly large, then we will observe that the slower modes will dominate.

8.4.9 From (7.15), we find $\boldsymbol{w}(n+1) = \boldsymbol{w}(n) - \mu\nabla e^2(n)$ (1), where n is the iteration number, and $\nabla = [(\partial/\partial w_0)\ (\partial/\partial w_1) \cdots (\partial/\partial w_{M-1})]^T$, $\boldsymbol{w}(n) = [w_0(n)\ w_1(n)\ \cdots\ w_{M-1}(n)]^T$. But

$$\begin{aligned} \partial e^2(n)/\partial w_i &= 2e(n)\partial e(n)/\partial w_i = 2e(n)\partial[d(n) - y(n)]/\partial w_i \\ &= -2e(n)\partial y(n)/\partial w_i = -2e(n)\partial\left[\sum_{i=0}^{M-1} w_i(n)x(n-i)\right]/\partial w_i(n) \\ &= -2e(n)x(n-i) \end{aligned}$$

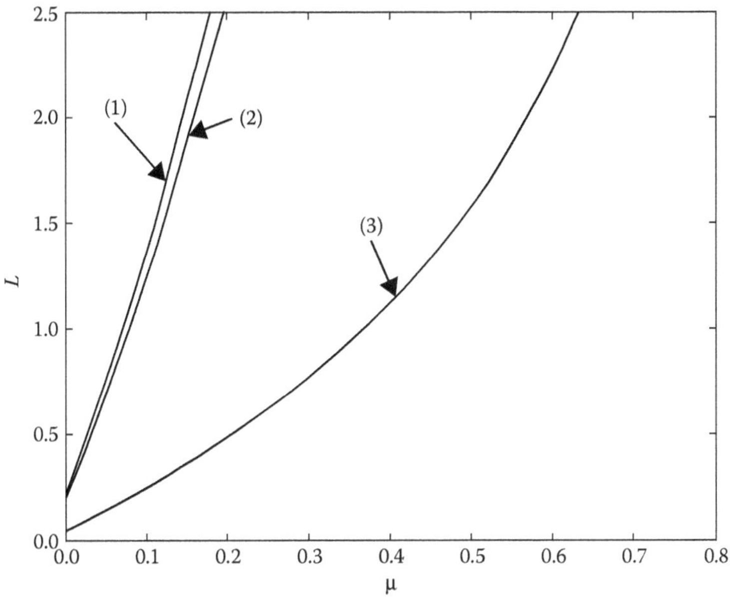

FIGURE P8.4.10

and hence, $\nabla e^2(n) = -2e(n)\boldsymbol{x}(n)$, where $\boldsymbol{x}(n) = [x(n)\ \ x(n-1)\ \ x(n-2) \cdots x(n-M+1)]^T$. Therefore, (1) becomes $\boldsymbol{w}(n+1) = \boldsymbol{w}(n) + 2\mu e(n)\boldsymbol{x}(n)$.

8.4.10 See Figure P8.4.10 for the three cases. We observe that for case (3), for example, the step-size parameter must be less than 0.35.

9 Variants of Least Mean-Square Algorithm

9.1 THE NORMALIZED LEAST MEAN-SQUARE ALGORITHM

Consider the conventional least mean-square (LMS) algorithm with the fixed step-size parameter μ replaced with a time-varying variable $\mu(n)$ as follows (we substitute 2μ with μ for simplicity):

$$\boldsymbol{w}(n+1) = \boldsymbol{w}(n) + \mu(n)e(n)\boldsymbol{x}(n) \tag{9.1}$$

Next, define ***a posteriori*** error, $e_{\mathrm{ps}}(n)$, as

$$e_{\mathrm{ps}}(n) = d(n) - \boldsymbol{w}^T(n+1)\boldsymbol{x}(n) \tag{9.2}$$

Substituting (9.1) in (9.2), and taking into consideration the error equation $e(n) = d(n) - \boldsymbol{w}^T(n)\boldsymbol{x}(n)$, we obtain

$$e_{\mathrm{ps}}(n) = \left[1 - \mu(n)\boldsymbol{x}^T(n)\boldsymbol{x}(n)\right]e(n) \qquad \boldsymbol{x}^T(n)\boldsymbol{x}(n) = \left\|\boldsymbol{x}(n)\right\|^2 = \sum_{i=0}^{M-1}\left|x(n-i)\right|^2 \tag{9.3}$$

Minimizing $e_{\mathrm{ps}}(n)$ with respect to $\mu(n)$ results in (see Problem 9.1.1)

$$\mu(n) = \frac{1}{\left\|\boldsymbol{x}(n)\right\|^2} \tag{9.4}$$

Substituting (9.4) in (9.1), we find

$$\boldsymbol{w}(n+1) = \boldsymbol{w}(n) + \frac{1}{\left\|\boldsymbol{x}(n)\right\|^2}e(n)\boldsymbol{x}(n) \tag{9.5}$$

However, the most common normalized LMS (NLMS) algorithm is

$$\boldsymbol{w}(n+1) = \boldsymbol{w}(n) + \frac{\mu}{\left\|\boldsymbol{x}(n)\right\|^2}e(n)\boldsymbol{x}(n) \tag{9.6}$$

Note that the NLMS is actually a variable step-size algorithm in which this step size is inversely proportional to the total instantaneous energy of the input signal $x(n)$, estimated over values within the tapped delay line length.

A modified version of the NLMS algorithm (sometimes called ε-NLMS) is given by

$$\begin{aligned} w(n+1) &= w(n) + \frac{\mu}{\varepsilon + \|x(n)\|^2} e(n)x(n) \\ &= w(n) + \frac{\mu}{\varepsilon + \|x(n)\|^2}\left[d(n) - w^T(n)x(n)\right]x(n) \end{aligned} \tag{9.7}$$

In the above equation, ε is a small positive number that was introduced to avoid division by zero at a time when the signal becomes zero or extremely small. Table 9.1 gives the LMS and NLMS formulas.

TABLE 9.1
Some Variants of the LMS Formulas

Algorithm	Recursion
LMS with constant step size	$w(n+1) = w(n) + \mu e(n)x(n)$; $w(n+1) = w(n) + \mu[d(n) - w^T(n)x(n)]x(n)$
LMS with time-varying step size	$w(n+1) = w(n) + \mu(n)[d(n) - w^T(n)x(n)]x(n)$
ε-NLMS	$w(n+1) = w(n) + \frac{\mu}{\varepsilon + \|x(n)\|^2}[d(n) - w^T(n)x(n)]x(n)$
ε-NLMS with power normalization	$w(n+1) = w(n) + \frac{\mu}{\varepsilon + p(n+1)}[d(n) - w^T(n)x(n)]x(n)$ $p(n+1) = ap(n) + (1-a)\lvert x(n+1)\rvert^2 \quad p(0) = 0$

TABLE 9.2
The Normalized Real and Complex LMS Algorithms

Real-Valued Functions		Complex-Valued Functions
Input		
Initialization vector	$w(n) = \mathbf{0}$	
Input vector	$x(n)$	
Desired output	$d(n)$	
Step-size parameter	μ	
Constant	ε	
Filter length	M	
Output		
Filter output	$y(n)$	
Coefficient vector	$w(n+1)$	
Procedure		
1. $y(n) = w^T(n)x(n) = w(n)x^T(n)$		1. $y(n) = w^H(n)\,x(n)$
2. $e(n) = d(n) - y(n)$		2. $e(n) = d(n) - w^H(n)\,x(n)$
3. $w(n+1) = w(n) + \frac{\mu}{\varepsilon + x^T(n)x(n)} e(n)x(n)$		3. $w(n+1) = w(n) + \frac{\mu}{\varepsilon + x^H(n)x(n)} e^*(n)x(n)$ (H = Hermitian, conjugate transpose)

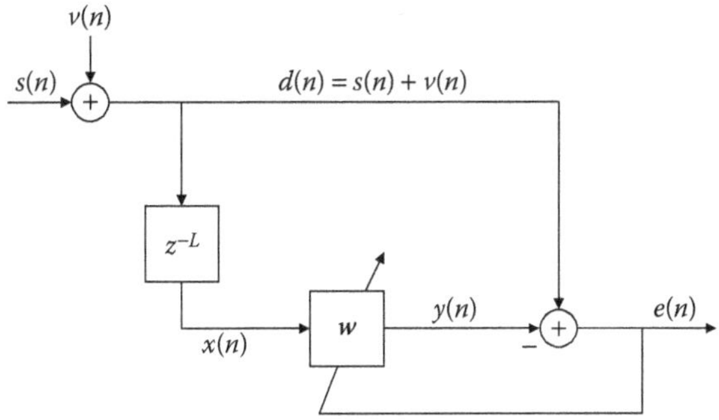

FIGURE 9.1

Example 9.1.1 (Noise Cancellation)

Compare the LMS and NLMS formulas for a line enhancer setup shown in Figure 9.1.

Solution: The two Book m-functions are given below.

Book m-Function for Unnormalized LMS

```
function[w,y,x,J]=lms_ex9_1_1_lms(s,mu,M)
        %s=sinusoidal signal;mu=step-size factor;
        %M=length of adaptive filter;
N=length(s);y=zeros(1,N);w=zeros(1,M);
e=zeros(1,N);
x=[0 s+rand(1,N)-0.5];
for n=M:N
        x1=x(n:-1:n-M+1);
        y(n)=w*x1';
        d(n+1)=x(n);
        e(n)=d(n+1)-y(n);
        w=w+2*mu*e(n)*x1;
end;
J=e.^2;
```

Book m-Function for NLMS

```
function[w,y,x,J,Jav]=lms_ex9_1_1_norm_lms(s,mu,M,c)
        %s=sinusoidal signal;mu=step-size factor;
        %M=length of adaptive filter;c=very small number;
N=length(s);y=zeros(1,N);w=zeros(1,M);
e=zeros(1,N);
d=[0s+rand(1,N)-0.5];
x=[d(1,2:N) 0];
for n=M:N
        x1=x(n:-1:n-M+1);
        y(n)=w*x1';
        e(n)=d(n)-y(n);
        w=w+((2*mu*e(n)*x1)/(c+x1*x1'));
end;
J=e.^2;
```

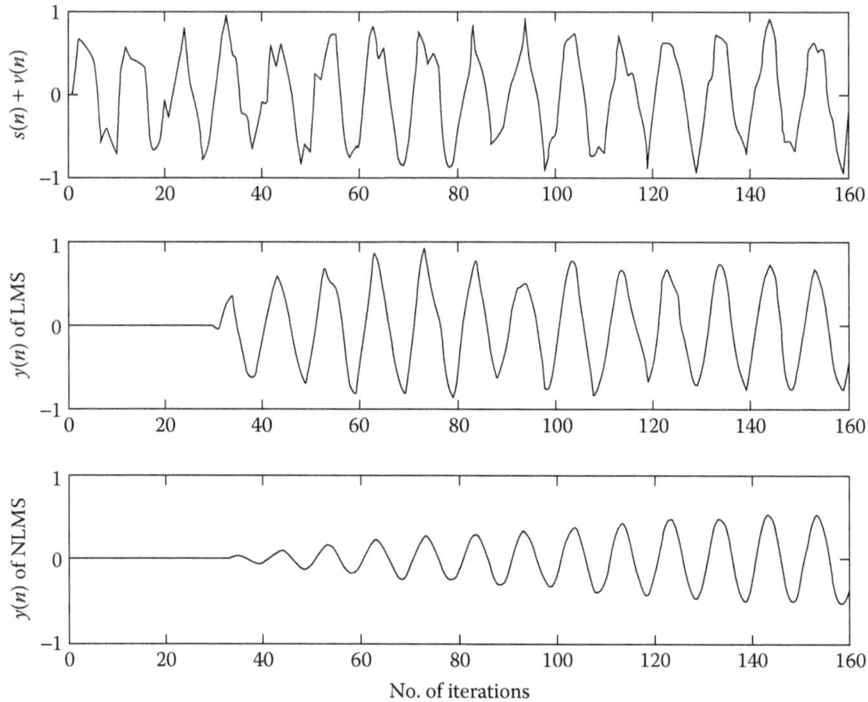

FIGURE 9.2

Figure 9.2 shows the output of the adaptive filter for the LMS and NLMS filters. The constants that were used were as follows: $\text{mu} = 0.01$, $M = 30$, $c = 0.0001$, $s = \sin(0.2\pi n)$. We observe that the NLMS filter output is closer to the sine signal than the LMS filter. The signal $v(n)$ identifies the white noise.

■

Example 9.1.2

If we desire to plot the learning function J, it is recommended to use an average process for a better presentation. The following Book m-function cancels the noise embedded in the signal.

Book m-Function for Noise Canceling in a Sine Function

```
function[w,jav,js]=lms_denoising_Jav1(N,mu,M,av,a)
es=zeros(1,N);
for m=1:av%av=integer equal to the number of desired
     %averaging;a=multiplier of the random noise,
     %a*randn(1,N);
     w=zeros(1,M);
```

```
        dn=sin(0.1*pi*[1:N])+a*randn(1,N);
        x=[0 0 dn(1,1:N-2)];%delay 2 units;
        for n=M:N
               x1=x(n:-1:n-M+1);
               y(n)=w*x1';
               e(n)=dn(n)-y(n);
               w=w+mu*e(n)*x1;
        end;
  es=es+e;
  y=zeros(1,N);
  end;
  jav=(es/av).^2;
  for n=1:N-5
        js(n)=(jav(n)+jav(n+1)+jav(n+2)+jav(n+3))/4;
  end;
```

Figure 9.3 was produced using the following constants: mu = 0.005, $M = 40$, av = 1 and 300, $a = 0.5$, and $N = 1500$. The signal was a sine function with a white Gaussian additive noise. The learning curves were smoothed for a better understanding of the variations of the two learning curves. The averaging produces a lower mean-square error as it was expected.

■

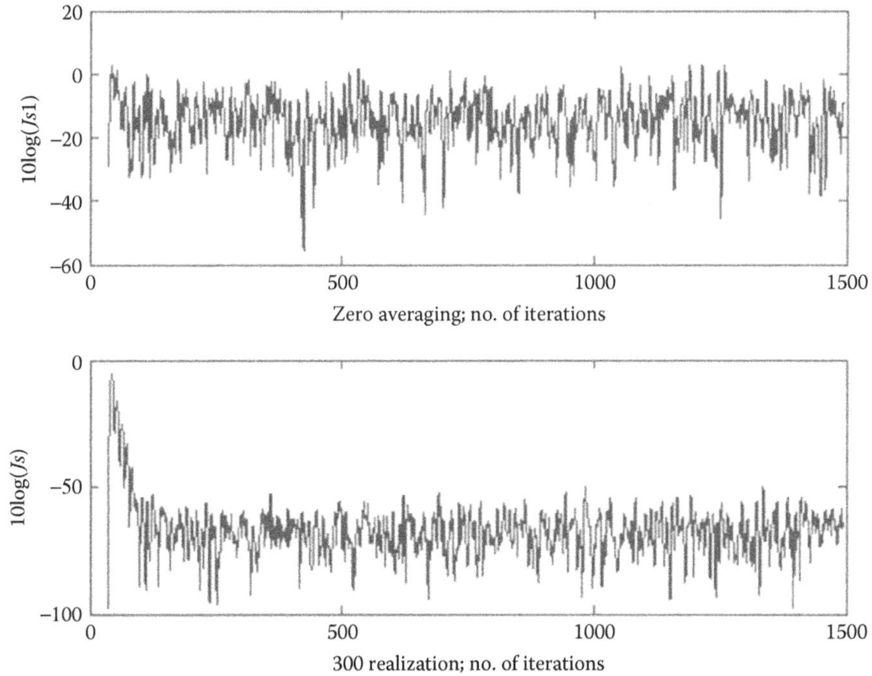

FIGURE 9.3

9.2 POWER NORMALIZED LMS

Observe that the NLMS algorithm can be written in the form (we have substituted $2\mu = \mu$ for convenience):

$$\mathbf{w}(n+1) = \mathbf{w}(n) + \frac{\mu/K}{\varepsilon/K + \|\mathbf{x}(n)\|^2/K} e(n)\mathbf{x}(n) \tag{9.8}$$

The form of the above equation suggests a method for approximating the norm square of the data. The method replaces $\|\mathbf{x}(n)\|^2/K$ by a variable $p(i)$ which is updated as follows:

$$p(n+1) = bp(n) + (1-b)|\mathbf{x}(n)|^2 \qquad p(0) = 0 \tag{9.9}$$

with a positive scalar b and values in the range $0 \le b < 1$. Hence, (9.8) takes the form:

$$\begin{aligned} \mathbf{w}(n+1) &= \mathbf{w}(n) + \frac{\mu}{\varepsilon + p(n+1)} e(n)\mathbf{x}(n) \\ &= \mathbf{w}(n) + \frac{\mu}{\varepsilon + p(n+1)}\left[d(n) - \mathbf{w}(n)\mathbf{x}^T(n)\right]\mathbf{x}(n) \end{aligned} \tag{9.10}$$

From the above equation, it is understood that the step size of power NLMS (PNLMS) in (9.10) is approximately M times smaller than the step size of the NLMS.

Example 9.2.1 (Channel Equalization)

It is desired to estimate a communication channel, shown in Figure 9.4, so that the noisy output signal from the channel space system is denoised by using a PNLMS.

Solution: At each instant time n, the measured output of the channel, $d(n)$, is compared with the output of the adaptive filter, $y(n)$, and an error signal $e(n) = d(n) - y(n)$, is generated. The error is then used to adjust the filter coefficients according to (9.10). In this case, the following constants and functions were used:

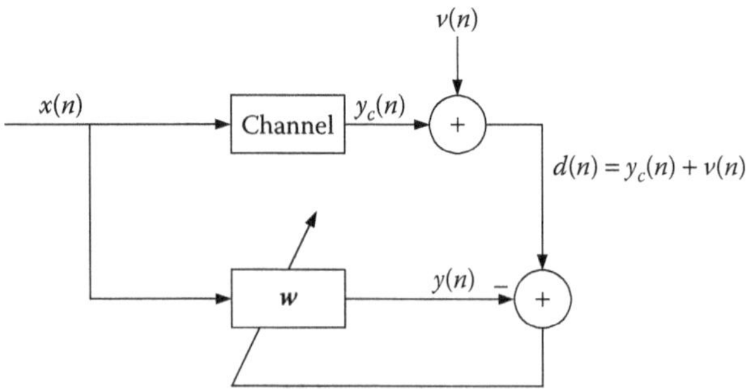

FIGURE 9.4

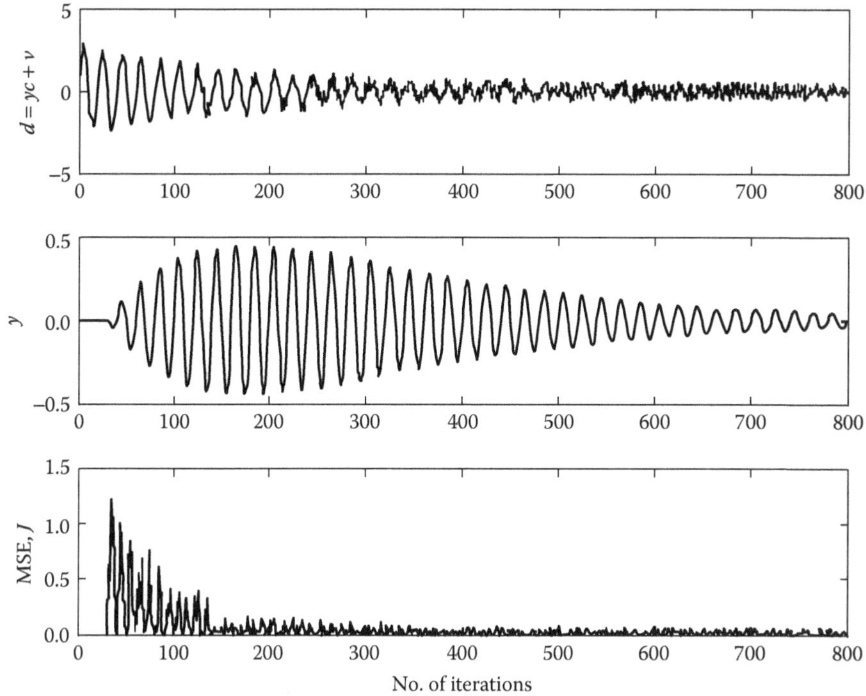

FIGURE 9.5

$$x(n) = 2\,0.995^n \sin(0.1\pi n), \mu = 0.004, M = 30,$$

$$b = 0.5, c = 0.0001, \text{Channel} = \text{FIR}\ [0.9 \quad 0.5 \quad -0.2]$$

The output of the adaptive filter will assume values close to the input to the communication channel. Therefore, from an input/output point of view, the adaptive filter behaves as the channel. The top of Figure 9.5 shows the output of the channel with additive white noise having zero mean value, $v(n) = \text{rand} - 0.5$. The middle figure presents the output of the adaptive filter and the third presents the MSE versus iteration. The Book m-function that was used to create the figure as follows:

Book m-Function for PNLMS

```
function[w,y,yc,x,J]=lms_power_norm_lms(mu,M,b,c)
      %mu=step-size factor;
      %M=length of adaptive filter;c=very small number;
      %yc=conv(x,[filter coefficients])=channel
      %output;d=yc+v;v=noise;0<b<1;
n=1:1500;p(M)=0;
x=2*0.995.^n.*sin(0.1*pi*n);
N=length(x);y=zeros(1,N);w=zeros(1,M);
yc=conv(x,[0.9 0.5 -0.2]);%conv(x,y)=MATLAB function;
for n=M:N
      x1=x(n:-1:n-M+1);
      p(n+1)=b*p(n)+(1-b)*x1*x1';
      y(n)=w*x1';
```

```
        e(n)=(yc(n)+rand-0.5)-y(n);
        w=w+(2*mu*e(n)*x1)/(c+p(n+1));
end;
J=(e/max(e)).^2;
```

We also observe that if b in (9.9) is zero, then the Book m-function above becomes the NLMS m-function.

■

Example 9.2.2 (Channel Equalization)

Figure 9.6a shows a baseband data transmission system equipped with an adaptive channel equalizer and a training system. The signal $\{s(n)\}$ transmitted through the communication channel is amplitude- or phase-modulated pulses. The communication channel distorts the signal, the most important one is the pulse spreading, and results in overlapping of pulses, thus creating the **intersymbol interference** phenomenon. The noise $v(n)$ further deteriorates the fidelity of the signal. It is ideally required that the output of the equalizer is the signal $s(n)$. Therefore, an initialization period is used during which the transmitter sends a sequence of training symbols that are known to the receiver (**training mode**). This approach is satisfactory if the channel does not change the characteristics rapidly in time. However, for slow changes the output from the channel can be treated as the desired signal for further adaptation of the equalizer so that its variations can be followed (**decision-directed mode**).

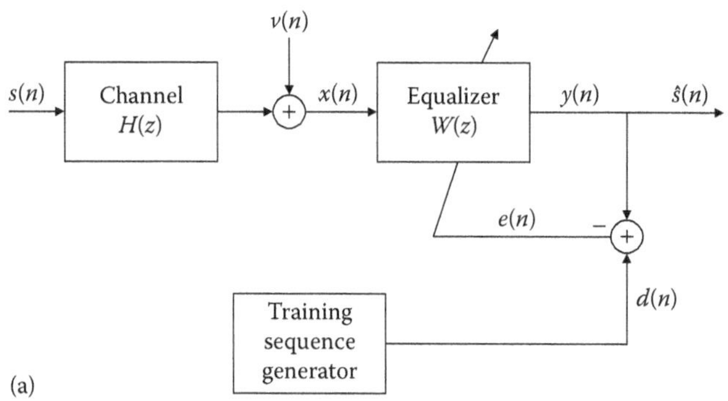

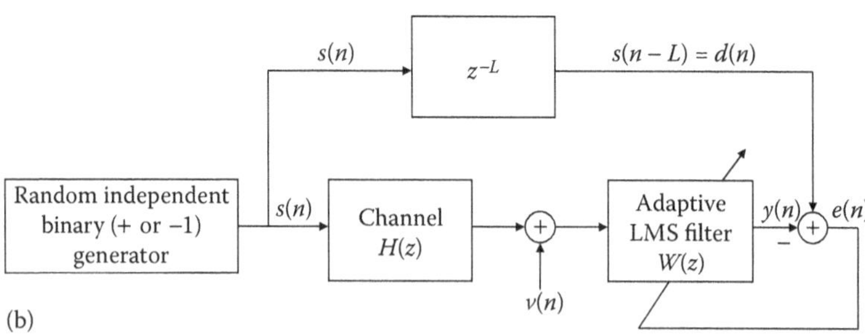

FIGURE 9.6

If the equalization filter is the inverse of the channel filter, $W(z)=1/H(z)$, the output will be that of the input to the channel, assuming, of course, that noise is small. To avoid singularities from the zero of the channel transfer function inside the unit circle, we select an equalizer such that $W(z)H(z)\cong z^{-L}$. This indicates that the output of the filter $W(z)$ is that of the input to the channel shifted by L units of time. Sometimes, more general filters of the form $Y(z)/H(z)$ are used, where $Y(z)\neq z^{-L}$. These systems are known as the partial-response signaling systems. In these cases, $Y(z)$ is selected such that the amplitude spectra are about equal over the range of frequencies of interest. The result of this choice is that $W(z)$ has a magnitude response of about one, thereby minimizing the noise enhancement.

Figure 9.6b shows a channel equalization problem at the training stage. The channel noise $v(n)$ is assumed to be white Gaussian with variance σ_v^2. The equalizer is an M-tap finite impulse response (FIR) filter and the desired output is assumed to be delayed replica of the signal $s(n), s(n-L)$. The signal $s(n)$ is white, has variance $\sigma_s^2=1$, has zero mean value, is made up of random +1 and −1 and is uncorrelated with $v(n)$. The channel transfer functions were assumed to be of FIR form. In this example, we used the following transfer functions:

$$H(z)=H_1(z)=1$$
$$H(z)=H_2(z)=0.9+0.4z^{-1}-0.2z^{-2} \tag{9.11}$$

The following Book m-function produced Figure 9.7.

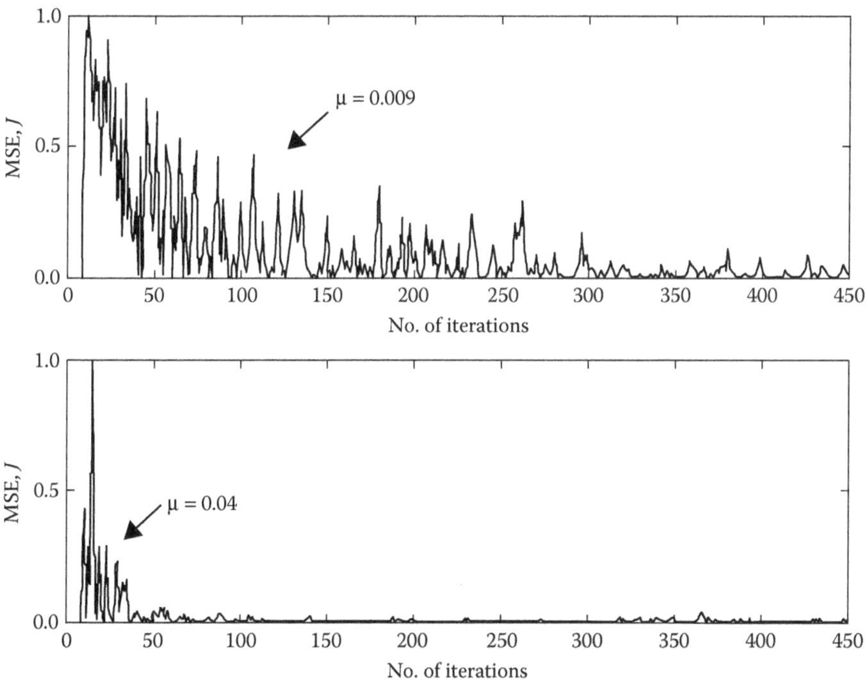

FIGURE 9.7

Book m-Function for Channel Equalization

```
function[Jav,wav,dn,e,x]=lms_equalizer_ex9_2_2(av,M,L,h,N,mu,c)
     %function[Jav,wav,dn,e,x]= lms_equalizer_ex9_2_2
     %(av,M,L,h,N,mu,c)
     %this function solves the example depicted
     %in Fig 9.7;av=number of times to average e(error or
     %learning curve); w(filter coefficient);N=length
     %of signal s;L=shift of the signal s to become dn;
     %h=assumed impulse response of the channel system;
     %mu=step-size factor; M=number of adaptive filter coefficients;
     %c=constant multiplier;
wn=zeros(1,M);
J=zeros(1,N);
for i=1:av
     for n=1:N
          v(n)=c*randn;
          s(n)=(rand-0.5);
          if s(n)<=0
               s(n)=-1;
          else
               s(n)=1;
          end;
     end;
dn=[zeros(1,L)s(:,1:N-L)];
ych=filter(h,1,s);
x=ych(1,1:N)+v;
[w,y,e,J,w1,Js]=lms1(x,dn,mu,M);
wn=wn+w;
J=J+e.^2;
end;
Jav=J/av;
wav=wn/av;
```

The constants used to create Figure 9.7 were as follows: $\text{av}=300$, $M=10$, $L=1$, $h=[0.9 \quad 0.4 \quad -0.2]$, $N=2000$, $c=0.05$.

■

9.3 SELF-CORRECTING LMS FILTER

We can arrange the standard LMS filter in a series form as shown in Figure 9.8. This book proposed configuration LMS filtering permits us to process the signal using filters with fewer coefficients, thus saving in computation time. Figure 9.9a shows the desired signal with noise. Figure 9.9b shows the output of the first stage. Figure 9.9c shows the output of the third stage of the self-correcting LMS filter (SCLMSF), $\{x1(n)\}$. Each stage LMS filter has four coefficients. We also used the following signals and constants: $s=\sin(0.1n\pi), v=[\text{rand}(1,2000)-0.5]$, $x(n)=dn(n+2)$(delay by 2 units), $I=3$, $mu=0.005$, $M=4$. Problem 9.3.1 solves for the SCNLMSF.

The output of the ith stage is related to the previous one as follows:

$$x_{i+1}(n)=x_i(n)*w_{i+1}(n)$$

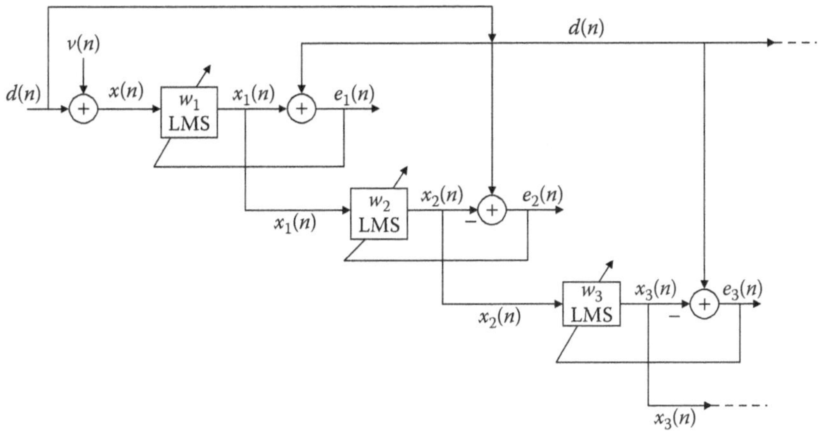

FIGURE 9.8

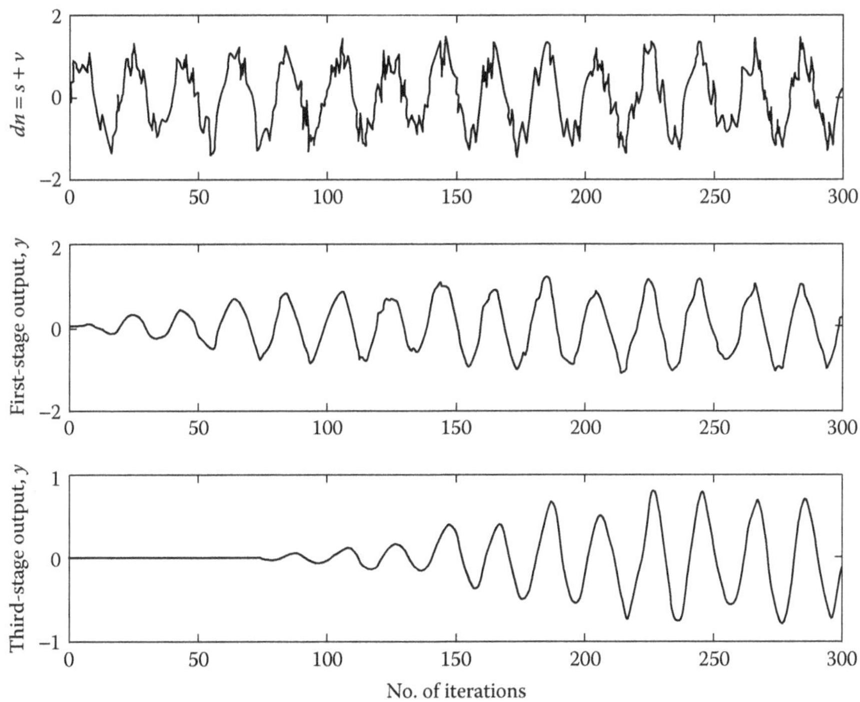

FIGURE 9.9

Book m-Function for SCLMS Algorithm

```
function[w,y,e,J]=lms_self_correcting_lms(x,dn,mu,M,I)
      %function[w,y,e,J]=lms_self_correcting_lms(x,dn,mu,M,I);
[w(1,:),y(1,:),e(1,:)]=lms1(x,dn,mu,M);
for i=2:I%I=number of iterations, I<8-10 is sufficient;
```

```
        [w(i,:),y(i,:),e(i,:)]=lms1(y(i-1,:),dn,mu,M);
end;
J=e.^2;
```

9.4 THE SIGN-ERROR LMS ALGORITHM

The sign-error algorithm is defined by

$$w(n+1) = w(n) + \mu \text{sign}[e(n)]x(n) \tag{9.12}$$

where:

$$\text{sign}(a) = \begin{cases} 1 & a > 0 \\ 0 & a = 0 \\ -1 & a < 0 \end{cases} \tag{9.13}$$

is the **signum** function. By introducing the signum function and setting μ to a value of power of 2, the hardware implementation is highly simplified (shift and add/subtract operations only).

Book m-Function for Sign-Error Algorithm

```
function[w,y,e,J,w1]=lms_sign_error(x,dn,mu,M)
        %function[w,y,e,J,w1]=lms_sign_error(x,dn,mu,M);
        %all quantities are real valued;
        %x=input data to the filter;dn=desired signal;
        %M=order of the filter;
        %mu=step size parameter;
        %x and dn must be of the same length;
N=length(x);
y=zeros(1,N);
w=zeros(1,M);%initialized filter coefficient vector;
for n=M:N
        x1=x(n:-1:n-M+1);%for each n the vector x1 is produced
                %of length M with elements from x in reverse order;
        y(n)=w*x1';
        e(n)=dn(n)-y(n);
        w=w+mu*sign(e(n))*x1;
        w1(n-M+1,:)=w(1,:);
end;
J=e.^2;
        %the columns of w1 depict the history of the filter
        %coefficients;
```

9.5 THE NLMS SIGN-ERROR ALGORITHM

The NLMS sign-error algorithm is

$$w(n+1) = w(n) + \mu \frac{\text{sign}[e(n)]x(n)}{\varepsilon + \|x(n)\|^2} \qquad \|x(n)\|^2 = x(n)^T x(n) \tag{9.14}$$

The Book m-function for the NLMS sign-error algorithm is given below. The m-file `sign()` is a MATLAB function (see MATLAB functions).

Book m-Function for NLMS Sign-Error Algorithm

```
function[w,y,e,J,w1]=lms_normalized_sign_error(x,dn,mu,M,ep)
      %function[w,y,e,J,w1]=lms_normalized_sign_error
      %(x,dn,mu,M,ep);
      %all quantities are real valued;
      %x=input data to the filter;dn=desired signal;
      %M=order of the filter;
      %mu=step size parameter;x and dn must be of the same
      %length;
      %ep=sm
N=length(x);
y=zeros(1,N);
w=zeros(1,M);%initialized filter coefficient vector
for n=M:N
      x1=x(n:-1:n-M+1);%for each n the vector x1 is produced
            %of length M with elements from x in reverse order;
      y(n)=w*x1';
      e(n)=dn(n)-y(n);
      w=w+2*mu*sign(e(n))*x1/(ep+x1*x1');
      w1(n-M+1,:)=w(1,:);
end;
J=e.^2;
      %the columns of w1 depict the history of the filter
      %coefficients;
```

Example 9.5.1

Compare the sign-error and NLMS sign-error algorithms for denoising a sinusoidal signal as shown in Figure 9.10.

Solution: The results are shown in Figure 9.11 using the following constants and functions: $\mu = 0.005$, $M = 15$, $\varepsilon = 0.0001$, $s = \sin(0.1\pi n)$, $x = \sin(0.1\pi n) + \text{rand}(n) - 0.5$, $x(n) = dn(n+1)$, $n = 1:1500$.

We observe that for the same inputs the NLMS sign-error algorithm is superior to unnormalized one.

■

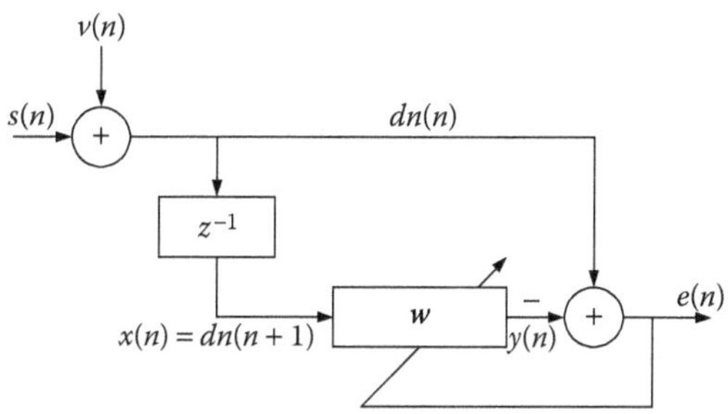

FIGURE 9.10

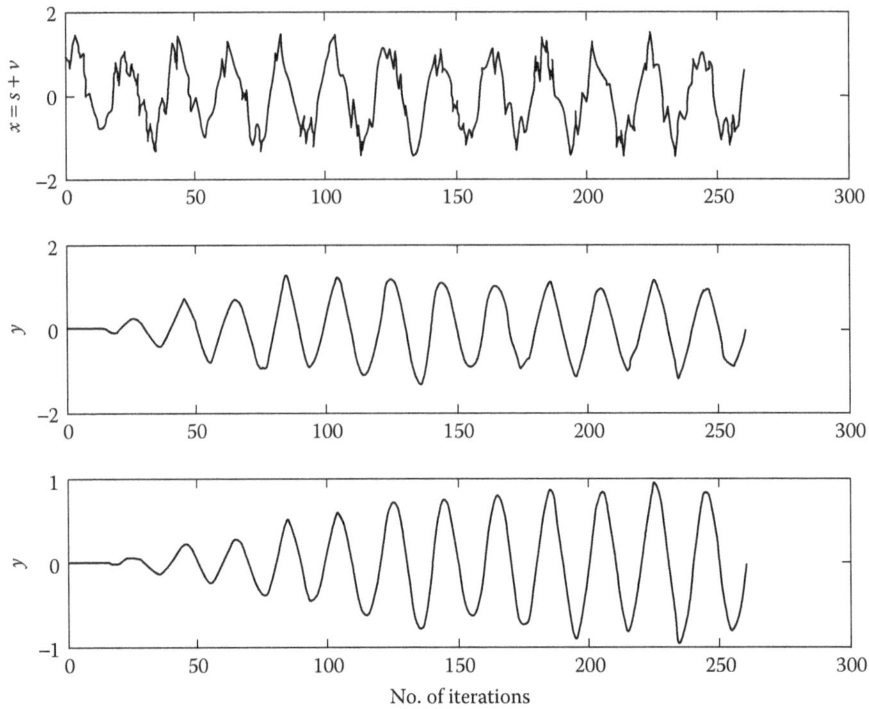

FIGURE 9.11

9.6 THE SIGN-REGRESSOR LMS ALGORITHM

The sign-regressor or data-sign algorithm is given by

$$w(n+1) = w(n) + \mu e(n)\text{sign}[x(n)] \tag{9.15}$$

where the sign function is applied to $x(n)$ on an element-by-element basis.

Book m-Function for Sign-Regressor Algorithm

```
function[w,y,e,J,w1]=lms_sign_regressor(x,dn,mu,M)
      %function[w,y,e,J,w1]=lms_sign_error(x,dn,mu,M);
      %all quantities are real valued;
      %x=input data to the filter;dn=desired signal;
      %M=order of the filter;
      %mu=step size parameter;
      %x and dn must be of the same length;
N=length(x);
y=zeros(1,N);
w=zeros(1,M);%initialized filter coefficient vector;
for n=M:N
      x1=x(n:-1:n-M+1);%for each n the vector x1 is produced
          %of length M with elements from x in reverse order;
      y(n)=w*x1';
```

```
        e(n)=dn(n)-y(n);
        w=w+mu*e(n)*sign(x1);
        w1(n-M+1,:)=w(1,:);
end;
J=e.^2;
        %the columns of w1 depict the history of the filter
        %coefficients;
```

9.7 SELF-CORRECTING SIGN-REGRESSOR LMS ALGORITHM

Book m-Function for Self-Correcting Sign-Regressor LMS Algorithm

```
function[w,y,e,J]=lms_self_correcting_sign_regressor_
        %lms(x,dn,mu,M,I)
        %function[w,y,e,J]=
        %lms_self_correcting_sign_regressor_lms(x,dn,mu,M,I);
        %x=input data to the filter;dn=desired signal;length(x)=
        %length(dn);
        %y=output of the filter, an Ixlength(x) matrix;J=error
        %function, an Ixlength(x) matrix;I=number of stages;
[w(1,:),y(1,:),e(1,:),J(1,:)]=lms_sign_regressor(x,dn,mu,M);
for i=2:I
        [w(i,:),y(i,:),e(i,:),J(i,:)]=lms1(y(i-1,:),dn,mu,M);
end;
J=e.^2;
```

Figure 9.12 was obtained using the following signals and constants: $s = \sin(0.2\pi n)$, $v = 1.2[\text{rand}(1,2000) - 0.5]$, $x = s + v$, delay by 2 units, $mu = 0.01$, $M = 2$, $I = 4$. The reader should observe the power of this Book proposed technique. The adaptive filters have only two coefficients, and by repeating this process only three times, it gives excellent results.

9.8 THE NORMALIZED SIGN-REGRESSOR LMS ALGORITHM

The normalized sign-regressor is given by

$$\boldsymbol{w}(n+1) = \boldsymbol{w}(n) + \mu \frac{e(n)\text{sign}[\boldsymbol{x}(n)]}{\varepsilon + \|\boldsymbol{x}(n)\|^2}, \quad \|\boldsymbol{x}(n)\|^2 = \boldsymbol{x}(n)^T \boldsymbol{x}(n) \tag{9.16}$$

Book m-Function for Normalized Sign-Regressor LMS Algorithm

```
function[w,y,e,J,w1]=lms_normalized_sign_regressor(x,dn,mu,M)
        %function[w,y,e,J,w1]=lms_sign_error(x,dn,mu,M);
        %all quantities are real valued;
        %x=input data to the filter;dn=desired signal;
        %M=order of the filter;
        %mu=step-size parameter;
        %x and dn must be of the same length;
N=length(x);
y=zeros(1,N);
```

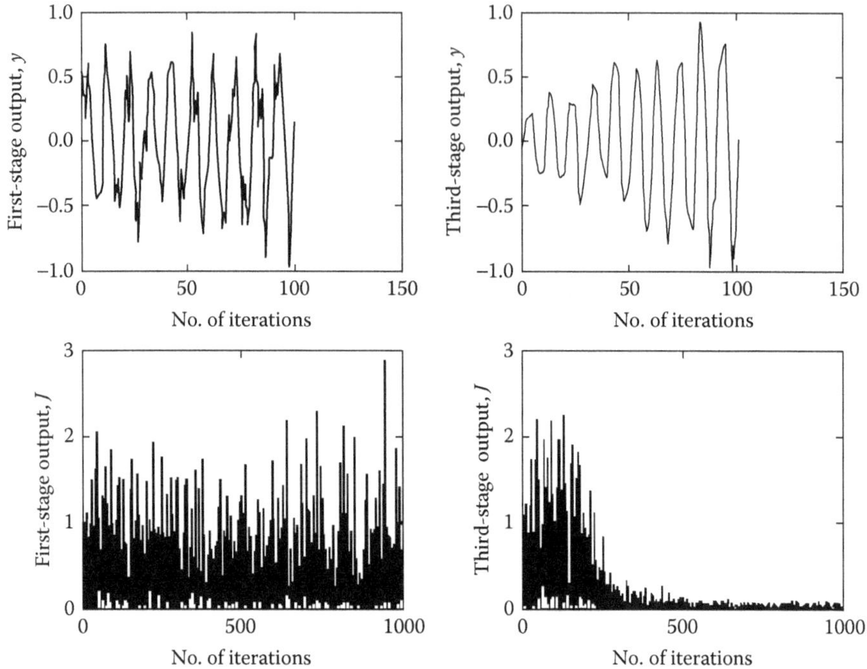

FIGURE 9.12

```
w=zeros(1,M);%initialized filter coefficient vector;
for n=M:N
      x1=x(n:-1:n-M+1);%for each n the vector x1 is produced
            %of length M with elements from x in reverse order;
      y(n)=w*x1';
      e(n)=dn(n)-y(n);
      w=w+mu*e(n)*sign(x1)./(0.0001+x1*x1');
      w1(n-M+1,:)=w(1,:);
end;
J=e.^2;
      %the columns of w1 depict the history of the filter
      %coefficients;
```

9.9 THE SIGN–SIGN LMS ALGORITHM

The sign–sign LMS algorithm is defined by

$$\boldsymbol{w}(n+1)=\boldsymbol{w}(n)+\mu\,\mathrm{sign}[e(n)]\mathrm{sign}[\boldsymbol{x}(n)] \tag{9.17}$$

Book m-Function for Sign–Sign LMS Algorithm

```
function[w,y,e,J,w1]=lms_sign_sign(x,dn,mu,M)
      %function[w,y,e,J,w1]=lms_sign_sign(x,dn,mu,M)
      %all quantities are real valued;
      %x=input data to the adaptive filter;
```

```
        %dn=desired signal;
        %M=order of the filter;
        %mu=step-size parameter;x and dn must be of
        %the same length;
N=length(x);
y=zeros(1,N);
w=zeros(1,M);%initialized filter coefficient vector
for n=M:N
        x1=x(n:-1:n-M+1);%for each n the vector x1 is produced
                %of length M with elements from x in reverse order;
        y(n)=w*x1';
        e(n)=dn(n)-y(n);
        w=w+2*mu*sign(e(n))*sign(x1);
        w1(n-M+1,:)=w(1,:);
end;
J=e.^2;
        %the columns of w1 depict the history of the filter
        %coefficients;
```

Book m-Function for Self-Correcting Sign–Sign LMS Algorithm

```
function[w,y,e,J]=lms_self_correcting_sign_sign_
        lms(x,dn,mu,M,I)
        %function[w,y,e,J]=lms_self_correcting_sign_sign_lms
        %(x,dn,mu,M,I);
        %x=input data to the filter;y=output
        %data from the filter,
        %y is an Ixlength(x) matrix; J=learning curves,
        %an Ixlength(x) matrix;mu=step-size parameter;
        %M=number of coefficients;
        %I=number of stages;w=an Ixlength(x) matrix of filter
        %coefficients; dn=desired signal;
[w(1,:),y(1,:),e(1,:),J(1,:)]=lms_sign_sign(x,dn,mu,M);
for i=2:I
        [w(i,:),y(i,:),e(i,:),J(i,:)]=lms1(y(i-1,:),dn,mu,M);
end;
J=e.^2;
```

9.10 THE NORMALIZED SIGN–SIGN LMS ALGORITHM

The sign–sign LMS algorithm is defined by

$$w(n+1) = w(n) + \mu \frac{\text{sign}[e(n)]\text{sign}[x(n)]}{\varepsilon + \|x(n)\|^2} \qquad \|x(n)\|^2 = x'(n)x(n) \tag{9.18}$$

Book m-Function for Normalized Sign–Sign LMS Algorithm

```
function[w,y,e,J,w1]=lms_normalized_sign_sign(x,dn,mu,M)
        %function[w,y,e,J,w1]=lms_normalized_sign_sign(x,dn,mu,M)
        %all quantities are real valued;
        %x=input data to the adaptive filter;
```

```
        %dn=desired signal;
        %M=order of the filter;
        %mu=step-size parameter;x and dn must be of
        %the same length;
N=length(x);
y=zeros(1,N);
w=zeros(1,M);%initialized filter coefficient vector;
for n=M:N
        x1=x(n:-1:n-M+1);%for each n the vector x1 is produced
                %of length M with elements from x in reverse order;
        y(n)=w*x1';
        e(n)=dn(n)-y(n);
        w=w+2*mu*sign(e(n))*sign(x1)./(0.0001+x1*x1');
        w1(n-M+1,:)=w(1,:);
end;
J=e.^2;
        %the columns of w1 depict the history of the filter
        %coefficients;
```

Figure 9.13 compares different types of sign-type LMS algorithm. The signal to be detected at the output of the LMS filter is a sine wave and it is shown in Figure 9.13a. The noisy signal is shown in Figure 9.13b. To recapture the sine wave from the noisy, $\{x(n)\}$, signal, we use the setup shown in Figure 9.1. The constants and functions that were used in this example are as follows:

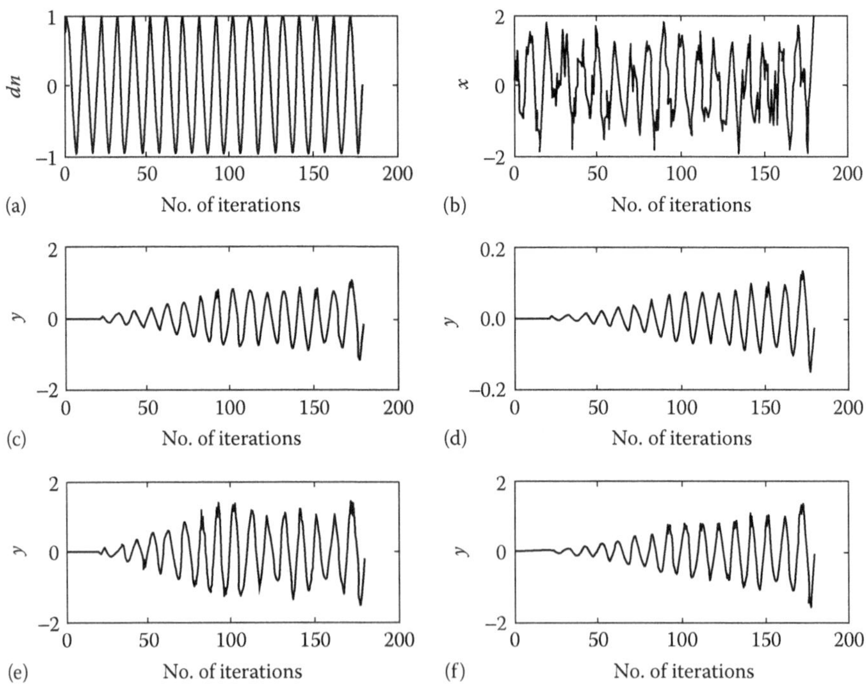

FIGURE 9.13

$$n = 1:2000;\ s = \sin(0.2n\pi);\ x = s + v = s + 0.5\text{randn}(1,2000)$$

$$\sigma_v^2 = 0.24;\ \mu = 0.002;\ M = 20;\ \text{delay} = 2\ (= z^{-2})$$

Figure 9.13c shows the output of the sign-regressor LMS filter. Figure 9.13d shows the output of a normalized sign-regressor LMS filter. Figure 9.13e shows the output of a sign–sign LMS filter. Figure 9.13f shows the output of a normalized sign–sign LMS filter. It is apparent that the normalized ones perform better.

Note: The self-correcting approach can be used with any one of the LMS algorithms and their normalized form.

9.11 VARIABLE STEP-SIZE LMS

The variable step-size LMS (VSLMS) algorithm was introduced to facilitate the conflicting requirements. A large step-size parameter is needed for fast convergence, whereas a small step-size parameter is needed to reduce the misadjustment factor. When the adaptation begins and $\boldsymbol{w}(n)$ is far from its optimum value, the step-size parameter should be large in order for the convergence to be rapid. However, as the filter value $\boldsymbol{w}(n)$ approaches the steady-state solution, the step-size parameter should decrease in order to reduce the excess MSE.

To accomplish the variation of the sep-size parameter, each filter coefficient is given a separate time-varying step-size parameter such that the LMS recursion algorithm takes the form:

$$w_i(n+1) = w_i(n) + \mu_i(n)e(n)x(n-i) \qquad i = 0, 1, \ldots, M-1 \tag{9.19}$$

where:

$w_i(n)$ is the ith coefficient of $\boldsymbol{w}(n)$ at iteration n

$\mu_i(n)$ is its associated step size

The step sizes are determined in an *ad hoc* manner, based on monitoring sign changes in the instantaneous gradient estimate $e(n)x(n-i)$. It was argued that successive changes in the sign of the gradient estimate indicates that the algorithm is close to its optimal solution, and hence, the step-size value must be decreased. The reverse is also true. The decision of decreasing the value of the step size by some factor c_1 is based on some number m_1 successive changes of $e(n)x(n-i)$. Increasing the step-size parameter by some factor c_2 is based on m_2 successive sign changes. The parameters m_1 and m_2 can be adjusted to optimize performance, as can the factors c_1 and c_2.

The set of update (9.19) may be written in the matrix form:

$$\boldsymbol{w}(n+1) = \boldsymbol{w}(n) + \mu(n)e(n)\boldsymbol{x}(n) \qquad \mu(n) = \begin{bmatrix} \mu_0(n) & & & \\ & & & \mathbf{0} \\ & \mu_1(n) & & \\ & & \vdots & \\ & \mathbf{0} & & \\ & & & \mu_{M-1}(n) \end{bmatrix} \tag{9.20}$$

TABLE 9.3
The VSLMS Algorithm

Input	Initial coefficient vector: $\boldsymbol{w}(0)$
	Input data vector: $\boldsymbol{x}(n) = [x(n)\, x(n-1) \cdots x(n-M+1)]^T$
	Gradient term: $g_0(n-1) = e(n-1)x(n-1), g_1(n-1) = e(n-1)x(n-1), \ldots,$ $g_{M-1}(n-1) = e(n-1)x(n-M)$
	Step-size parameter: $\mu_0(n-1), \mu_1(n-1), \ldots, \mu_{M-1}(n-1)$
	a = small positive constant
	μ_{max} = positive constant
Outputs	Desired output: $d(n)$
	Filter output: $y(n)$
	Filter update: $\boldsymbol{w}(n+1)$
	Gradient term: $g_0(n), g_1(n), \ldots, g_{M-1}(n)$
	Update step-size parameter: $\mu_0(n), \mu_1(n), \ldots, \mu_{M-1}(n)$
Execution	1. $y(n) = \boldsymbol{w}^T(n)\, \boldsymbol{x}(n)$
	2. $e(n) = d(n) - y(n)$
	3. Weights and step-size parameter adaptation
	For $i = 0, 1, 2, \ldots, M-1$
	$g_i(n) = e(n) \times (n-i)$
	$\mu_i(n) = \mu_i(n-1) + \sigma \text{sign}(g_i(n))\text{sign}(g_i(n))\%\sigma$ = small positive step-size parameter;
	if $\mu_i(n) > \mu_{max}, \mu_i(n) = \mu_{max}$
	if $\mu_i(n) < \mu_{min}, \mu_i = \mu_{min}$
	$w_i(n+1) = w_i(n) + \mu_i(n) g_i(n)$
	end

The VSLMS algorithm is given in Table 9.3.

The following proposed Book m-function can be used to take care, for example, of a noisy function whose added noise is varying. Figure 9.14 shows the results using the constant value of the step-size value and a varying one. The constants and functions used were as follows:

$$N = 2000, n = 1:2000, dn = \sin(0.2\pi n) + 1.5\, 0.988^n \,\text{randn}(1{,}2000),$$

$$x(n) = dn(n+2), \text{delay by 2 units}, mu = 0.05, \text{ for the constant step-size case}$$

Book m-Function for Varying Variance of the Signal That Adjusts Appropriately the Value of μ

```
function[w,y,e,J,w1]=lms_varying_mu(x,dn,M,N)
w=zeros(1,M);%N=length of x and dn signals;
for n=M:N
       x1=x(n:-1:n-M+1);
       if var(x1)>=0.6
              mu=0.0005;
```

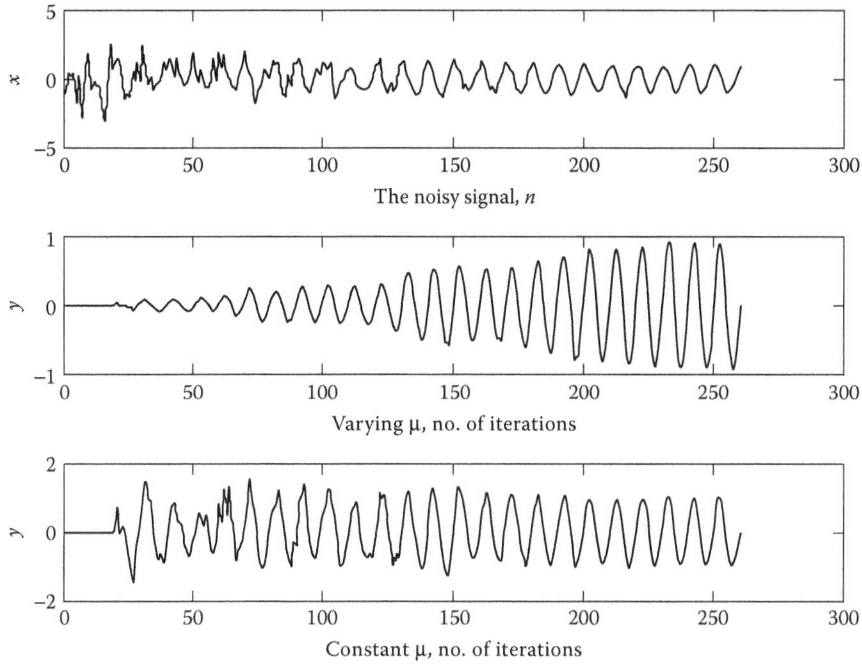

FIGURE 9.14

```
        elseif 0.6<var(x1)<0.2
               mu=0.005;
        elseif var(x1)<=0.2
               mu=0.05;
        end
        y(n)=w*x1';
        e(n)=dn(n)-y(n);
        w=w+mu*e(n)*x1;
        w1(n-M+1,:)=w(1,:);
end;
J=e.^2;
```

9.12 THE LEAKY LMS ALGORITHM

Let us assume that $\{\boldsymbol{x}(n)\}$ and $\{d(n)\}$ are jointly wide-sense stationary processes that determine when the coefficients $\boldsymbol{w}(n)$ converge in the mean to $\boldsymbol{w}^o = \boldsymbol{R}_x^{-1}\boldsymbol{p}_{dx}$. That is,

$$\lim_{n\to\infty} E\{\boldsymbol{w}(n)\} = \boldsymbol{w}^o = \boldsymbol{R}_x^{-1}\boldsymbol{p}_{dx} \tag{9.21}$$

We start by taking the expectation of both sides of the LMS recursion as follows:

$$E\{\boldsymbol{w}(n+1)\} = E\{\boldsymbol{w}(n)\} + \mu E\{d(n)\boldsymbol{x}(n)\} - \mu E\{\boldsymbol{x}(n)\boldsymbol{x}^T(n)\boldsymbol{w}(n)\} \tag{9.22}$$

where:

$y(n) = \boldsymbol{x}^T(n)\boldsymbol{w}(n)$

Assuming that $\boldsymbol{x}(n)$ and $\boldsymbol{w}(n)$ are statistically independent (**independence** assumption), (9.22) becomes

$$\begin{aligned} E\{\boldsymbol{w}(n+1)\} &= E\{\boldsymbol{w}(n)\} + \mu E\{d(n)\boldsymbol{x}(n)\} - \mu E\{\boldsymbol{x}(n)\boldsymbol{x}^T(n)\}E\{\boldsymbol{w}(n)\} \\ &= \left(\boldsymbol{I} - \mu \boldsymbol{R}_x\right) E\left\{\boldsymbol{w}(n)\right\} + \mu\, \boldsymbol{p}_{dx}(n) \end{aligned} \tag{9.23}$$

which is similar to the equations of the steepest-descent method [see (7.18)], where we have set $\mu' = \mu$ for simplicity. The difference in the above equation is the presence of the ensemble average symbol. This suggests that the steepest descent method is applicable to ensemble average $E\{\boldsymbol{w}(n+1)\}$. Rewriting (7.23) in matrix form, we obtain

$$\xi'(n) = (\boldsymbol{I} - \mu \boldsymbol{\Lambda})^n \xi'(0) \quad k = 0, 1, \ldots, M-1 \quad n = 1, 2, 3, \ldots \tag{9.24}$$

We observe that $\boldsymbol{w}(n) \to \boldsymbol{w}^o$ if $\xi(n) = \boldsymbol{w}(n) - \boldsymbol{w}^o \to 0$ as $n \to \infty$ or when $\xi'(n) = \boldsymbol{Q}^T \xi$ converges to zero. The kth row of (9.24) is

$$\xi_k'(n) = (1 - \mu\lambda_k)^n \xi_k'(0) \tag{9.25}$$

which indicates that $\xi_k'(n) \to 0$ if

$$\left|1 - \mu\lambda_k\right| < 1 \ \text{ or } \ -1 < 1 - \mu\lambda_k < 1 \ \text{ or } \ 0 < \mu < \frac{2}{\lambda_k} \tag{9.26}$$

To have more restrictive condition, we can use the inequality

$$0 < \mu < \frac{1}{\lambda_{\max}} \tag{9.27}$$

If $\lambda_k = 0$, (9.25) indicates that no convergence takes place as n approaches infinity. Since it is possible for these undamped modes to become unstable, it is important for the stabilization of the LMS algorithm to force these modes to zero. One way to avoid this difficulty is to introduce a leakage coefficient γ into the LMS algorithm as follows:

$$\boldsymbol{w}(n+1) = (1 - \mu\gamma)\boldsymbol{w}(n) + \mu e(n)\boldsymbol{x}(n) \tag{9.28}$$

where:

$0 < \gamma << 1$

The effect of introducing the **leakage coefficient** γ is to force any undamped modes to become zero and to force the filter coefficients to zero if either $e(n)$ or $\boldsymbol{x}(n)$ is zero. [The homogeneous equation $w_i(n+1) = (1 - 2\mu\gamma)w_i(n)$ has the solution $w_i(n) = A(1 - \mu\gamma)^n$, where A is a constant.]

We write (9.28) in the form: $[e(n) = d(n) - \boldsymbol{x}^T(n)\boldsymbol{w}(n)]$.

$$\begin{aligned} w(n+1) &= (1-\mu\gamma)w(n) + \mu[d(n) - x^T(n)w(n)]x(n) \\ &= w(n) - \mu\gamma w(n) + \mu d(n)x(n) - x^T(n)w(n)x(n) \\ &= [I - \mu(x(n)x^T(n) + \gamma I)]w(n) + \mu d(n)x(n) \end{aligned} \tag{9.29}$$

We set $x^T(n)w(n)x(n) = x(n)x^T(n)w(n)$ because $x^T(n)w(n)$ is a number. By taking the expected value of both sides of the above equation and using the independence assumption, we obtain

$$\begin{aligned} E\{w(n+1)\} &= [I - \mu E(\{x(n)x^T(n)\} + \gamma I)]E\{w(n)\} + \mu E\{d(n)x(n)\} \\ &= [I - \mu(R_x + \gamma I)]E\{w(n)\} + \mu p_{dx}(n) \end{aligned} \tag{9.30}$$

Comparing the above equation with (9.23), we observe that the autocorrelation matrix R_x of the LMS algorithm has been replaced with $R_x + \gamma I$. Since the eigenvalues of $R_x + \gamma I$ are $\lambda_k + \gamma$ and $\lambda_k \geq 0$, all the modes of the leaky LMS algorithm will be decayed to zero. Furthermore, the constraint for the step-size parameter becomes

$$0 < \mu < \frac{1}{\lambda_{\max} + \gamma} \tag{9.31}$$

As $n \to \infty, w(n+1) \cong w(n)$, and hence, (9.30) becomes

$$\lim_{n\to\infty} E\{w(n)\} = \left[R_x + \gamma I\right]^{-1} p_{dx} \tag{9.32}$$

which indicates that the leakage coefficient produces a bias into the steady-state solution $R_x^{-1} p_{dx}$. For another way to produce the leaky LMS algorithm, see Problem 9.11.1.

Book m-Function for Leaky LMS Algorithm

```
function[w,y,e,J,w1]=lms_leaky_lms(x,dn,mu,gama,M)
      %function[w,y,e,J,w1]=lms_leaky_lms(x,dn,mu,gama,M);
      %all signals are real valued;x=input to filter;
      %y=output from the filter;dn=desired signal;
      %mu=step-size factor;gama=gamma factor<<1;
      %M=number of filter coefficients;w1=matrix whose M
      %rows give the history of each filter coefficient;
N=length(x);
y=zeros(1,N);
w=zeros(1,M);
for n=M:N
      x1=x(n:-1:n-M+1);
      y(n)=w*x1';
      e(n)=dn(n)-y(n);
      w=(1-mu*gama)*w+mu*e(n)*x1;
      w1(n-M+1,:)=w(1,:);
end;
J=e.^2;
```

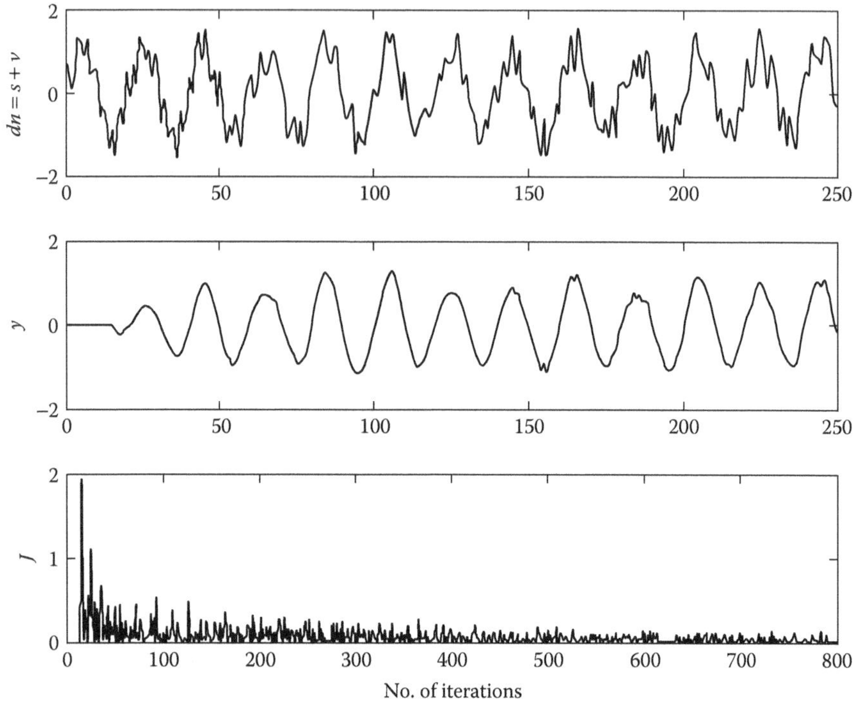

FIGURE 9.15

Figure 9.15 was produced with the following constants and functions: $s(n)=\sin(0.1\pi n)$, $v(n)=1.2[\text{rand}(1,2000)-0.5]$, $dn=s+v$, delay for x was 2, $dn(n+2)=x(n)$, $mu=0.005$, $M=15$, $\gamma=0.01$.

For the normalized leaky LMS case, see Problem 9.11.2.

9.13 THE LINEARLY CONSTRAINED LMS ALGORITHM

In all previous analyses of the Wiener filtering problem, the steepest descent method, the Newton's method, and the LMS algorithm, no constraint was imposed on the solution of minimizing the MSE. However, in some applications, there might be some mandatory constraints that must be taken into consideration in solving optimization problems. For example, the problem of minimizing the average output power of a filter, while the frequency response must remain constant at specific frequencies. In this section, we discuss the filtering problem of minimizing the MSE subject to a general constraint.

The error between the desired signal and the output of the filter is

$$e(n)=d(n)-\boldsymbol{w}^T(n)\boldsymbol{x}(n) \tag{9.33}$$

We then minimize this error in the mean-square sense subject to the constraint:

$$\boldsymbol{c}^T\boldsymbol{w}=a \tag{9.34}$$

where:

a is a constant

$\boldsymbol{c}$ is a constant vector

Using the Lagrange multiplier method (see also Appendix 4), we write

$$J_c = E\{e^2(n)\} + \lambda(\boldsymbol{c}^T \boldsymbol{w} - a) \tag{9.35}$$

where:

λ is the Lagrange multiplier

Therefore, the following relations must be satisfied simultaneously:

$$\nabla_w J_c = 0 \qquad \frac{\partial J_c}{\partial \lambda} = 0 \tag{9.36}$$

The second term produces the constraint (9.34). Next, we substitute the error $e(n)$ in (9.35) to obtain (see Problem 9.12.1)

$$J_c = J_{\min} + \xi^T \boldsymbol{R}_x \xi + \lambda(\boldsymbol{c}^T \xi - a') \tag{9.37}$$

where:

$$\xi(n) = \boldsymbol{w}(n) - \boldsymbol{w}^o, \quad \boldsymbol{w}^o = \boldsymbol{R}_x^{-1} \boldsymbol{p}_{dx}, \quad \boldsymbol{R}_x = E\left\{\boldsymbol{x}(n)\boldsymbol{x}^T(n)\right\} \tag{9.38}$$

and

$$\boldsymbol{p}_{dx} = E\{d(n)\boldsymbol{x}^T(n)\} \quad a' = a - \boldsymbol{c}^T \boldsymbol{w}^o \tag{9.39}$$

The solution has now changed to $\nabla_\xi J_c = 0$ and $\partial J_c / \partial \lambda = 0$. Hence, from (9.37), we obtain (see also Appendix 2)

$$\nabla_\xi J_c = \begin{bmatrix} \dfrac{\partial J_c}{\partial \xi_1} \\ \vdots \\ \dfrac{\partial J_c}{\partial \xi_M} \end{bmatrix} = \begin{bmatrix} 2\xi_1 r_1 + 2\xi_2 r_2 + & \cdots & +2\xi_M r_M \\ \vdots & & \vdots \\ 2\xi_1 r_M + 2\xi_2 r_{M-1} + & \cdots & +2\xi_M r_1 \end{bmatrix} + \lambda \begin{bmatrix} c_1 \\ \vdots \\ c_M \end{bmatrix} = \boldsymbol{0} \tag{9.40}$$

or in a matrix form

$$2\boldsymbol{R}_x \xi_c^o + \lambda \boldsymbol{c} = \boldsymbol{0} \tag{9.41}$$

where:

ξ_c^o is the constraint optimum of the vector ξ

In addition, the constraint gives the relation:

$$\frac{\partial J_c}{\partial \lambda} = \boldsymbol{c}^T \xi_c^o - a' = 0 \tag{9.42}$$

Solving the system of the last two equations for λ and ξ_c^o, we obtain

$$\lambda = -\frac{2a'}{\boldsymbol{c}^T \boldsymbol{R}_x^{-1} \boldsymbol{c}} \qquad \xi_c^o = \frac{a' \boldsymbol{R}_x^{-1} \boldsymbol{c}}{\boldsymbol{c}^T \boldsymbol{R}_x^{-1} \boldsymbol{c}} \tag{9.43}$$

Substituting the value of λ in (9.37), we obtain the minimum value of J_c as

$$J_c = J_{\min} + \frac{a'^2}{\boldsymbol{c}^T \boldsymbol{R}_x^{-1} \boldsymbol{c}} \tag{9.44}$$

But $\boldsymbol{w}(n) = \xi(n) + \boldsymbol{w}^o$, and hence, using (9.43) we obtain the equation:

$$\boldsymbol{w}_c^o = \boldsymbol{w}^o + \frac{a' \boldsymbol{R}_x^{-1} \boldsymbol{c}}{\boldsymbol{c}^T \boldsymbol{R}_x^{-1} \boldsymbol{c}} \tag{9.45}$$

Note: The second term of (9.44) is the excess MSE produced by the constraint.

To obtain the recursion relation subject to constraint $\boldsymbol{c}^T \boldsymbol{w} = a$, we must proceed in two steps:

$$\text{Step 1: } \boldsymbol{w}'(n) = \boldsymbol{w}(n) + \mu e(n) \boldsymbol{x}(n) \tag{9.46}$$

$$\text{Step 2: } \boldsymbol{w}(n+1) = \boldsymbol{w}'(n) + \eta(n) \tag{9.47}$$

where $\eta(n)$ is chosen so that $\boldsymbol{c}^T \boldsymbol{w}(n+1) = a$ while $\eta^T(n)\eta(n)$ is minimized. In other words, we choose the vector $\eta(n)$ so that the constraint $\boldsymbol{c}^T \boldsymbol{w} = a$ after Step 2 while the perturbation introduced by $\eta(n)$ is minimized. The problem can be solved using the Lagrange multiplier method that gives

$$\eta(n) = \frac{a - \boldsymbol{c}^T \boldsymbol{w}'(n)}{\boldsymbol{c}^T \boldsymbol{c}} \boldsymbol{c} \tag{9.48}$$

Thus, we obtain the final form of (9.47) as

$$\boldsymbol{w}(n+1) = \boldsymbol{w}'(n) + \frac{a - \boldsymbol{c}^T \boldsymbol{w}'(n)}{\boldsymbol{c}^T \boldsymbol{c}} \boldsymbol{c} \tag{9.49}$$

The constraint algorithm is given in Table 9.4. Figure 9.16 shows the results of a constrained LMS filter with the following data: $s = \sin(0.1\pi n)$, $v = \text{noise} = 2(\text{rand} - 0.5)$, $dn = s + v$, x delayed by 2 from dn, $\boldsymbol{c} = \text{ones}(1,32)$, $a = 0.8$, $\mu = 0.005$, $M = 32$.

Note: Solving the constrained optimization problem using the Lagrange multiplier method, the NLMS can be obtained (see Problem 9.12.2).

9.14 THE LEAST MEAN FOURTH ALGORITHM

The optimal weight vector $\boldsymbol{w}^o$ that solves $\min_{\boldsymbol{w}} E\left|d(n) - \boldsymbol{x}^T(n)\boldsymbol{w}(n)\right|^4$ (see Sayed 2003) can be approximated iteratively by the recursion:

$$\begin{aligned} &\boldsymbol{w}(n+1) = \boldsymbol{w}(n) + \mu \boldsymbol{x}(n) e(n) e(n)^2 \\ &\boldsymbol{w}(n+1) = \boldsymbol{w}(n) + \mu \boldsymbol{x}^*(n) e(n) \left|e(n)\right|^2 \quad \text{(complex case)} \end{aligned} \tag{9.50}$$

For the application of the least mean fourth (LMF) algorithm, see Problem 9.13.1.

TABLE 9.4
The Linearly Constrained LMS Algorithm

Input	Initial coefficient vector, $\boldsymbol{w}(0) = \mathbf{0}$
	Input data, $\boldsymbol{x}(n)$
	Desired output, $d(n)$
	Constant vector, $\boldsymbol{c}$
	Constraint constant, a
Output	Filter output, $y(n)$
Procedure	$y(n) = \boldsymbol{w}^T(n)\boldsymbol{x}(n)$
	$e(n) = d(n) - y(n)$
	$\boldsymbol{w}'(n) = \boldsymbol{w}(n) + \mu e(n)\boldsymbol{x}(n)$
	$\boldsymbol{w}(n+1) = \boldsymbol{w}'(n) + \dfrac{a - \boldsymbol{c}^T\boldsymbol{w}'(n)}{\boldsymbol{c}^T\boldsymbol{c}}\boldsymbol{c}$

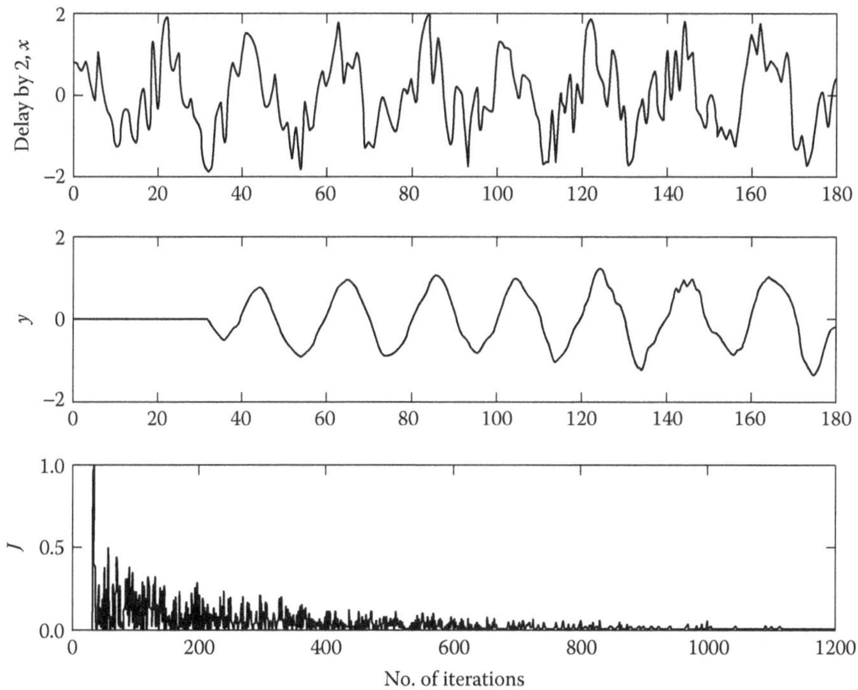

FIGURE 9.16

9.15 THE LEAST MEAN MIXED NORM LMS ALGORITHM

The optimal weight vector $\boldsymbol{w}^o$ that solves $\min_{\boldsymbol{w}} E[\delta\,|e|^2 + 1/2(1-\delta)\,|e|^4]$, $e = d - \boldsymbol{x}^T\boldsymbol{w}$ for some constant $0 \le \delta \le 1$, can be approximated iteratively using the recursion (see Sayed 2003):

$$\mathbf{w}(n+1) = \mathbf{w}(n) + \mu \mathbf{x}(n) e(n)[\delta + (1-\delta)|e(n)|^2] \tag{9.51}$$

For the application of the least mean mixed norm (LMMN) algorithm and its comparison with the normalized one, see Problem 9.14.1.

9.16 SHORT-LENGTH SIGNAL OF THE LMS ALGORITHM

The proposed Book program that alleviates the problem of using a short-length signal present is given below. To test the proposed approach, we proceed to identify an unknown system as shown in Figure 9.17.

Book Program for Short-Time Signals

```
>>M=8;mu=0.03;
>>v=rand(1,200)-0.5;%length(v)=200;
>>x=conv(v,[0.90.2-0.2]);
>>d=conv(x,[42-0.40.2-0.1]);
>>x=x(1,1:200);
>>d=d(1,1:200);
>>ve=[vvvvvvvv];%extended signal;
>>xe=conv(ve,[0.90.2-0.2]);
>>dn=conv(xe,[42-0.40.2-0.1]);
>>xe=xe(1,1:1600);
>>de=dn(1,1:1600);
>>[w,y,e,J,w1]=lms_short_signal(x,d,mu,M);
>>[we,ye,ee,Je,w1e]=lms_short_signal(xe,de,mu,M);
```

Book m-Function for Short-Time Signals

```
function[w,y,e,J,w1]=lms_short_signal(x,dn,mu,M)
N=length(x);w=0.02*(rand(1,M)-0.5);
for n=M:N
      x1=x(n:-1:n-M+1);
      y(n)=w*x1';
      e(n)=dn(n)-y(n);
      w=w+mu*e(n)*x1;
```

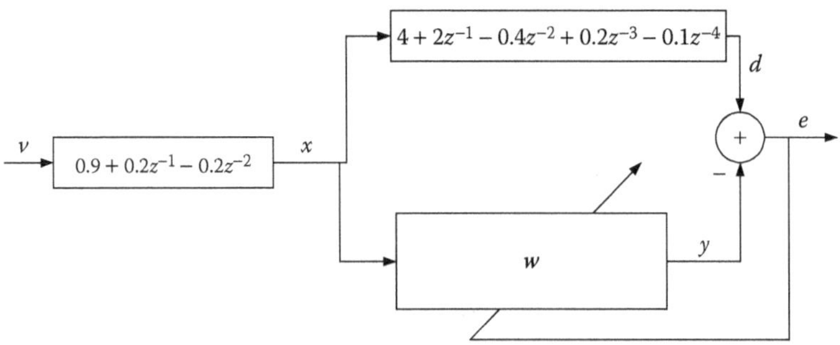

FIGURE 9.17

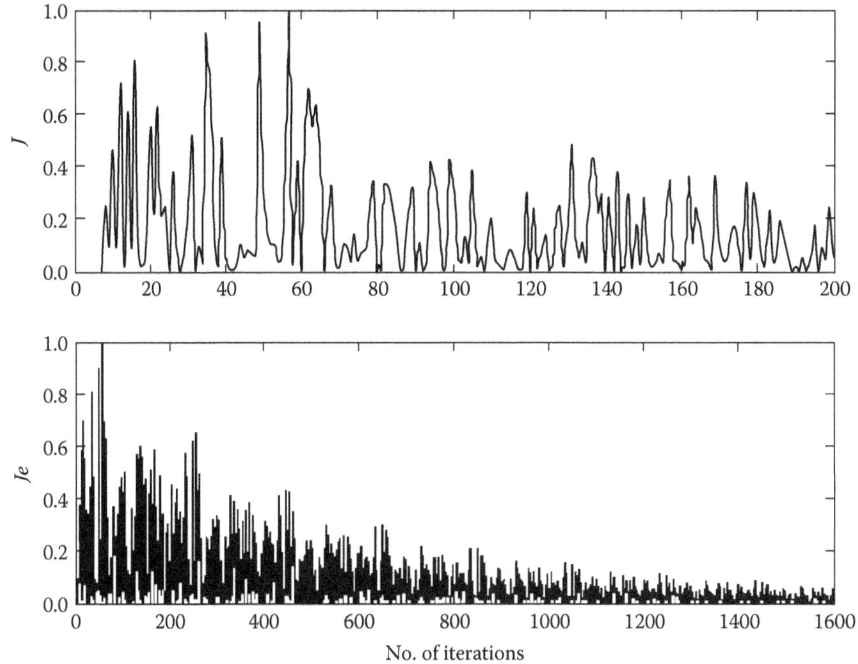

FIGURE 9.18

```
w1(n-M+1,:)=w(1,:);
end;
J=e.^2;
```

The results of the short signal with 200 time units are as follows:

$$1.5904 \quad 0.9112 \quad -0.2863 \quad -0.0321 \quad 0.1554 \quad -0.0754 \quad -0.0357 \quad 0.2486$$

The results of the extended signal with 1600 time units are as follows:

$$3.8982 \quad 2.0002 \quad -0.3960 \quad 0.1534 \quad -0.0385 \quad -0.0379 \quad 0.0087 \quad 0.0519$$

Figure 9.18 shows the learning curve for the two cases. It is interesting to note that this approach is successful for stationary random input signals.

9.17 THE TRANSFORM DOMAIN LMS ALGORITHM

The implementation of the LMS filter in the frequency domain can be accomplished simply by taking the discrete Fourier transform (DFT) of both the input data, $\{x(n)\}$, and the desired signal, $\{d(n)\}$. The advantage of doing this is due to the fast processing of the signal using the fast Fourier transform (FFT) algorithm. However, this procedure requires a block-processing strategy, which results in storing a number of incoming data in buffers, and thus, some delay is unavoidable.

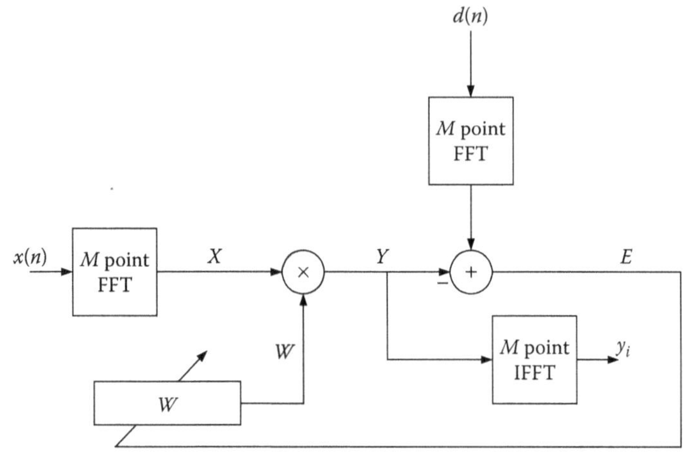

FIGURE 9.19

The block diagram approach for the transform domain LMS (TDLMS) algorithm is shown in Figure 9.19. The signals are processed by a block-by-block format, that is, $\{x(n)\}$ and $\{d(n)\}$ are sequenced into blocks of length M so that

$$x_i(n) = x(iM + n),\ d_i(n) = d(iM + n) \quad n = 0, 1, \ldots, M - 1, \quad i = 0, 1, 2, \ldots \tag{9.52}$$

The values of the ith block of the signals $\{x_i(n)\}$ and $\{d_i(n)\}$ are Fourier transforms using the DFT to find $X_i(k)$ and $D_i(k)$, respectively. Because of the DFT properties, the sequences $X_i(k)$ and $D_i(k)$ have M complex elements corresponding to frequency indices ("bins") $k = 0, 1, 2, \ldots, M -$. Hence,

$$X_i(k) = DFT\{x_i(n)\} = \sum_{n=0}^{M-1} x_i(n) e^{-j\left(\frac{2\pi nk}{M}\right)} \quad k = 0, 1, \ldots, M - 1 \tag{9.53}$$

$$D_i(k) = DFT\{d_i(n)\} = \sum_{n=0}^{M-1} d_i(n) e^{-j\left(\frac{2\pi nk}{M}\right)} \quad k = 0, 1, \ldots, M - 1 \tag{9.54}$$

During the ith block processing, the output of each frequency bin of the adaptive filter is computed as follows:

$$Y_i(k) = W_{i,k} X_i(k) \quad k = 0, 1, 2, \ldots, M - 1 \tag{9.55}$$

where:

$W_{i,k}$ is the kth frequency bin corresponding to the ith update (corresponding to the ith block data)

The error in the frequency domain is

$$E_i(k) = D_i(k) - Y_i(k) \quad k = 0, 1, 2, \ldots, M - 1 \tag{9.56}$$

The system output is given by

$$y_i(n) = y(iM+n) = IDFT\{Y_i(k)\} = \frac{1}{M}\sum_{k=0}^{M-1} Y_i(k)e^{j\left(\frac{2\pi nk}{M}\right)} \qquad n = 0, 1, \ldots, M-1 \tag{9.57}$$

To update the filter coefficients we use, by analogy of the LMS recursion, the following recursion is used:

$$\boldsymbol{W}_{i+1} = \boldsymbol{W}_i + \mu \boldsymbol{E}_i \bullet \boldsymbol{X}_i^* \tag{9.58}$$

where:

$$\boldsymbol{W}_{i+1} = [W_{i+1,0} \quad W_{i+1,1} \quad \cdots \quad W_{i+1,M-1}]^T$$

$$\boldsymbol{W}_i = [W_{i,0} \quad W_{i,1} \quad \cdots \quad W_{i,M-1}]^T$$

$$\boldsymbol{E}_i = [E_i(0) \quad E_i(1) \quad \cdots \quad E_i(M-1)]^T$$

$$\boldsymbol{X}_i^* = [X_i^*(0) \quad X_i^*(1) \quad \cdots \quad X_i^*(M-1)]^T$$

The dot ($\bullet$) in (9.58) implies element-by-element multiplication and the asterisk (*) stands for complex conjugate. If we set $\boldsymbol{X}_i^*$ in the form:

$$\boldsymbol{X}_i = \text{diag}\{X_i(0) \quad X_i(1) \quad \cdots \quad X_i(M-1)\}$$

$$= \begin{bmatrix} X_i(0) & 0 & \cdots & 0 \\ 0 & X_i(1) & \cdots & 0 \\ \vdots & \vdots & \cdots & 0 \\ 0 & 0 & \cdots & X_i(M-1) \end{bmatrix} \tag{9.59}$$

then (9.58) becomes

$$\boldsymbol{W}_{i+1} = \boldsymbol{W}_i + \mu \boldsymbol{X}_i^* \boldsymbol{E}_i \tag{9.60}$$

Therefore, Equations 9.52 through 9.58 constitute the frequency domain of the LMS algorithm. The Book m-function gives the adaptive filter coefficients after *I* blocks (or iterations) is given below.

Book m-Function for the Frequency Domain of the LMS

```
function[A]=lms_FT_lms(x,d,M,I,mu)
      %function[A]=lms_FT_lms(x,d,M,I,mu);
wk=zeros(1,M);%M=number of filter coefficients;
for i=0:I%I=number of iterations (or blocks);
      if I*M>length(x)-1
            ('error:I*M<length(x)-1')
      end;
      x1=x(M*(i+1):-1:i*M+1);
      d1=d(M*(i+1):-1:i*M+1);
      xk=fft(x1);
```

```
        dk=fft(d1);
        yk=wk.*xk;
        ek=dk-yk;
        wk=wk+mu*ek.*conj(xk);
        A(i+1,:)=wk;
end;
        %all the rows of A are the wk's at an increase order
        %of iterations(blocks); to filter the data, wk must be
        %inverted in the time domain, convolve with the data x
        %and then plot the real part of the output y, e.g.,
        %wn4=the forth iteration =ifft(A(4,:)),
        %yn4=filter(wn4/4,1,x)for even M;
```

Example 9.16.1

The following Book m-file (or script file) produces the output of Figure 9.20. To produce the figure in the Command window, we just write: `ex9 _ 16 _ 1`.

Book m-File (Script File)

```
%ex9_16_1
M=40;I=12;mu=0.01;
n=0:999;
```

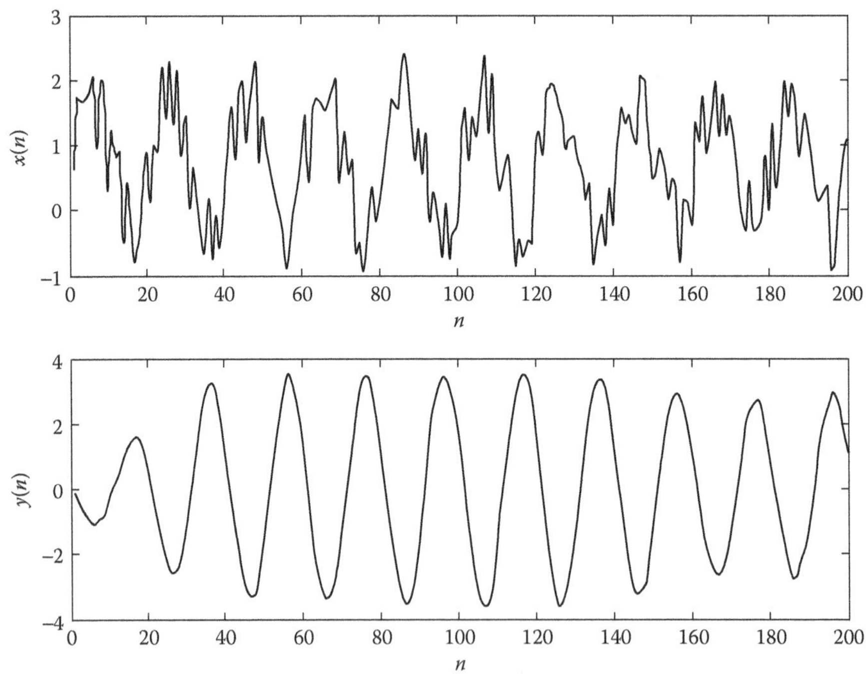

FIGURE 9.20

```
d=sin(0.1*pi*n);v=1.5*rand(1,1000);
x=d+v;
[A]=lms_FT_lms(x,d,M,I,mu);
wn10=ifft(A(10,:))/M;%inverse FT of row 10 of
     %matrix A;
yn10=filter(wn10/I,1,x)/M;
subplot(2,1,1);plot(x(1,1:200),'k');
xlabel('n');ylabel('x(n)');
subplot(2,1,2);plot(real(yn10(1,1:200)),'k');
xlabel('n');ylabel('y(n)');
```

■

9.17.1 Convergence

Let the signals $\{x(n)\}$ and $\{y(n)\}$ be jointly WSS and the initial filter coefficient be zero, $\mathbf{W}_o = \mathbf{0}$. Based on (9.60) and (9.56) (see Problem. 9.16.1), we obtain

$$W_{i+1,k} = \left(1 - \mu\left|X_i\right|^2\right)W_{i,k} + \mu D_i(k)X_i^*(k) \tag{9.61}$$

The expected value of (9.61), assuming that $W_{i,k}$ and $X_i(k)$ are statistically independent, is given by

$$E\left\{W_{i+1,k}\right\} = \left(1 - \mu E\left\{\left|X_i(k)\right|^2\right\}\right)E\left\{W_{i,k}\right\} + \mu E\left\{D_i(k)X_i^*(k)\right\} \tag{9.62}$$

Because $x(n)$ and $x(n)$ are stationary, their statistical characteristics do not change from block to block, and therefore, the ensembles $E\{|X_i(k)|^2\}$ and $E\{D_i(k)X_i^*(k)\}$ are independent of i but depend on k. Taking the z-transform of (9.62) with respect to i of the dependent variable $W_{i,k}$, we find the relation (see Table 3.3.1 and Problem 9.16.2):

$$W_k(z) = -\mu E\left\{\left|X_i(k)\right|^2\right\}\frac{W_k(z)}{z-1} + \frac{\mu E\left\{D_i(k)X_i^*(k)\right\}}{z-1} \tag{9.63}$$

Applying the final value theorem (see Table 3.2 and Problem 9.16.2), we obtain the steady-state value for the filter coefficients:

$$E\left\{W_k^{\infty}\right\} = \frac{E\left\{\{D_i(k)X_i^*\}\right\}}{E\left\{\left|X_i(k)\right|^2\right\}} \tag{9.64}$$

Let the mean filter coefficient error $E_i(k)$ be defined by

$$E_i(k) = E\{W_{i,k}\} - E\{W_k^{\infty}\} \tag{9.65}$$

Then, using (9.64) and (9.62), we find (see Problem 9.16.3)

$$E_{i+1} = \left(1 - \mu E\left\{\left|X_i\right|^2\right\}\right)E_i(k) \quad k = 0, 1, 2, 3, \ldots, M-1 \tag{9.66}$$

9.18 THE ERROR NORMALIZED STEP-SIZE LMS ALGORITHM

The variable error normalized step size (ENSS) depends on the optimization problem as follows:

Define **a posteriori** error $e_p(n)$ as

$$e_p(n) = d(n) - \mathbf{w}^T(n+1)\mathbf{x}(n) \tag{9.67}$$

Then, we can define the following constraint optimization problem (Sayed 2003):

$$\min_{w} \left\|\mathbf{w}(n) - \mathbf{w}(n-1)\right\|^2 \text{ subject to } e_p(n) = \left[1 - \mu\frac{\left\|\mathbf{x}(n)\right\|^2}{1+\mu\left\|\mathbf{e}_L(n)\right\|^2}\right]e(n) \tag{9.68}$$

where:

$$\left\|\mathbf{e}(n)\right\|^2 = \sum_{i=0}^{L-1}\left|e(n-i)\right|^2 \tag{9.69}$$

By defining the difference as

$$\Delta\mathbf{w}(n) = \mathbf{w}(n+1) - \mathbf{w}(n) \tag{9.70}$$

and multiplying both sides of (9.70) by $\mathbf{x}^T(n)$ from left, we obtain

$$\mathbf{x}^T(n)\Delta\mathbf{w}(n) = \mathbf{x}^T(n)\mathbf{w}(n+1) - \mathbf{x}^T(n)\mathbf{w}(n) \tag{9.71}$$

By adding and subtracting $d(n)$ from the right-hand side of (9.71), we have

$$\mathbf{x}^T(n)\Delta\mathbf{w}(n) = e(n) - e_p(n) \tag{9.72}$$

Substituting (9.68) in (9.72), we obtain

$$\mathbf{x}^T(n)\Delta\mathbf{w}(n) = \mu e(n)\frac{\left\|\mathbf{x}(n)\right\|^2}{1+\mu\left\|\mathbf{e}_L(n)\right\|^2} \tag{9.73}$$

There are infinite many solutions to (9.73). The solution that results in a minimum value of $\Delta\mathbf{w}(n)$ in the Euclidean second norm sense is the one taken directly from (9.73) (Sayed 2003), that is,

$$\Delta\mathbf{w}(n) = \mu e(n)\frac{\mathbf{x}(n)}{1+\mu\left\|\mathbf{e}_L(n)\right\|^2} \tag{9.74}$$

$$\left[\mathbf{x}^T(n)\mathbf{x}(n) = \left\|\mathbf{x}(n)\right\|^T\right]$$

Equation 9.74 can be proved by assuming that $[\Delta\mathbf{w}(n)+z]$ is any other solution, where z is a column vector. The Euclidian vector norm of this solution can be written as

$$\begin{aligned}\|\Delta w(n)+z\|^2 &= (\Delta w(n)+z)^T(\Delta w(n)+z)\\ &= \|\Delta w(n)\|^2+\|z\|^2+\Delta w^T(n)z+z^T\Delta w(n)\end{aligned} \tag{9.75}$$

Substituting (9.74) in (9.75) yields

$$\|\Delta w(n)+z\|^2 = \|\Delta w(n)\|^2+\|z\|^2+\mu e(n)\frac{x^T(n)z}{1+\mu\|e_L(n)\|^2}+\mu e(n)\frac{z^T x(n)}{1+\mu\|e_L(n)\|^2} \tag{9.76}$$

Since $\Delta w(n)$ is a solution, we conclude from (9.74) that $x^T(n)z = 0 = z^T x(n)$, and hence, (9.76) becomes

$$\|\Delta w(n)+z\|^2 = \|\Delta w(n)\|^2+\|z\|^2 \tag{9.77}$$

The right-hand side of (9.77) is larger than $\|\Delta w(n)\|^2$ for any nontrivial vector z. Thus, the minimum value of $\|\Delta w(n)+z\|^2$ is $\|\Delta w(n)\|^2$ (i.e., when $z=\mathbf{0}$), which proves (9.74). Now, substituting (9.74) in (9.70), we obtain the ENSS LMS algorithm:

$$w(n+1) = w(n)+\frac{\mu x(n)}{1+\mu\|e_L(n)\|^2}x(n)e(n) \tag{9.78}$$

The fractional quantity in (9.78) represents the variable step size of the algorithm, that is,

$$\mu(n) = \frac{\mu}{1+\mu\|e_L(n)\|^2} \quad \text{or} \quad \mu(n) = \frac{1}{1+\dfrac{1}{\mu}\|e_L(n)\|^2} \tag{9.79}$$

The above equation shows that $\mu(n)$ is an increasing function of the step size μ. In a stationary environment, the best choice of L is $L = n$, which in turn makes $\mu(n)$ a monotonic decreasing function of μ. Clearly, large values of μ should be used in this case to increase $\mu(n)$, thus obtaining the fast rate of convergence mainly at the early stages of adaptation, where the error value is large. In stationary environments, the length of the error vector should be constant (L = constant value) such that the algorithm will adapt to statistical input data. In this case, as n increases, $\|e_L(n)\|^2$ decreases and $\mu(n)$ increases to a maximum value μ, which is the step size of the LMS algorithm.

The proposed Book algorithm performs better than many other algorithms that have been examined in stationary environments for both small and large filter lengths.

Example 9.17.1

Compare the ENSS LMS algorithm with the LMS and NLMS algorithms. To check the algorithms, we used the identification of an unknown system as shown in Figure 9.21. The results are shown in Figure 9.22.

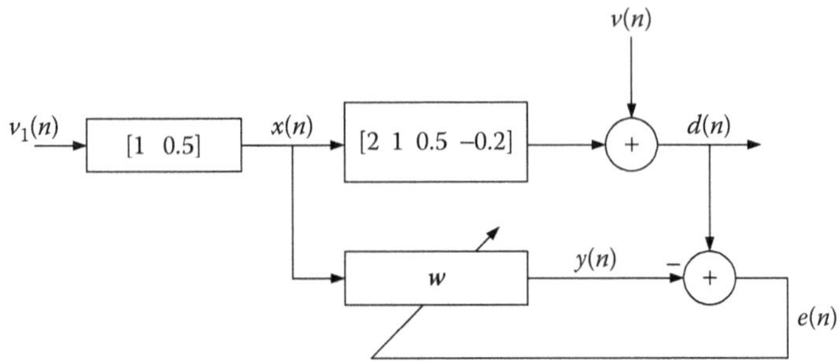

FIGURE 9.21

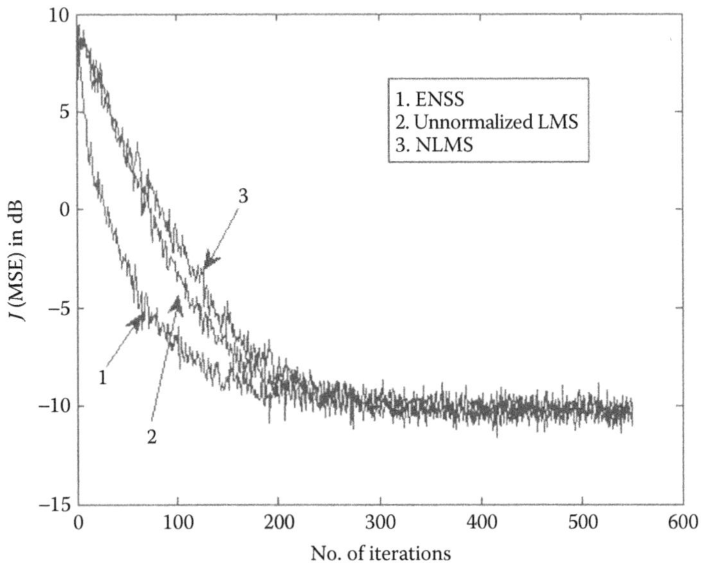

FIGURE 9.22

Solution: The following Book m-functions and program were used.

Book m-Functions for ENSS Application

```
function[J,w]=lms_error_normalized_SS(h,avn,N,mu1,M)
J=zeros(1,N);
for i=1:avn
      x=filter([1 0.5],1,randn(1,N));y=zeros(1,N);
      w=zeros(1,M);e=zeros(1,N);X=zeros(1,M);
      d=filter(h,1,x);v=0.3*randn(1,N);
      D=0;
      %x=input to the unknown filter h;
      %y=output from the adaptive filter;
      %avn=number of averaging;d=output from
      %the unknown filter plus system noise;
```

```
        %M=number of adaptive filter coefficients;
    for k=1:N
        X=[x(k) X(1:M-1)];
        den=X*X'+0.0001;
        y=w*X';
        e(k)=d(k)+v(k)-y;
        D=D+e(k)^2;
        den1=D*mu1+1;
        w=w+(mu1/den1)*e(k)*X;
        J(k)=J(k)+abs(e(k))^2;
    end
end
J=J/avn;

function[J,w]=lms_average_normalized_lms(h,avn,N,mu1,M)
J=zeros(1,N);
for i=1:avn
    x=filter([1 0.5],1,randn(1,N));y=zeros(1,N);
    w=zeros(1,M);e=zeros(1,N);X=zeros(1,M);
    d=filter(h,1,x);v=0.3*randn(1,N);
        %J=learning curve(MSE);h=unknownfilter;
        %w=adaptive filter coefficients;v=internal
        %system noise;avn=number of times been
        %averaged;M=number of adaptive filter
        %coefficients;N=number of data;
    for k=1:N
        X=[x(k) X(1:M-1)];
        den=X*X'+0.0001;
        y=w*X';
        e(k)=d(k)+v(k)-y;
        w=w+(mu1/den)*e(k)*X;
        J(k)=J(k)+abs(e(k))^2;
    end
end
J=J/avn;

function[J,w]=lms_average_unnormalized_lms(h,avn,N,mu1,M)
J=zeros(1,N);
    %for explanation see the above two functions;
for i=1:avn
    x=filter([1 0.5],1,randn(1,N));y=zeros(1,N);
    w=zeros(1,M);e=zeros(1,N);X=zeros(1,M);
    d=filter(h,1,x);v=0.3*randn(1,N);
    for k=1:N
        X=[x(k) X(1:M-1)];
        y=w*X';
        e(k)=d(k)+v(k)-y;
        w=w+mu1*e(k)*X;
        J(k)=J(k)+abs(e(k))^2;
    end
end
J=J/avn;
```

Book MATLAB Program

```
>>[J,w]=lms_error_normalized_SS([210.5-0.2],200,1000, … 0.8, 6);
>>[J1,w1]=lms_average_normalized_lms([210.5-0.2], 200, … 1000,
      0.06, 6);
>>[J2,w2]=lms_average_unnormalized_lms([210.5-0.2], 200, …
      1000, 0.01,6);
>>plot(10*log10(J(1,1:550)),'k'); hold on; >>plot(10*log10(J1(1,
      1:550)),'k');
>>hold on; plot(10*log10(J2(1,1:550)),'k');xlabel('No. of
      iterations');
>>ylabel('J (MSE) in dB');
```

■

9.19 THE ROBUST VARIABLE STEP-SIZE LMS ALGORITHM

Based on regularization Newton's recursion (Sayed 2003), we write

$$\boldsymbol{w}(n+1) = \boldsymbol{w}(n) + \mu(n)[\varepsilon(n)\boldsymbol{I} + \boldsymbol{R}_x]^{-1}[\boldsymbol{p} - \boldsymbol{R}_x\boldsymbol{w}(n)] \tag{9.80}$$

where:

n is the iteration number
$\boldsymbol{w}$ is an $M \times 1$ vector of adaptive filter weights
$\varepsilon(n)$ is an iteration-dependent regularization parameter
$\mu(n)$ is an iteration-dependent step size
I is an $M \times M$ identity matrix
$\boldsymbol{p}(n) = E\{d(n)\boldsymbol{x}(n)\}$ is the cross-correlation vector between the desired signal $d(n)$ and the input signal $\{\boldsymbol{x}(n)\}$
$\boldsymbol{R}_x(n) = E\{\boldsymbol{x}(n)\boldsymbol{x}^T(n)\}$ is the autocorrelation matrix of $\boldsymbol{x}(n)$

Writing (9.80) in the LMS form by replacing $\boldsymbol{p}$ and $\boldsymbol{R}_x$ by their instantaneous approximations $[d(n)\boldsymbol{x}(n)]$ and $[\boldsymbol{x}(n)\boldsymbol{x}^T(n)]$, respectively, with the appropriate proposed weights, we obtain

$$\boldsymbol{w}(n+1) = \boldsymbol{w}(n) + \mu\left\|\boldsymbol{e}_L(n)\right\|^2 \times [\alpha\left\|\boldsymbol{e}_L(n)\right\|^2 \boldsymbol{I} + \gamma\boldsymbol{x}(n)\boldsymbol{x}^T(n)]^{-1}\boldsymbol{x}(n)e(n) \tag{9.81}$$

where:

μ is a positive constant step size
α and γ are positive constants
$e(n)$ as the system output error is defined by

$$e(n) = d(n) - \boldsymbol{x}^T(n)\boldsymbol{w}(n) \tag{9.82}$$

$$\left\|\boldsymbol{e}(n)\right\|^2 = \sum_{i=0}^{n-1}\left|e(n-i)\right|^2 \tag{9.83}$$

$$\|e_L(n)\|^2 = \sum_{i=0}^{L-1} |e(n-i)|^2 \tag{9.84}$$

Equation 9.83 is the square norm of the error vector, $e(n)$, estimated over its entire updated length n, and (9.84) is the square norm of the error vector, $e(n)$, estimated over the last L values.

Expanding (9.81) and applying the matrix inversion formula (see Appendix 2),

$$[A + BCD]^{-1} = A^{-1} - A^{-1}B[C^{-1} + DA^{-1}B]^{-1} DA^{-1} \tag{9.85}$$

with $A = \alpha\|e(n)\|^2 I$, $B = x(n)$, $C = \gamma$, and $D = x^T(n)$, we obtain

$$\begin{aligned}\left[\alpha\|e(n)\|^2 I + \gamma x(n)x^T(n)\right]^{-1} &= \alpha^{-1}\|e(n)\|^{-2} I \\ &\quad - \alpha^{-1}\|e(n)\|^{-2} Ix(n)\frac{x^T(n)\alpha^{-1}\|e(n)\|^{-2}}{\gamma^{-1} + x^T(n)\alpha^{-1}\|e(n)\|^{-2} x(n)}\end{aligned} \tag{9.86}$$

Multiplying both sides of (9.86) by $x(n)$ from the right and rearranging the equation, we have (see Problem 9.18.1)

$$\left[\alpha\|e(n)\|^2 I + \gamma x(n)x^T(n)\right]^{-1} x(n) = \frac{x(n)}{\alpha\|e(n)\|^2 + \gamma\|x(n)\|^2} \tag{9.87}$$

Substituting (9.87) in (9.81), we obtain a new proposed robust variable step size (RVSS):

$$w(n+1) = w(n) + \frac{\mu\|e_L(n)\|^2}{\alpha\|e(n)\|^2 + (1-\alpha)\|x(n)\|^2} x(n)e(n) \tag{9.88}$$

where we replaced γ by $(1-\alpha) \geq 0$ in (9.88) without loss of generality. The fractional quantity in (9.88) may be viewed as a time-varying step size, $\mu(n)$, of the proposed RVSS algorithm. Clearly, $\mu(n)$ is controlled by normalized of both error and input data.

The parameters α, L, and μ are appropriately chosen to achieve the best trade-off between the rate of convergence and low final mean-square error. A small constant ε could be added to the denominator of (9.88) to prevent instability of the algorithm if the denominator becomes too small. The quantity $\|e_L(n)\|^2$ is large at the beginning of adaptation, and it decreases as n increases, while $\|x(n)\|^2$ fluctuates depending on the recent values of the input signal. However, $\|e(n)\|^2$ is an increasing function of n since $e(n)$ is a vector of increasing length. To compute (9.83) with minimum computational complexity, the error value produced in the first iteration is squared and stored. The error value in the second iteration is squared and added to the previous stored value. Then, the result is stored in order to be used in the next iteration and so on.

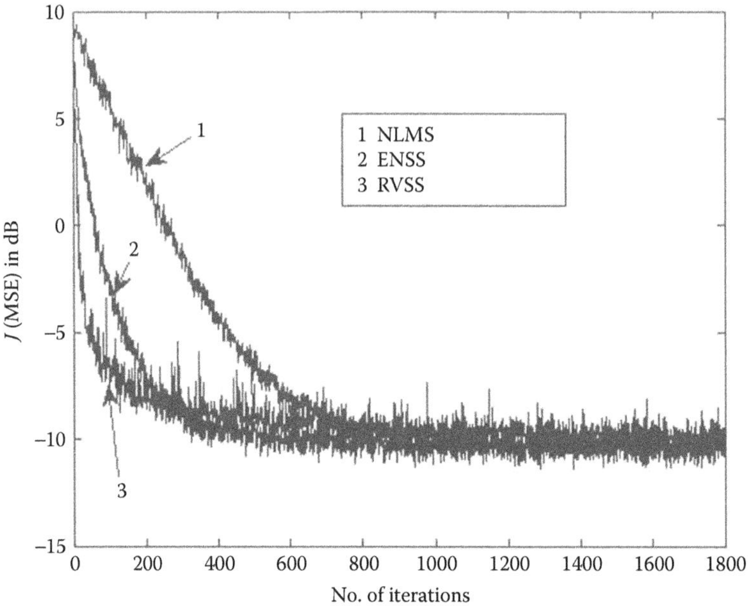

FIGURE 9.23

Example 9.18.1

Use the unknown system shown in Figure 9.21, study the RVSS algorithm with respect to normalized and ENSS algorithms. Figure 9.23 shows the desired results.

Solution: The following Book m-function and program were used.

Book m-Function for RVSS Algorithm

```
function[J,w]=lms_robust_variable_SS(avn,N,mu1,h,M)
J=zeros(1,N);
     %M>length(h);J=MSE-learning curve;avn=number of averaging
     %the learning curve; N=number of inputs;M=number
     %of adaptive filter coefficients; h=unknown system;
     %mu1=constant step-size;v=internal system noise;
for i=1:avn
     y=zeros(1,N);w=zeros(1,M);e=zeros(1,N);
     X=zeros(1,M);D=zeros(1,M);
     x=filter([1 0.5],1,0.3*randn(1,N));v=0.1*randn(1,N);
     for k=1:N
         d=filter(h,1,x)+v;
         X=[x(k) X(1,1:M-1)];
         den=X*X'+0.0001;
         y=w*X';
         e(k)=d(k)-y;
         D=[e(k) D(1,1:M-1)];
         denx=D*D';
```

```
            a=0.5;
            mu=(mu1*denx)/((a*den+(1-a)*e*e'));
            w=w+mu*e(k)*X;
            J(k)=J(k)+abs(e(k))^2;
        end;
end;
J=J/avn;
```

Book MATLAB Program

```
>>[J,w]=lms_robust_variable_SS(300,2000,0.07,[2 1 0.5 ...
     -0.2],6);
>>[J1,w1]=lms_error_normalized_SS([2 1 0.5 - ...
     0.2],300,2000,0.15,6);
>>[J2,w2]=lms_average_normalized_lms([2 1 0.5 - ...
     0.2],300,2000,0.02,6);
>>plot(10*log10(J(1,1:1800)),'k');
>>hold on;plot(10*log10(J1(1,1:1800)),'k');
>>hold on;plot(10*log10(J2(1,1:1800)),'k');xlabel('No. of ...
     iterations');
>>ylabel(J, MSE in dB');
```

The reader will observe that RVSSs adjust faster than the other algorithms.

■

Example 9.18.2

Use NLMS to study the results for an abrupt change in the unknown system, from h becomes $-h$, which must be identified. The unknown system is shown in Figure 9.21.

Solution: The following Book m-functions and program were used to produce Figures 9.24 and 9.25. The value of $\boldsymbol{w}$ that was found with the constants used was 1.9988 0.9973 0.5573 −0.2052 0.0055 0.0519.

```
function[J,mukav]=lms_abrupt_change_normalized_ ...
     lms(avn,N,M,mu1)
     %N must be even;muk=gives the variation of step-size;
     %N=number of data;M>=length(h);avn=number of averaging;
     %v=internal system noise;mu1=step-size constant;J=MSE;
J=zeros(1,N);muk=zeros(1,N);
for i=1:avn
     h=[2 1 0.5 -0.2];
     y=zeros(1,N);w=zeros(1,M);e=zeros(1,N);
     X=zeros(1,M);D=zeros(1,M);
     x=filter([1 0.5],1,0.3*randn(1,N));
     v=0.1*randn(1,N);
     for k=1:N
         d=filter(h,1,x)+v;
         X=[x(k) X(1:M-1)];
         den=X*X'+0.0001;
         y=w*X';
```

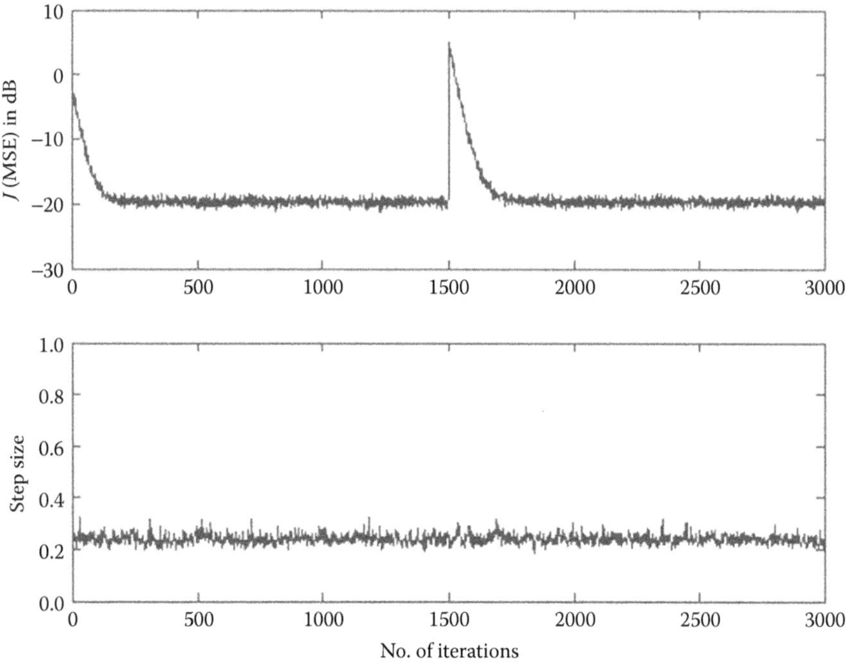

FIGURE 9.24

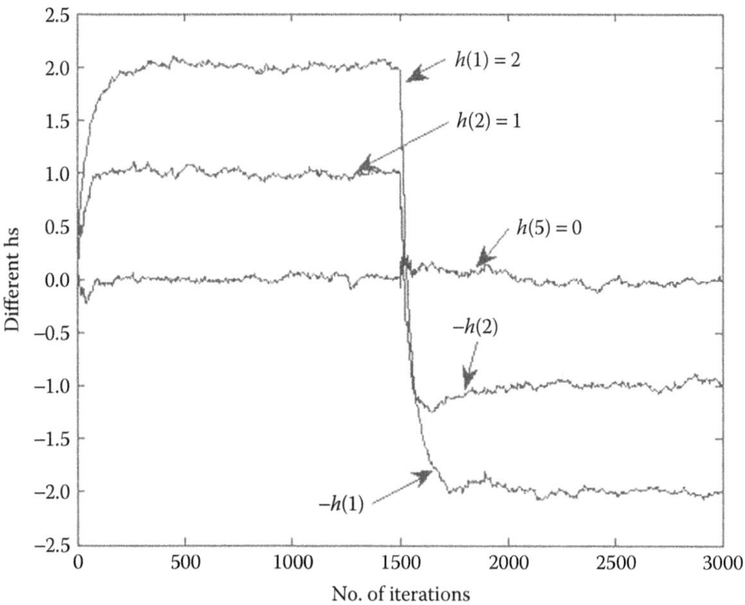

FIGURE 9.25

```
            e(k)=d(k)-y;
               if k==N/2;%N must be even;
                     h=-h;
               end;
               w=w+(mu1/den)*e(k)*X;
               muk(k)=muk(k)+(mu1/den);
               J(k)=J(k)+abs(e(k))^2;
         end;
end;
mukav=muk/avn;
J=J/avn;

function[w1,w]=lms_abrupt_change_normalized_lms_w(N,M,mu1)
     %N must be even;muk=gives the variation of step-size;
     %N=number of data;M>=length(h);avn=number of averaging;
     %v=internal system noise;mu1=step-size constant;J=MSE;
h=[2 1 0.5 -0.2];
y=zeros(1,N);w=zeros(1,M);e=zeros(1,N);
X=zeros(1,M);D=zeros(1,M);
x=filter([1 0.5],1,0.3*randn(1,N));
v=0.1*randn(1,N);
for k=1:N
     d=filter(h,1,x)+v;
     X=[x(k) X(1:M-1)];
     den=X*X'+0.0001;
     y=w*X';
     e(k)=d(k)-y;
     if k==N/2;%N must be even;
           h=-h;
     end;
     w=w+(mu1/den)*e(k)*X;
     w1(k,:)=w(1,:);
end;
```

Book MATLAB Program

```
>>[J,mukav]=lms_abrupt_change_normalized_lms(200,3000,6,0.1);
>>[w1,w]=lms_abrupt_change_normalized_lms_w(3000,6,0.1);
```

For Figure 9.24, we used the following commands:

```
>>subplot(2,1,1);plot(10*log10(J),'k');
>>xlabel('No. of iterations');
>>ylabel('J, MSE in dB');subplot(2,1,2);plot(mukav,'k');
>>axis([03000 0 1]);
>>ylabel('Step-size');xlabel('No. of iterations');
```

For Figure 9.25, we used the following commands:

```
>>plot(w1(:,1),'k');hold on;plot(w1(:,2),'k');
>>hold on;plot(w1(:,5),'k');
>>xlabel('No. of iterations');ylabel('Different hs');
```

■

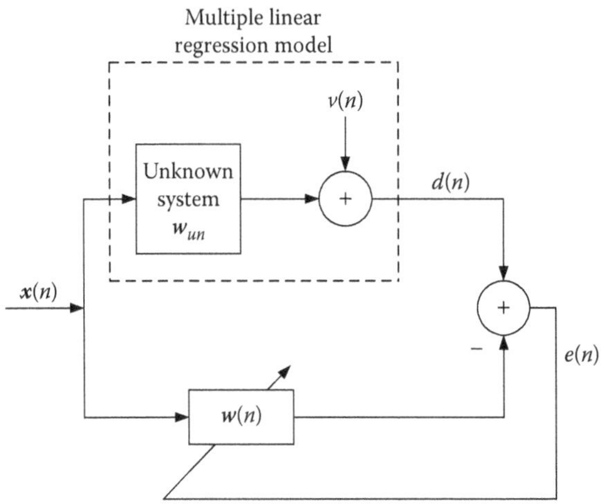

FIGURE 9.26

9.20 THE MODIFIED LMS ALGORITHM

From Figure 9.26, the desire signal is the output of the linear regression model system, and it is given by (Haykin 1996)

$$d(n) = \boldsymbol{w}_{\text{un}}^T \boldsymbol{x}(n) + v(n) \tag{9.89}$$

where:

$\boldsymbol{w}_{un}$ is the coefficient vector of the unknown system
$\boldsymbol{x}(n)$ is the input vector (regressor)
$v(n)$ is a nonmeasurable white noise with variance σ_v^2

The LMS FIR filter is chosen to minimize the following index of performance:

$$J(\boldsymbol{w},K) = E\left\{e^{2K}(n)\right\} \quad K = 1, 2, 3 \ldots \tag{9.90}$$

From the above equation and using the instantaneous gradient vector, we find the **modified** LMS filter (Haykin 1996):

$$\boldsymbol{w}(n+1) = w(n) + \mu K \boldsymbol{x}(n) e^{2K-1}(n) \tag{9.91}$$

The following Book m-function simulates the above equation:

Book m-Function for Modified LMS Filter

```
function[w,e,Jav]=lms_modified_lms(mu,av,a,b,N,M,K)
for m=1:av%av=integer equal to the number of desired
      %averaging;a=multiplier of the random noise,
      %a*randn(1,N);
    w=zeros(1,M);
```

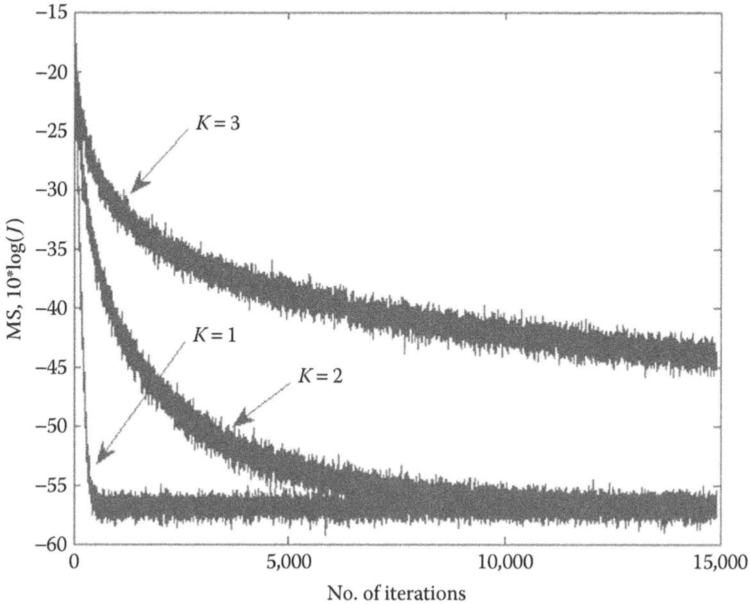

FIGURE 9.27

```
        x=a*(rand(1,N)-0.5);
        d=filter([0.9 0.4-0.2],1,x)+b*(rand(1,N)-0.5);
              %multiplier of multilinear regressor model;
        for n=M:N
              x1=x(n:-1:n-M+1);
              y(n)=w*x1';
              e(n,m)=d(n)-y(n);%e(n,m) becomes a Nxav
                   %matrix;
              w=w+mu*K*e(n,m)^(2*K-1)*x1;
        end
        y=zeros(1,N);
end
Jav=sum((e.^2),2)/av;%e.^2 squares each matrix element;
```

Figure 9.27 shows the results with the following inputs: $mu = 0.04$, av = 300 (ensemble number), $a = 1.5$, $b = 0.2$, $N = 15{,}000$, $M = 6$. It is observed that the modified LMS filter is not as efficient as the standard LMS filter.

■

9.21 MOMENTUM LMS

Another LMS type filter has been proposed, known as **momentum LMS**, which is given by

$$\boldsymbol{w}(n+1) = \boldsymbol{w}(n) + (1-g)[\boldsymbol{w}(n) - \boldsymbol{w}(n-1)] + ag\mu e(n)\boldsymbol{x}(n) \tag{9.92}$$

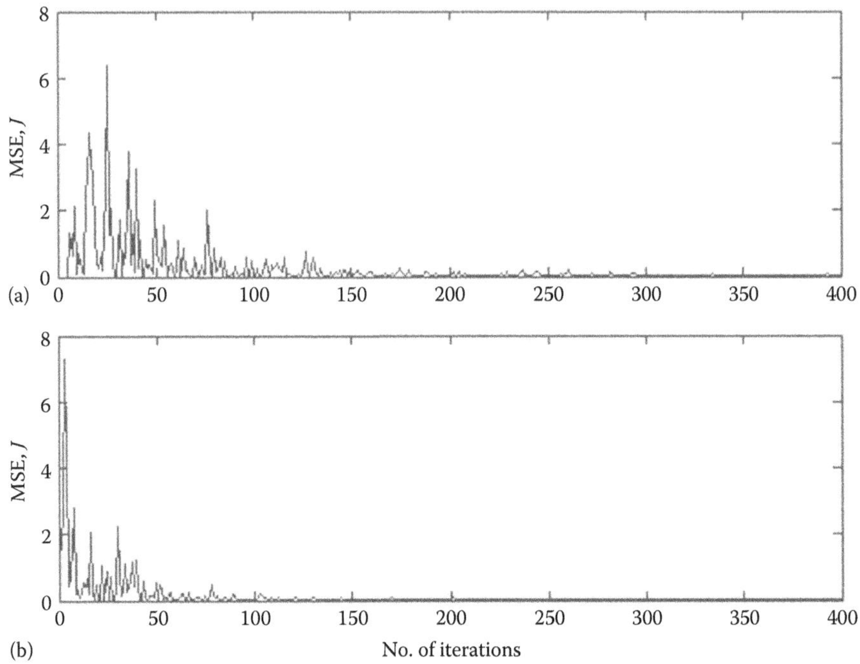

FIGURE 9.28

where:

$$0 < g < 0$$
$$a > 1$$

The following Book m-file was used to produce Figure 9.28.

Book m-Function for Momentum LMS

```
function[w,y,e,J,w1]=lms_momentum_lms(x,d,mu,g,a,M)
      %0<g<1,
      %function[w,y,e,J,w1]=lms_momentum_lms(x,dn,mu,M);
      %all quantities are real valued;
      %x=input data to the filter; d=desired signal;
      %M=order of the filter;
      %mu=step-size factor; x and d must be
      %of the same length, in the program they are
      %reversed;
      %w1=a matrix of dimensions: length(x)xM,
      %each column represents the variation of
      %each filter coefficient;
N=length(x);w(2,1:M)=zeros(1,M);w1=zeros(1,M);
w(1,1:M)=zeros(1,M);
xr=fliplr(x);dr=fliplr(d);
```

```
for n=1:N-M
        x1=xr(n:1:M+n-1);%for each n the vector x1 is
                         %of length M;
        y(n)=w(n+1,1:M)*x1';
        e(n)=dr(n)-y(n);
        w(n+2,:)=w(n+1,:)+(1-g)*(w(n+1,:)-w(n,:))+g*a*mu*e(n)*x1;
        w1(n,:)=w(1,:);
end;
J=e.^2;%J is the learning curve of the adaptation;
```

Figure 9.28a is the result of the standard LMS filter with the constants: mu = 0.06, $M = 6$. Figure 9.28b is the result of the momentum LMS with the following constants: mu = 0.06, $M = 6$, $g = 0.4$, $a = 4$. For comparison, we used the system identification problem with an FIR system with coefficients $[2 \;\; 4 \;\; -1]$. The coefficients after 1900 iterations were found to be (1) 2.0020 3.9968 −0.9960 −0.0039 0.0031 −0.0019 and (2) 2.0000 4.0000 −1.0000 −0.0000 0.0000 −0.0000.

9.22 THE BLOCK LMS ALGORITHM

We can subdivide a set of data in L sections and process the data sequentially. The following Book m-function processes the data in a block-by-block format.

Book m-Function for Block LMS Processing

```
function[w,y,e,J,w1]=lms_block_lms(x,d,mu,M,L)
        %all quantities are real valued;
        %x=input data to the filter; d=desired signal;
        %M=order of the filter;
        %mu=step-size factor; x and d must be
        %of the same length, in the program they are
        %reversed;N/L must be an integer;
        %w1=a matrix of dimensions: length(x)x(M+1),
        %each column represents the variation of
        %each filter coefficient;
N=length(x);w(2,1:M+1)=zeros(1,M+1);w1=zeros(1,M+1);
w(1,1:M+1)=zeros(1,M+1);
xr=fliplr(x);dr=fliplr(d);
for m=1:L-1
        if m*(N/L)+1>N
        end;
        for n=((m-1)*(N/L)+1):1:m*(N/L)
                x1=xr(n:1:M+n);%for each n the vector x1 is of
                length M+1;
                y(n)=w(n+1,1:M+1)*x1';
                e(n)=dr(n)-y(n);
                w(n+2,:)=w(n+1,:)+(mu/L)*e(n)*x1;
                w1(n,:)=w(1,:);
        end;
end;
J=e.^2;%J is the learning curve of the adaptation;
```

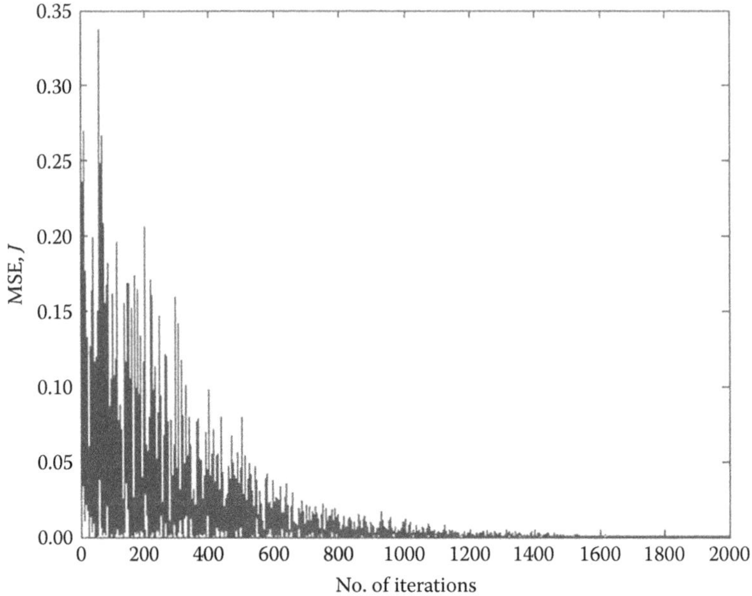

FIGURE 9.29

Using the identification of an FIR system with impulse response $h = [0.9\ \ 0.4\ \ -0.1]$, with a zero-mean white noise as input we obtain the learning curve shown in Figure 9.29. The constants used were as follows: $N = 12000$, $L = 4$, $mu = 0.08$, $d = \text{Filter}([0.9\ \ 0.4\ \ -0.1],1,x)$. The values of the adaptive filter were as follows: at the 3000th iteration $w(3000,:)$ = 0.8945 0.3975 −0.0991 0.0003 −0.0001 0.0003 0.0002; at the 6000th iteration $w(6000,:)$ = 0.9000 0.4000 −0.1000 −0.0000 −0.0000 −0.0000 0.0000.

9.23 THE COMPLEX LMS ALGORITHM

The complex LMS algorithm is given in Table 9.5. Figure 9.30 shows the two learning curves: one for unnormalized and the other for normalized complex LMS adaptive filters. It shows that after the first about 100 iterations, the normalized filter performs better. The input for both the unknown system and the adaptive filter was $x = \text{randn}(1,12000) + j0.2\text{filter}([0.9\ \ 0.2],1,\text{rand}(1,12000) - 0.5)$. The output from the unknown filter was $d = \text{Filter}([0.9\ \ 0.4\ \ -0.1],1,x)$. To obtain the unknown filter, we write real(w) = 0.9092 0.4052 −0.1044 −0.0053 −0.0011 −0.0069.

Book m-File for Normalized Complex LMS Filter

```
function[w,y,e,J,w1,Js]=lms_complex_normalized_lms(x,dn,mu,M)
      %x=input data to the filter; dn=desired signal;
      %M=order of the filter;
      %mu=step-size factor; x and dn must be
```

```
        %of the same length;
        %Js=smooths the learning curve;
        %w1=a matrix of dimensions: length(x)xM,
        %each column represents the variation of
        %each filter coefficient;
N=length(x);w=zeros(1,M);w1=zeros(1,M);
for n=M:N
```

TABLE 9.5
The Complex LMS Algorithm

Unnormalized

$\boldsymbol{x}(0) = \boldsymbol{w}(0) = [0 \quad 0 \quad 0 \cdots 0]^T$

For $n \geq 0$

$e(n) = d(k) - \boldsymbol{w}^H(n)\boldsymbol{x}(n)$

$\boldsymbol{w}(n+1) = \boldsymbol{w}(n) + \mu e^*(n)\boldsymbol{x}(n)$

Normalized

For $n \geq 0$

$e(n) = d(k) - \boldsymbol{w}^H(n)\boldsymbol{x}(n)$

$\boldsymbol{w}(n+1) = \boldsymbol{w}(n) + \mu e^*(n)\boldsymbol{x}(n)/[0.0001 + \boldsymbol{x}(n)\boldsymbol{x}^H(n)]$

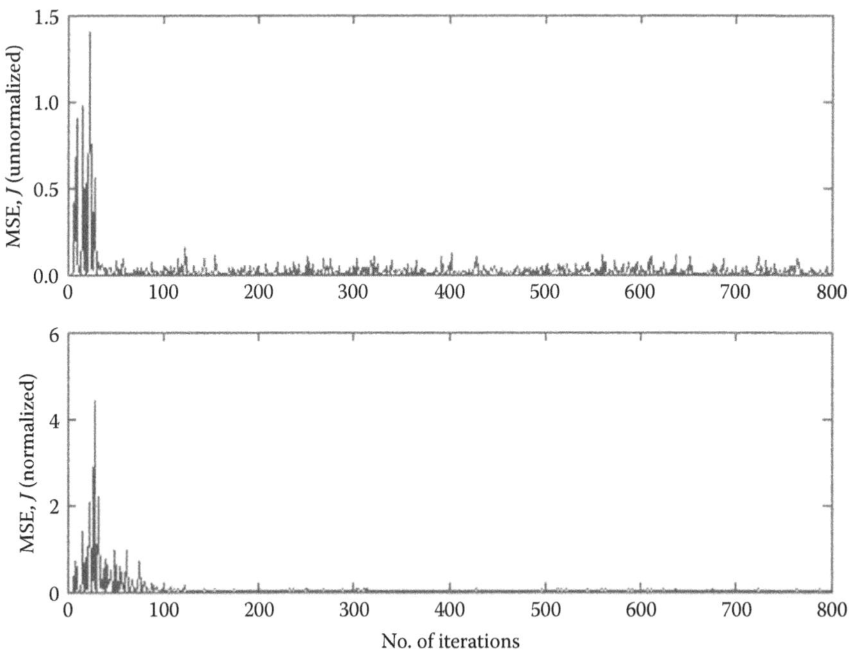

FIGURE 9.30

```
        x1=x(n:-1:n-M+1);%for each n the vector x1 is
              %of length M with elements from x in
              %reverse order;
        y(n)=conj(w)*x1';
        e(n)=dn(n)-y(n);
        w=w+2*mu*conj(e(n))*x1/(0.00001+x1*x1');
        w1(n-M+1,:)=w(1,:);
end;
J=conj(e).*e;%J is the learning curve of the adaptation;
for n=1:length(x)-5
        Js(n)=(J(n)+J(n+1)+J(n+2))/3;
end;
```

Book m-File for Unnormalized Complex LMS Filter

```
function[w,y,e,J,w1,Js]=lms_complex_lms(x,dn,mu,M)
        %x=input data to the filter; dn=desired signal;
        %M=order of the filter;
        %mu=step-size factor; x and dn must be
        %of the same length;
        %Js=smooths the learning curve;
        %w1=a matrix of dimensions: length(x)xM,
        %each column represents the variation of
        %each filter coefficient;
N=length(x);w=zeros(1,M);w1=zeros(1,M);
for n=M:N
        x1=x(n:-1:n-M+1);%for each n the vector x1 is
              %of length M with elements from x in
              %reverse order;
        y(n)=conj(w)*x1';
        e(n)=dn(n)-y(n);
        w=w+2*mu*conj(e(n))*x1;
        w1(n-M+1,:)=w(1,:);
end;
J=conj(e).*e;%J is the learning curve of the adaptation;
for n=1:length(x)-5
        Js(n)=(J(n)+J(n+1)+J(n+2))/3;
end;
```

9.24 THE AFFINE LMS ALGORITHM

The algorithm is defined as follows (see Chapter 10):

First we create the signal matrix, which is of the form:

$$\boldsymbol{X}(n)=\begin{bmatrix} x(n) & x(n-1) & \cdots & x(n-N+1) & x(n-N) \\ x(n-1) & x(n-2) & \cdots & x(n-N) & x(n-N-1) \\ \vdots & \vdots & & \vdots & \vdots \\ x(n-M) & x(n-M-1) & \cdots & x(n-N-M+1) & x(n-N-M) \end{bmatrix} \tag{9.93}$$

The output vector from the adaptive filter is given by

$$\boldsymbol{y}(n) = \boldsymbol{X}^T(n)\boldsymbol{w}(n) = \begin{bmatrix} y_0(n) \\ y_1(n) \\ \vdots \\ y_N(n) \end{bmatrix} \tag{9.94}$$

The desired and error signals are given by

$$\boldsymbol{d}(n) = \begin{bmatrix} d(n) \\ d(n-1) \\ \vdots \\ d(n-N) \end{bmatrix}, \ \boldsymbol{e}(n) = \begin{bmatrix} e_0(n) \\ e_1(n) \\ \vdots \\ e_N(n) \end{bmatrix} = \begin{bmatrix} d(n) - y_0(n) \\ d(n-1) - y_1(n) \\ \vdots \\ d(n-N) - y_N(n) \end{bmatrix} = \boldsymbol{d}(n) - \boldsymbol{y}(n) \tag{9.95}$$

The affine algorithm is given in Table 9.6.

The $\boldsymbol{w}$ has dimensions $M \times (N - M + 1)$. Therefore, we observe the evolution of $\boldsymbol{w}$ toward its exact value. The following Book m-file was used to identify the unknown system: $h = \begin{bmatrix} 0.9 & 0.4 & -0.1 \end{bmatrix}$. In this case, we used the following constants: $N = 200$, $M = 5$, $\mu = 0.05$, $g = 0.2$, $x = \text{filter}([0.9\ \ 0.2], 1, \text{rand}(1,200) - 0.5)$, $d = \text{filter}\left([0.9\ \ 0.4\ \ -0.1], 1, x\right)$.

Book m-File for System Identification

```
function[w,y,e,J]=lms_affine_lms(x,d,mu,M,g)
      %all quantities are real valued;0<g<1
      %x=input data to the filter; dn=desired signal;
      %M=order of the filter;
      %mu=step-size factor; x and dn must be
      %of the same length;
      %Js=smooths the learning curve;
      %w1=a matrix of dimensions: length(x)xM,
      %each column represents the variation of
      %each filter coefficient;
```

TABLE 9.6
The Affine Projection Algorithm

Initialization

$\boldsymbol{x}(0) = \boldsymbol{w}(0) = [0 \ \ 0 \ \cdots \ 0]^T$

γ = small constant

For $n \geq 0$

$\boldsymbol{e}(n) = \boldsymbol{d}(n) - \boldsymbol{X}^T(n)\boldsymbol{w}(n)$

$\boldsymbol{w}(n+1) = \boldsymbol{w}(n) + \mu \boldsymbol{X}(n)(\boldsymbol{X}^T(n)\boldsymbol{X}(n) + \gamma \boldsymbol{I})^{-1}\boldsymbol{e}(n)$

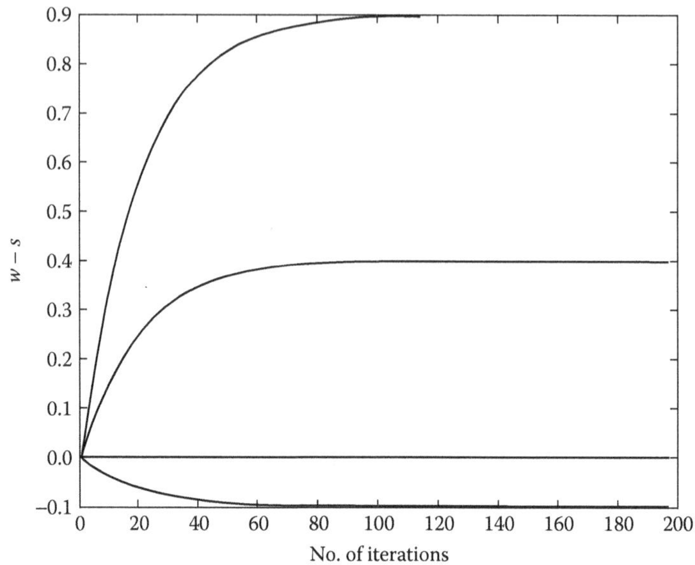

FIGURE 9.31

```
N=length(x);w=zeros(1,M)';
xr=fliplr(x)';dr=fliplr(d)';
for k=1:M
        X(k,:)=xr(k:1:N-M+k);%Mx(N-M+1)
end;
for n=1:N-M+1
        y=X'*w(:,n);%((N-M+1)xM)x(Mx1)=(N-M+1)x1
        e=dr(1:(N-M+1))-y;
        w(:,n+1)=w(:,n)+mu*X*inv(X'*X+g*eye(N-M+1))*e;
end;
J=e.^2;%J is the learning curve of the adaptation;
```

The evolving of the first four out of five coefficients versus the iteration number is shown in Figure 9.31.

9.25 THE COMPLEX AFFINE LMS ALGORITHM

The complex affine LMS algorithm is presented in Table 9.7.

We have tested the algorithm given in the table by using the following constants and vectors to identify an unknown system with $h = [0.9\ \ 0.4\ \ -0.1]$. The input vectors and constants were as follows:

$$x = \text{filter}([0.9\ \ 0.2], 1, 0.5\,\text{randn}(1,200)) + j[\text{rand}(1,200) - 0.5],$$

$$d = \text{filter}([0.9\ \ 0.4\ \ -0.1], 1, x)\mu = 0.05, \gamma = 0.2, M = 5$$

TABLE 9.7
The Complex Affine Algorithm

$\boldsymbol{x}(0) = \boldsymbol{w}(0) = [0 \;\; 0 \;\cdots\; 0]^T$

γ = small constant

For $n \geq 0$

$\boldsymbol{e}^*(n) = \boldsymbol{d}^*(n) - \boldsymbol{X}^H(n)\boldsymbol{w}(n)$

$\boldsymbol{w}(n+1) = \boldsymbol{w}(n) + \mu\boldsymbol{X}(n)(\boldsymbol{X}^H(n)\boldsymbol{X}(n) + \gamma\boldsymbol{I})^{-1}\boldsymbol{e}^*(n)$

Based on the above constants and vectors, we obtained the following approximations:

$$\text{real}(\boldsymbol{w}(:,20)) = 20\text{th iteration} = [0.5590 \;\; 0.2487 \;\; -0.0621 \;\; 0.0000 \;\; -0.0000]^T$$

$$\text{real}(\boldsymbol{w}(:,100)) = 100\text{th iteration} = [0.8943 \;\; 0.3975 \;\; -0.0994 \;\; 0.0000 \;\; -0.0000]^T$$

$$\text{real}(\boldsymbol{w}(:,150)) = 150\text{th iteration} = [0.8996 \;\; 0.3998 \;\; -0.1000 \;\; 0.0000 \;\; -0.0000]^T$$

To obtain the above results, we used the following Book m-function:

Book m-Function for Complex Affine Algorithm

```
function[w,y,e,J]=lms_complex_affine_lms(x,d,mu,M,g)
        %all quantities are real valued;0<g<1
        %x=input data to the filter; d=desired signal;
        %M=order of the filter;
        %mu=step-size factor; x and d must be
        %of the same length;
N=length(x);w=zeros(1,M)';
xr=fliplr(x)';dr=conj(fliplr(d)');
for k=1:M
        X(k,:)=xr(k:1:N-M+k);%Mx(N-M+1)
end;
for n=1:N-M+1
        y=X'*w(:,n);%((N-M+1)xM)x(Mx1)=(N-M+1)x1
        e=dr(1:(N-M+1))-y;
        w(:,n+1)=w(:,n)+mu*X*inv(X'*X+g*eye(N-M+1))*e;
end;
J=e.*conj(e);%J is the learning curve of the adaptation;
```

PROBLEMS

9.1.1 Verify (9.3) and (9.4).

9.1.2 Use NLMS to identify the system shown in Figure 8.5. Plot the filter coefficients for two different step-size values.

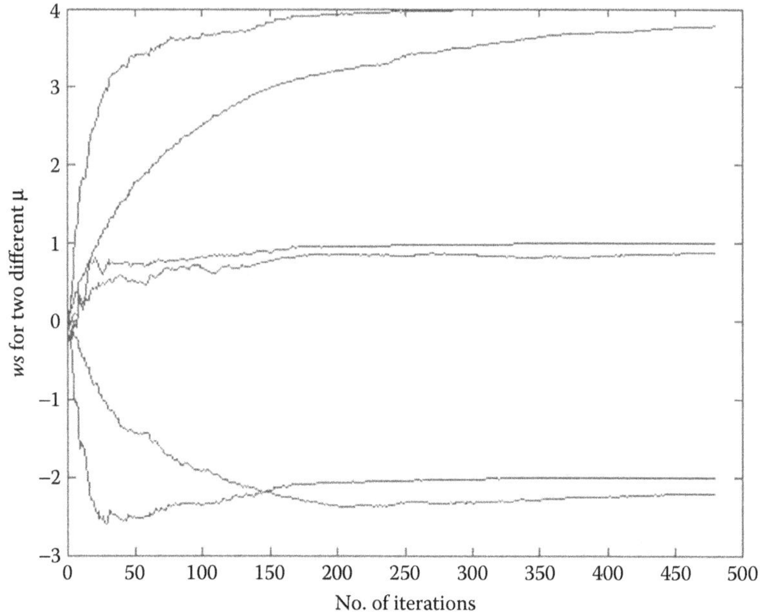

FIGURE P9.1.4.1

9.1.3 Repeat Example 9.1.2 but with an NLMS algorithm.

9.1.4 Plot the average learning function for identifying the system shown in Figure P9.1.4.1. Plot also the average learning curve for an NLMS. Compare with the nonaverage situation.

9.3.1 Find the output of the first and second stages of an SCNLMSF that is shown in Figure 9.8.

9.11.1 Develop the leaky LMS algorithm by minimizing the modified mean-square error:

$$J(n) = e(n)^2 + \gamma \mathbf{w}^T(n)\mathbf{w}(n)$$

9.11.2 Compare the normalized and unnormalized leaky LMS algorithms.

9.12.1 Verify (9.37).

9.12.2 We can derive the NLMS subject to a constraint: Minimize $\min_{\mathbf{w}} \|\mathbf{w}(n+1) - \mathbf{w}(n)\|^2 = \min_{\mathbf{w}}\{[\mathbf{w}(n+1) - \mathbf{w}(n)]^T[\mathbf{w}(n+1) - \mathbf{w}(n)]\}$ subject to the constraint $d(n) = \mathbf{w}^T(n+1)\mathbf{x}(n)$.

9.13.1 Study the application of the LMF algorithm by making changes to different constants and input signals with varying noise strengths.

9.14.1 Study the application of the LMMN LMS algorithm by varying constants and input signals and with varying noise strengths. Compare the results between the standard and the normalized algorithm.

9.16.1 Verify (9.61).

9.16.2 Verify (9.63).

9.16.3 Verify (9.66).
9.18.1 Verify (9.87).
9.18.2 Repeat Examples 9.18.1 and 9.18.2 using uncorrelated input signals.
9.18.3 Repeat Examples 9.18.1 and 9.18.2 using the simple LMS algorithm.

HINTS–SOLUTIONS–SUGGESTIONS

9.1.1

$$\boldsymbol{w}(n+1) = \boldsymbol{w}(n) + \mu(n)e(n)\boldsymbol{x}(n) \tag{1}$$

$$e_{ps}(n) = d(n) - \boldsymbol{w}^T(n+1)\boldsymbol{x}(n) \tag{2}$$

Substituting (2) in (1), we get

$$\begin{aligned} e_{ps}(n) &= d(n) - [\boldsymbol{w}^T(n) + \mu(n)e(n)\boldsymbol{x}^T(n)]\boldsymbol{x}(n) \\ &= d(n) - \boldsymbol{w}^T(n)\boldsymbol{x}(n) - \mu(n)e(n)\boldsymbol{x}^T(n)\boldsymbol{x}(n) \\ &= e(n) - \mu(n)e(n)\left\|\boldsymbol{x}(n)\right\|^2 \end{aligned}$$

$$\frac{\partial e_{ps}^2(n)}{\partial \mu(n)} = 2\left[e(n) - \mu(n)e(n)\left\|\boldsymbol{x}(n)\right\|^2\right] = 0 \Rightarrow \mu(n) = \frac{1}{\left\|\boldsymbol{x}(n)\right\|^2}$$

9.1.2 From Figure P9.1.2, we observe that with $\mu = 0.1$ the adaptive coefficients approach the desired ones within 50–100 iterations. However, for $\mu = 0.02$, they need more than 400 iterations to approach the right value. Also observe that with the smaller step-size value, the curves are smoother. The Book m-function used to obtain Figure P9.1.2 is as follows:

```
function[w,w1,y,x,J]=lms_power_norm_lms_pr9_1_2(mu,M,b,c)
    %mu=step-size factor;
    %M=length of adaptive filter;c=very small number;
    %yc=conv(x,[filter coefficients])=channel
```

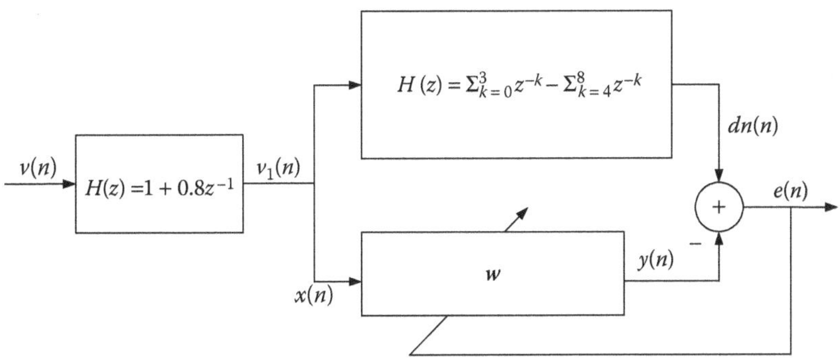

FIGURE P9.1.2

```
        %output;d=yc+v;v=noise;0<b<1;
n=1:1500;p(M)=0;
x=conv([rand(1,1500)-0.5],[0.9 -1 0.2]);
N=length(x);y=zeros(1,N);w=zeros(1,M);
d=conv([1 -2 4],x);
for n=M:N
        x1=x(n:-1:n-M+1);
        p(n+1)=b*p(n)+(1-b)*x1*x1';
        y(n)=w*x1';
        e(n)=d(n)-y(n);
        w=w+(2*mu*e(n)*x1)/(c+p(n+1));
        w1(n-M+1,:)=w(1,:);
end;
J=(e/max(e)).^2;
```

The unknown FIR system had the following coefficients: 1, –2, 4. Also the white noise was passed through an FIR filter with coefficients: 0.9, –1, 0.2.

9.1.3 The Book m-function for averaging LMS and NLMS denoising algorithms is given as follows:

```
function[w,y,e,Jav]=lms_denoising_norm_lms_Jav(N,mu,M,av,a,ep)
for m=1:av%av=integer equal to the number of desired
        %averaging;a=multiplier of the random noise,
        %a*randn(1,N);ep=a very small number, e.g., ep=0.0001;
   w=zeros(1,M);
   dn=sin(0.2*pi*[1:N])+a*randn(1,N);
   x=[0 0 dn(1,1:N-2)];%delay 2 units;
   for n=M:N
        x1=x(n:-1:n-M+1);
        y(n)=w*x1';
        e(n,m)=dn(n)-y(n);%e(n,m) becomes a Nxav
             %matrix;
        w=w+mu*e(n,m)*x1/(ep+x1*x1');
   end;
   y=zeros(1,N);
end;
Jav=sum((e.^2),2)/av;%e.^2 squares each matrix element;
```

9.1.4 The constants and functions used were as follows: $\text{mu}=0.01$, $M=15$, $\text{av}=200$, $v=0.5*\text{randn}(1,4000)$. The plots of J's were logarithmical of the form $10*\log(Jav)$ versus iteration numbers. A typical result of the adaptive filter coefficients was as follows:

Columns 1 through 10

0.9724 1.0513 0.9279 1.0897 0.8992 −0.8913 −1.1115

−0.8920 −1.1000 −0.9129

Columns 11 through 15

−0.0718 0.0588 −0.0440 0.0308 −0.0166

The two Book m-functions are given below.

Book m-Function Averaging *J* with NLMS

```
function[w,y,e,Jav]=lms_normalized_system_ident_Jav(mu,M,av)
for m=1:av
    w=zeros(1,M);
    v=0.5*randn(1,4000);
    v1=conv(v,[1 0.9]);
    dn=conv(v1,[1 1 1 1 1 -1 -1 -1 -1 -1]);
    dn=dn(1,1:4000);
    x=v1(1,1:4000);
    N=length(x);
    for n=M:N
        x1=x(n:-1:n-M+1);
        y(n)=w*x1';
        e(n,m)=dn(n)-y(n);
        w=w+mu*e(n,m)*x1/(0.0001+x1*x1');
    end;
    y=zeros(1,N);
end;
Jav=sum((e.^2),2)/av;
```

Book m-Function Averaging *J* with Unnormalized LMS

```
function[w,y,e,Jav]=lms_system_ident_Jav(mu,M,av)
for m=1:av
      w=zeros(1,M);
      v=0.5*randn(1,4000);
      v1=conv(v,[1 0.9]);
      dn=conv(v1,[1 1 1 1 1 -1 -1 -1 -1 -1]);
      dn=dn(1,1:4000);
      x=v1(1,1:4000);
      N=length(x);
      for n=M:N
          x1=x(n:-1:n-M+1);
          y(n)=w*x1';
          e(n,m)=dn(n)-y(n);
          w=w+mu*e(n,m)*x1;
      end;
      y=zeros(1,N);
end;
Jav=sum((e.^2),2)/av;
```

Figure P9.1.4.2 shows the resulting average learning curves.

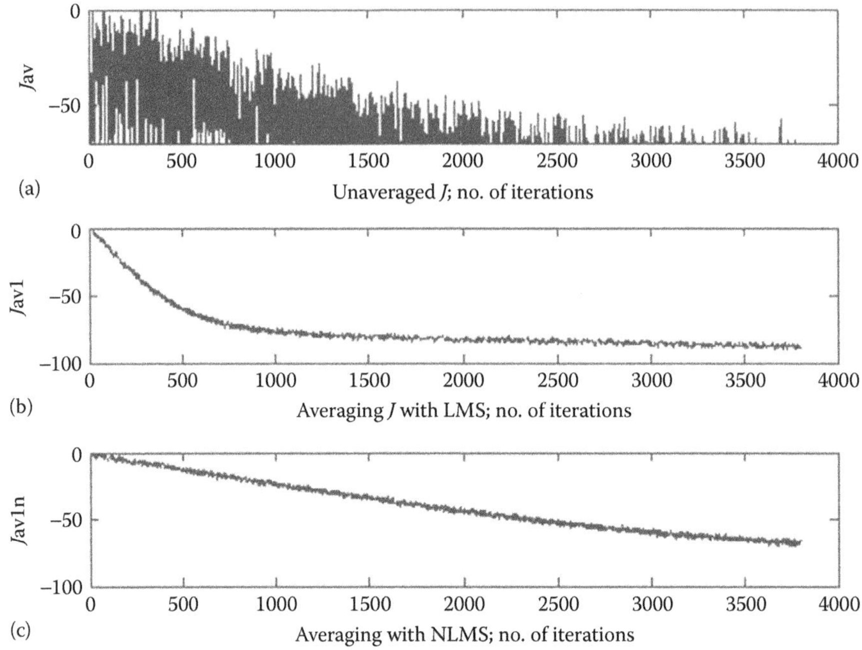

FIGURE P9.1.4.2

9.3.1 The results are shown in Figure of P9.3.1. The following constants and signals were used: $\mu = 0.01$, $M = 6$, $dn(n) = 0.995^n \cos(0.1\pi n)$, $x(n) = 0.995^n \cos(0.1\pi n) + 0.2\text{randn}(1,1500)$, $n = 1:1500$.

9.11.1 Taking the gradient of $J(n)$, we obtain

$$\nabla_w J(n) = -e(n)\boldsymbol{x}(n) + \gamma \boldsymbol{w}(n) \Rightarrow \boldsymbol{w}(n+1) = \boldsymbol{w}(n) - \mu \nabla_w J(n)$$

(steepest descent algorithm)

$$= \boldsymbol{w}(n) - \mu[-e(n)\boldsymbol{x}(n) + \gamma \boldsymbol{w}(n)] = (1 - \mu\gamma)\boldsymbol{w}(n) + \mu e(n)\boldsymbol{x}(n)$$

9.11.2 The normalized leaky LMS algorithm is given as follows:

```
function[w,y,e,J,w1]=lms_normalized_leaky_lms(x,dn,mu,gama,M)
  %function[w,y,e,J,w1]=lms_normalized_leaky_
  %lms(x,dn,mu,gama,M);
  %all signals are real valued;x=input to filter;
  %y=output from the filter;dn=desired signal;
  %mu=step-size factor;gama=gamma factor<<1;
  %M=number of filter coefficients;w1=matrix whose M
  %rows give the history of each filter coefficient;
N=length(x);
y=zeros(1,N);
w=zeros(1,M);
for n=M:N
  x1=x(n:-1:n-M+1);
  y(n)=w*x1';
```

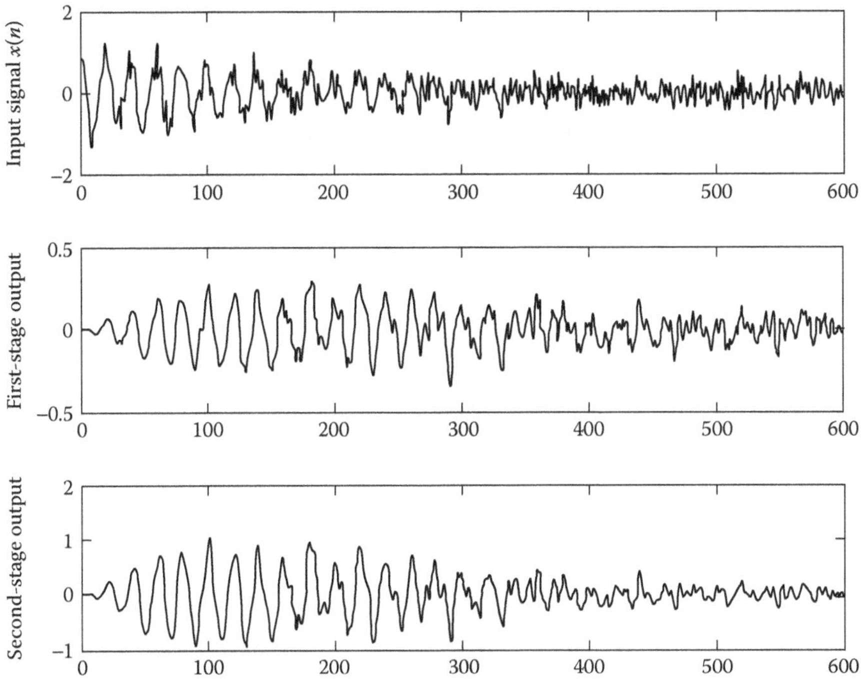

FIGURE P9.3.1

```
    e(n)=dn(n)-y(n);
    w=(1-mu*gama)*w+mu*e(n)*x1/(0.0001+x1*x1');
    w1(n-M+1,:)=w(1,:);
end;
J=e.^2;
```

The reader will find that the normalized leaky LMS algorithm performs better.

9.12.1

$$J = E\{[d(n)-\mathbf{w}^T(n)\mathbf{x}(n)][d(n)-\mathbf{x}^T(n)\mathbf{w}(n)]+\lambda(\mathbf{c}^T\mathbf{w}(n)-a)\}$$

$$= E\{d^2(n)-d(n)\mathbf{w}^T(n)\mathbf{x}(n)-d(n)\mathbf{x}^T(n)\mathbf{w}(n)+\mathbf{w}^T(n)\mathbf{x}(n)\mathbf{x}^T(n)\mathbf{w}(n)$$

$$+\lambda\mathbf{c}^T\mathbf{w}(n)-\lambda a\}$$

$$= \sigma_d^2-\mathbf{w}^T(n)E\{d(n)\mathbf{x}(n)\}-\mathbf{w}E\{d(n)\mathbf{x}(n)\}+\mathbf{w}^T(n)E\{\mathbf{x}(n)\mathbf{x}^o(n)\}\,\mathbf{w}(n)$$

$$+\lambda\mathbf{c}^T\mathbf{w}(n)-\lambda a$$

$$= \sigma_d^2-2\mathbf{w}^T(n)\mathbf{p}_{dx}+\mathbf{w}^T(n)\mathbf{R}_x\mathbf{w}(n)+\lambda\mathbf{c}^T\mathbf{w}(n)-\lambda a$$

$$= \sigma_d^2-2\mathbf{w}^T(n)\mathbf{p}_{dx}+[\mathbf{w}(n)-\mathbf{w}^o]^T\mathbf{R}_x[\mathbf{w}(n)-\mathbf{w}^o]+\mathbf{w}^{oT}\mathbf{R}_x\mathbf{w}(n)$$

$$+\mathbf{w}^T(n)\mathbf{R}_x\mathbf{w}^o+\mathbf{w}^{oT}\mathbf{R}_x\mathbf{w}^o+\lambda\{\mathbf{c}^T[\mathbf{w}(n)-\mathbf{w}^o]-(a-\mathbf{c}^T\mathbf{w}^o)\}$$

$$= \sigma_d^2-\mathbf{w}^{oT}\mathbf{p}_{dx}+\xi^T\mathbf{R}_x\xi+\lambda(\mathbf{c}^T\xi-a')$$

where:

$\boldsymbol{R}_x \boldsymbol{w}^o = \boldsymbol{p}_{dx}$, $[\boldsymbol{w}^{oT} \boldsymbol{R}_x \boldsymbol{w}(n)]^T = \boldsymbol{w}^T(n) \boldsymbol{R}_x \boldsymbol{w}^o$ ($\boldsymbol{R}_x =$ symmetric)

9.12.2 We write the cost function as follows:

$$J(n) = [\boldsymbol{w}(n+1) - \boldsymbol{w}(n)]^T [\boldsymbol{w}(n+1) - \boldsymbol{w}(n)] + \lambda[d(n) - \boldsymbol{w}(n+1)\boldsymbol{x}(n)] \quad (1)$$

Differentiate (1) with respect to $\boldsymbol{w}(n+1)$ (see also Appendix 2) to obtain

$$\frac{\partial J(n)}{\partial \boldsymbol{w}(n+1)} = 2[\boldsymbol{w}(n+1) - \boldsymbol{w}(n)] - \lambda \boldsymbol{x}(n) = \boldsymbol{0} \Rightarrow \boldsymbol{w}(n+1) = \boldsymbol{w}(n) + \frac{1}{2}\lambda \boldsymbol{x}(n) \quad (2)$$

Substitute (2) in the constraint $d(n) = \boldsymbol{w}^T(n+1)\boldsymbol{x}(n)$ to obtain

$$d(n) = \left[\boldsymbol{w}(n) + \frac{1}{2}\lambda \boldsymbol{x}(n)\right]^T \boldsymbol{x}(n) = \boldsymbol{w}^T(n)\boldsymbol{x}(n) + \frac{1}{2}\lambda \left\|\boldsymbol{x}(n)\right\|^2 \quad (3)$$

But $e(n) = d(n) - \boldsymbol{w}^T(n)\boldsymbol{x}(n)$ and (3) becomes

$$\lambda = \frac{2e(n)}{\left\|\boldsymbol{x}(n)\right\|^2} \quad (4)$$

Substituting (4) in (2), we obtain

$$\boldsymbol{w}(n+1) = \boldsymbol{w}(n) + \frac{1}{\left\|\boldsymbol{x}(n)\right\|^2} e(n)\boldsymbol{x}(n) \quad (5)$$

From (5), we obtain the final form by introducing the step-size factor μ to control the change in weight vector.

9.13.1 The Book m-function FIR LMF algorithm is given as follows:

```
function[w,y,e,J,w1]=lms_least_mean_fourth(x,dn,mu,M)
N=length(x);w=zeros(1,M);
for n=M:N
     x1=x(n:-1:n-M+1);
     y(n)=w*x1';
     e(n)=dn(n)-y(n);
     w=w+mu*x1*e(n)^3;
     w1(n-M+1,:)=w(1,:);
end;
J=e.^2;
```

9.14.1 The Book m-function for the LMMN LMS algorithm is given below:

```
function[w,y,e,J,w1]=lms_least_mean_mixed_norm(x,dn,mu,M,delta)
N=length(x);  w=zeros(1,M);%0<delta<1;
```

```
for n=M:N
     x1=x(n:-1:n-M+1);
     y(n)=w*x1';
     e(n)=dn(n)-y(n);
     w=w+mu*x1*e(n)*(delta+(1-delta)*e(n)^2);
     w1(n-M+1,:)=w(1,:);
end;
J=e.^2;
```

To obtain the normalized form of the above algorithm, divide the last term of line 7 with the expression $(\varepsilon + x1 * x1')$.

9.16.1 The kth value is

$$W_{i+1,k} = W_{i,k} + \mu X_i^* E_i = W_{i,k} + \mu X_i^* [D_i(k) - Y_i(k)] = W_{i,k} + \mu X_i^* [D_i(k) - W_{i,k} X_i(k)]$$

$$= W_{i,k} + \mu X_i^* D_i(k) - \mu W_{i,k} |X_i|^2 = \left(1 - \mu |X_i|^2\right) W_{i,k} + \mu X_i^* D_i(k)$$

9.16.2 The z-transform and the ensemble operator are linear operations and can be interchanged. Therefore, the z-transform is

$$zW_k(z) - zW_{0,k} = \left(1 - \mu E\left\{|X_i(k)|^2\right\}\right) W_k(z) + \left[\frac{\mu E\left\{D_i(k) X_i^*(k)\right\}}{\left(1 - z^{-1}\right)}\right] \quad (1)$$

In the above equation, $W_{0,k} = 0$ since it was assumed that the initial conditions have zero values and $W_k(z) = \sum_{i=0}^{\infty} E\left\{W_{i,k}\right\} z^{-1}$. Multiplying (1) by z^{-1} and $(z - 1)$, and applying the final value theorem (see Table 3.3.1), we obtain

$$E\{W_k^{\infty}\} = \lim_{z \to 1} [(z - 1) W_k(z)]$$

$$= \lim_{z \to 1} \frac{\mu E\{D_i(k) X_i^*(k)\}}{1 - \left(1 - \mu E\left\{|X_i(k)|^2\right\}\right) z^{-1}} = \frac{E\{D_i(k) X_i^*(k)\}}{E\left\{|X_i(k)|^2\right\}} \quad (2)$$

9.16.3

$$E_{i+1}(k) = E\{W_{i+1,k}\} - E\{W_k^{\infty}\}$$

$$= (1 - 2\mu E\{|X_i(k)|^2\}) E\{W_{i,k}\} + 2\mu E\{D_i(k) X_i^*(k)\} - \frac{E\{D_i(k) X_i^*(k)\}}{E\{|X_i(k)|^2\}}$$

$$= (1 - 2\mu E\{|X_i(k)|^2\}) E\{W_{i,k}\} - (1 - 2\mu E\{|X_i(k)|^2\}) \frac{E\{D_i(k) X_i^*(k)\}}{E\{|X_i(k)|^2\}}$$

$$= (1 - 2\mu E\{|X_i(k)|^2\}) E_i(k)$$

9.18.1

$$\left[\alpha\left\|\boldsymbol{e}(n)\right\|^{2}\boldsymbol{I}+\gamma\boldsymbol{x}(n)\boldsymbol{x}^{T}(n)\right]^{-1}\boldsymbol{x}(n)=\alpha^{-1}\left\|\boldsymbol{e}(n)\right\|^{-2}\boldsymbol{I}\boldsymbol{x}(n)-\alpha^{-1}\left\|\boldsymbol{e}(n)\right\|^{-2}\boldsymbol{I}\boldsymbol{x}(n)$$

$$\frac{\boldsymbol{x}^{T}(n)\alpha^{-1}\left\|\boldsymbol{e}(n)\right\|^{-2}\boldsymbol{x}(n)}{\gamma^{-1}+\boldsymbol{x}^{T}(n)\alpha^{-1}\left\|\boldsymbol{e}(n)\right\|^{-2}\boldsymbol{x}(n)}$$

$$=\frac{\gamma^{-1}\alpha^{-1}\left\|\boldsymbol{e}(n)\right\|^{-2}\boldsymbol{I}\boldsymbol{x}(n)+\alpha^{-1}\left\|\boldsymbol{e}(n)\right\|^{-2}\boldsymbol{I}\boldsymbol{x}(n)\boldsymbol{x}^{T}(n)\alpha^{-1}\left\|\boldsymbol{e}(n)\right\|^{-2}\boldsymbol{x}(n)}{\gamma^{-1}+\boldsymbol{x}^{T}(n)\alpha^{-1}\left\|\boldsymbol{e}(n)\right\|^{-2}\boldsymbol{x}(n)}$$

$$\frac{-\alpha^{-1}\left\|\boldsymbol{e}(n)\right\|^{-2}\boldsymbol{I}\boldsymbol{x}(n)\boldsymbol{x}^{T}(n)\alpha^{-1}\left\|\boldsymbol{e}(n)\right\|^{-2}\boldsymbol{x}(n)}{}$$

$$=\frac{\left\|\boldsymbol{e}(n)\right\|^{-2}\boldsymbol{x}(n)}{\alpha+\boldsymbol{x}^{T}(n)\gamma\left\|\boldsymbol{e}(n)\right\|^{-2}\boldsymbol{x}(n)}=\frac{\boldsymbol{x}(n)}{\alpha\left\|\boldsymbol{e}(n)\right\|^{2}+\gamma\boldsymbol{x}^{T}(n)\boldsymbol{x}(n)}$$

Appendix 1
Suggestions and Explanations for MATLAB Use

A1.1 SUGGESTIONS AND EXPLANATIONS FOR MATLAB USE

It is suggested that before start using the text, the reader, who does not have a lot of experience with MATLAB, goes over this appendix and tries to execute the presented material in MATLAB.

A1.1.1 Creating a Directory

It was found by the author that it is less confusing if for a particular project we create our own directory where our own developed MATLAB m-files are stored. However, any time we need anyone of these files, we must include the directory in the MATLAB path. Let us assume that we have the following directory path: c:\ap\sig-syt\ssmatlab. The following two approaches can be used:

```
>>cd 'c:\ap\sig-syst\ssmatlab'
```

or

```
>>path(path,'c:\ap\sig-syst\ssmatlab')%remember to introduce
       %the path any time you start new MATLAB operations;
       %the % symbol is necessary for the MATLAB to ignore the
       %explanations;
```

The MATLAB files are included in the "ssmatlab" directory.

A1.1.2 Help

In case we know the name of a MATLAB function and we would like to know how to use it, we write the following command in the Command window:

```
>>help sin
```

or

```
>>help exp
```

etc.

In case we want to look for a key word, we write

```
>>look for filter
```

A1.1.3 Save and Load

When we have created many variables in the Command window and we would like to save two of them in a particular directory and in a particular file, we proceed as follows:

```
>>cd 'c:\ap\matlabdata'
>>save data1 x dt%it saves in the matlabdata directory the
      %file data1 having the two variables x and dt;
```

Let us assume now that we want to bring these two variables in the working space to use them. We first change the directory, as we did earlier, and then we write in the Command window

```
>>load data1
```

Then, the two variables will appear in the working space ready to be used.

A1.1.4 MATLAB as Calculator

```
>>pi^pi-10;
>>cos(pi/4);
>>ans*ans;%the result will be (√2/2)×(√2/2)=1/2 because the
      %first output is eliminated, only the last output is
      %kept in the form of ans;
```

A1.1.5 Variable Names

```
>>x = [1 2 3 4 5];
>>dt = 0.1;
>>cos(pi*dt);%since no assignment takes place there is no
      %variable;
```

A1.1.6 Complex Numbers

```
>>z = 3+j*4;%note the multiplication sign;
>>zs = z*z;%or z^2 will give you the same results;
>>rz = real(z);iz = imag(z);%will give rz = 3, and iz = 4;
>>az = angle(z);abz = abs(z);%will give az = 0.9273 rad,
      %and abz = 5;
>>x = exp(-z)+4;%x = 3.9675+j0.0377;
```

A1.1.7 Array Indexing

```
>>x = 2:1:6;%x is an array of the numbers {2, 3, 4, 5, 6};
>>y = 2:-1:-2:%y is an array of the numbers {2, 1, 0, -1, -2};
>>z = [1 3 y];%z is an array of the numbers {1, 3, 2, 1, 0,
      %-1, -2}; note the required space between array numbers;
```

```
>>xt2=2*x;%xt2 is an array of numbers of x each one multiplied
      %by 2;
>>xty=x.*y;%xty is an array of numbers which are the result of
      %multiplication of corresponding elements, that is,
      %{4, 3, 0, -5, -12};
```

A1.1.8 Extracting and Inserting Numbers in Arrays

```
>>x=2:1:6;
>>y=[x zeros(1,3)];%y is an array of the numbers {2, 3, 4, 5,
      %6, 0, 0, 0};
>>z=y(1,3:7);%1 stands for row 1 which y is and 3:7 instructs
      %to keep columns 3 through 7 the result is the array {4, 5,
      %6, 0, 0};
>>lx=length(x);%lx is the number equal to the number of columns
      %of the row vector x, that is lx = 5;
>>x(1,2:4)=4.5*(1:3);%this assignment substitutes the elements
      %of x at column positions 2,3 and 4 with the numbers
      %4.5*[1 2 3] = 4.5, 9, and 13.5,note the columns of
      %2:4 and 1:3 are the same;
>>x(1,2:2:length(x))=pi;%substitutes the columns 2 and 4 of x
      %with the value of pi, hence the array is {2, 3.1416,
      %4,3.1416 6};
```

A1.1.9 Vectorization

```
>>n=0:0.2:1;
>>s=sin(0.2*pi*n);%the result of these two commands gives the
      %signal s (sine function) at times (values of n)
      %0, 0.2, 0.4, 0.6, 0.4, 1;
```

This approach is preferable since MATLAB executes the vectorization approach faster rather than the loop approach, which is given as follows:

```
>>s=[];%initializes all values of vector s to zero;
>>for n=0:5%note that the index must be integer;
>>s(n+1)=sin(0.2*pi*n*0.2);%since we want values of s
      %every 0.2 seconds we must multiply n by 0.2;
      %note also that for n = 0 the variable becomes
      %s(1) and this is because the array in MATLAB
      %always starts counting columns from 1;
>>end;
```

The results are identical with the previous one.

A1.1.10 Windowing

The following windows are used for correcting, to some extent, the effects of truncating a vector.

@bartlett—Bartlett window
@barthannwin—Modified Bartlett–Hanning window
@blackman—Blackman window
@blackmanharris—Minimum four-term Blackman–Harris window
@bohmanwin—Bohman window
@chebwin—Chebyshev window
@flattopwin—Flat-top window
@gausswin—Gaussian window
@hamming—Hamming window
@hann—Hann window
@kaiser—Kaiser window
@nuttallwin—Nuttall-defined minimum four-term Blackman–Harris window
@parzenwin—Parzen (de la Vallée-Poussin) window
@rectwin—Rectangular window
@tukeywin—Tukey window
@triang—Triangular window

An example for windowing is given as follows:

```
N=65;
w=window(@blackmanharris,N);
w1=window(@hamming,N);
w2=window(@gausswin,N,2.5);
plot(1:N,[w,w1,w2]); axis([1 N 0 1]);
legend('Blackman-Harris','Hamming','Gaussian');
```

A1.1.11 Matrices

If `a` and `b` are matrices such that `a` is a 2×3 matrix and `b` is a 3×3 matrix, then `c = a*b` is a 2×3 matrix.

`>>a=[1 2; 4 6];%a is a 2x2 matrix` $\begin{bmatrix} 1 & 2 \\ 4 & 6 \end{bmatrix}$`;`

`>>b=a';%b is a transposed 2x2 matrix of a and is` $\begin{bmatrix} 1 & 4 \\ 2 & 6 \end{bmatrix}$`;`

```
>>da=det(a);%da is a number equal to the determinant of a,
      %da = -2;
>>c=a(:);%c is a vector which is made up of the columns of a,
      %c = [1 4 2 6];
>>ia=inv(a);%ia is a matrix which is the inverse of a;
```

```
>>sa1=sum(a,1);%sa1 is a row vector made up of the sum of the
       %rows,sa1 = [5 8];
>>sa2=sum(a,2);%sa2 is a column vector made up by the sum of
       %the columns,sa2 = [3 10]';
```

A1.1.12 PRODUCING A PERIODIC FUNCTION

```
>>x=[1 2 3 4];
>>xm=x'*ones(1,5);%xm is 4x5 matrix and each of its column is x';
>>xp=xm(:)';%xp is a row vector, xp = [x x x x x];
```

A1.1.13 SCRIPT FILES

Script files are m-files in which we introduce their names in the Command window and receive the results. We must, however, have the directory that includes the file in the MATLAB search directories. You can modify the file in any desired way and get new results. Suppose that any time we ask for the file pexp.m, the magnitude and the angle of the exponential function $e^{j\omega}$ are plotted. To accomplish this, we first go to the Command window and open a new m-file. In the window, we type the file as shown below. As soon as we finish typing, we click on *Save as* and save the file in, say, c:\ap\ssmatlab. If we want to see the results, in the Command window we just write pexp and hit the Enter key.

Script File pexp.m

```
>>w=0:pi/500:pi-pi/500;%they are 500 at pi/500 appart;
>>x=exp(j*w);ax = abs(x);anx = angle(x);
>>subplot(2,1,1);plot(w,ax,'k');%'k' means plot line in black;
>>xlabel('\omega rad/s');ylabel('Magnitude');
>>subplot(2,1,2);plot(w,anx,'k');
>>xlabel('\omega rad/s');ylabel('Angle');
```

If we have the function $2e^{j\omega}/(e^{j\omega}-0.5)$ and want to plot the results as earlier, we substitute in the script file the function x with the function: `x=2*exp(j*w)./...(exp(j*w)-0.5);`

In the above MATLAB expression, note the dot before the slash. This instructs MATLAB to operate at each value of w separately and, thus, gives results at each frequency point.

A1.1.14 FUNCTIONS

We present here an example of how to write functions. The reader should also study the functions that are presented throughout the book. In the Fourier series, for example, we plot the functions of the form:

$$s(t)=\sum_{n=0}^{N} A_n \cos(n\omega_0 t)$$

and plot this sum of cosines, each one having a different amplitude and frequency. Let $A = [1 \quad 0.6 \quad 0.4 \quad 0.1]$, $\omega_0 = 2$, and $0 \le t \le 4$. We approach this solution by vectorizing the summation. The MATLAB function is of the form:

```
function[s]=sumofcos(A,N,w0,rangeoft)
n=0:N-1;
s=A*cos(w0*n'*rangeoft);
        %when we want to use this function at the command window to
        %find s we write for example:
        %>>A = [1 0.6 0.4 0.1];N = 4;w0 = 2;rangeoft = 0:0.05:6;
        %>>[s] = sumofcos(A,N,w0,rangeoft);
        %at the Enter key click the vector s is one of the
        %variables in the command window and it can be plotted
        %at the wishes of the reader; we
        %must secure that the directory in which sumofcos function
        %exists is in the MATLAB
        %path; After you type the function in the Editing window,
        %you save as.. in the directory, for example,
        %c:\ap\ssmatlab and filename: sumofcos.m.
```

It is recommended that the reader set small numbers for $N (N = 4)$ and the range of $t(0:0.2:1)$ and produce first the matrix `cos(w0*n'*t)` and then see the result `A*cos(w0*n'*t)`.

A1.1.15 Complex Expressions

We can produce the results by writing, for example,

```
>>x=[1 3 1 5 3 4 5 8];
>>plot(abs(fft(x,256)),'r');%will plot in red color the
        %spectrum of the vector x of 256 points;
```

A1.1.16 Axes

```
>>axis([xmin xmax ymin ymax]);%sets the max and min values of
        %the axes;
>>grid on;%turns on grid lines in the graph;
```

A1.1.17 2D Graphics

To plot a sine and a cosine signal

```
>>x=linspace(0,2*pi,40);%produces 40 equal spaced points
        %between 0 and 2π;
>>y=sin(x);plot(x,y,'r');%will plot the sine signal with red
        %color;
```

```
>>y1=cos(x);plot(x,y1,'g');%will plot the cosine signal with
      %green color;
```

For other color lines

```
'y'=yellow,'c'=cyan,'b'=blue,'w'=white,'k'=black
```

Type of lines

```
'g:'=green dotted line,'r-  '=red dashed line,'k-  x'=
black dashed line with x's,'k-.'=black dash-dot line,'+'=
plus sign,'ko'=black circles
```

Add Greek letters

```
\omega=will produce Greek lowercase omega,\Omega=will
produce Greek uppercase omega. The same is true for the rest of
the Greek letters. For example, if we want to write the frequency
in a figure under the x-axis, in the Command window we write:
>>xlabel('\omega rad/s');
For an omega with a subscript 01, we write:
>>xlabel('\omega_{01} rad/s');
```

Add grid lines

```
>>grid on;%this is done after the command plot;
```

Adjusting axes

```
>>axis square;%sets the current plot to be square rather
      %than the default rectangle;
>>axis off;%turn off all axis labeling, grid, and tick marks
      %marks;leave the title and any labels placed
      %by the 'text' and 'gtext' commands;
>>axis on;%turn on axis labeling, tick marks, and grid;
>>axis([xmin xmax ymin ymax]);%set the maximum and
      %minimum values of the axes using values given
      %in the row vector;
```

Subplots (Example)

```
>>n=0:100;x=sin(n*pi*n);y=cos(n*pi*n);z=x.*y;w=x+y;
      %subplot(2,2,1);plot(n,x);subplot(2,2,2);plot(n,y);
      %subplot(2,2,3);plot(n,z);subplot(2,2,4);plot(n,w);
```

Log plotting

```
>>semilogx(x);%will plot the vector x in log scale in x-axis
      %and linear scale in y-axis;
```

```
>>semilogy(x);%will plot the vector x in log scale in y-
      %direction and linear scale in the x-axis;
>>loglog(x);%will plot the vector x in log scale both axes;
```

Histogram

```
>>x=randn(1,1000);hist(x,40);colormap([0 0 0]);%will plot a
      %Gaussian histogram of 40 bars in white;if instead we
      %entered the vector [1 1 1] the bars would be
      %black;the vector [1 0 0] will give red and the
      %vector [0.5 0.5 0.5] will give gray;
>>x=-3.0:0.05:3;y=exp(-.*x);bar(x,y);colormap([.5.5.5]);
      %will produce bar figure of the bell curve with gray
      %color;
>>stairs(x,y,'k');%will produce a stair-like black curve;
```

Add words

```
>>gtext('the word');
```

At the return, a crosshair will appear in the figure. Move the center at the point in the figure where the word must start and click.

Add legend

```
>>plot(x1,y1,'+',x2,y2,'*');%there will be two curves in the
      %graph;
>>legend('Function 1','Function 2');
```

The following rectangle will appear in the figure:

+ Function 1
* Function 2

A1.1.18 3D Plots

A1.1.18.1 Mesh-Type Figures

If, for example, we desire to plot the function $f(x) = \mathrm{e}^{-(x^2+y^2)}$ in the ranges $-2 \le x \le 2$ and $-2 \le y \le 2$, we proceed as follows:

```
>>x=-2:0.1:2;y=-2:0.1:2;[X,Y]=meshgrid(x,y);
>>f=exp(-(X.*X+Y.*Y));mesh(X,Y,f);colormap([0  0  0]);
```

The above commands will produce a mesh-type 3D figure with black lines.

A1.2 GENERAL PURPOSE COMMANDS

A1.2.1 Managing Commands and Functions

`help`	Online help for MATLAB functions and m-files, for example, `>>help plot`
`path`	Shows the path to MATLAB directories which are available at the command window

A1.2.2 Managing Variables and Workplace

`clear`	Removes all the variables and items in the memory. Let us assume that the memory contains the variables `x`, `y`, `z`. Then, write `>>clear x z;` only `y` will remain in the memory.
`length`	A number that gives the length of a vector. `>>x=[1 3 2 5];` then `>>length(x);` will give the number 4. If we write `>>y=length(x);`, then we get the value of variable `y` equal to `4`.
`size`	Shows the array dimensions. `>>x=[1 3 2 5];` then `size(x)` will give the numbers 1 4, which means 1 row and 4 columns. Let us write `>>x=[1 2; 3 5; 6 4];` then `size(x)` will give the numbers 3 2, which means that x is a matrix of 3 × 4 dimensions
`who`	Produces a list of the variables in the memory.
`format`	Used for display as follows: `>>format short,pi;` will produce the number 1.1416, `>>format long,pi;` will produce the number 3.14159265358979, `>>format long,single(pi);` will produce the number 3.1415927.

A1.2.3 Operators and Special Characters

+	Plus
-	Minus
*	Number and matrix multiplications
.*	Array multiplication, for example, `>>x=[1 2 3];y=[2 3 4];z=x.*y;`. Hence, `z = [2 6 12]`
.^	Array power, for example, `>>x=[2 3 4];y=x.^3;`. Hence, `y=[8 27 64]`. `>>x=[2 4;1 5]; y=x.^2;`. Hence, `y=[4 16;1 25]`
/	Right division
./	Array division, for example, `>>x=[2 4 6];y=[4 4 12];z=x./y;`. Hence, `z=[0.5 1 0.5]`
:	Colon, for example, `>>x=[1 3 6 2 7 8];y=x(1,3:6);`. Hence, `y=[6 2 7 8]`
.	Decimal point
...	Continuation, for example, `>> x=[1 4 6 7 8 9 ...` `>>2 5 8 1];` The vector x is interpreted by MATLAB as a row vector having 10 elements
%	Comments, for example, `>>x=[1 4 2 6];%this is a vector.` The MATLAB ignores "this is a vector"
'	Transpose of a matrix or vector, for example, `>>x=[2 6 3];y=x';` will have $y = \begin{bmatrix} 2 \\ 6 \\ 3 \end{bmatrix}$

(*Continued*)

&	Logical AND
|	Logical OR
~	Logical NOT
`xor`	Logical exclusive (XOR)

A1.2.4 Control Flow

`for` Repeat operations a specific number of times.

```
>>for n=0:3
>>x(n+1)=sin(n*pi*0.1);%observe the n+1, if the
      %+1 was not there, x(0)is not defined by MATLAB
>>end;
```

Then, x = [0 0.3090 0.5878 0.8090]

```
>>for n=0:2
>>      for m=0:1
>>            x(n+1,m+1)=n^2+m^2;
>>      end;
>>end;
```

Then, $x = \begin{bmatrix} 0 & 1 \\ 1 & 2 \\ 4 & 5 \end{bmatrix}$

`while` Repeats statements an indefinite times of times

```
>>a=1;num=0;
>>while (1+a)<=2 & (1+a)>=1.0001
>>  a=s/2;
>>  num=num+1;
>>end;
```

We obtain a = 0.0001 and num = 14

`if` Executes statements conditionally

`if` *expression*

commands evaluated if true

`else`

commands evaluated if false

`end;`

If there are more than one alternative, the `if-else-end` statement takes the form:

```
elseif
     if expression1
          commands evaluated if expression 1 is true
     elseif expression2
          commands evaluated if expression 2 is true
     elseif....
          .
          .
          .
     else
          commands evaluated if no other expression is true
     end;
```

A1.3 ELEMENTARY MATRICES AND MATRIX MANIPULATION

A1.3.1 Elementary Matrices and Arrays

`eye(n,n)`	Identity matrix (its diagonal elements are 1 and all the others are 0)
`linspace`	`linspace(x1,x2)` generates 100 equally spaced points between `x1` and `x2` `linspace(x1,x2,N)` generates `N` equally spaced points between `x1` and `x2`
`ones`	`ones(1,5)` generates a row vector with its elements only ones `ones(2,4)` generates a 2×4 matrix with all its elements ones
`rand`	Uniformly distributed random numbers. `>>x=rand(1,5);` where `x` is a row vector of 5 elements of random numbers. `>>x=rand(2,3);` where `x` is a 2×3 matrix whose elements are random numbers
`randn`	Normally distributed random numbers. Applications are similar to `rand`
`zeros`	Creates arrays and matrices of all zeros. `>>x=zeros(1,4);` where `x` is a row vector of 4 elements all with zero value. `>>x=zeros(3,4);` where `x` is a 3×4 matrix with all of its elements zeros
: (colon)	Regularly spaced vector. `>>x = [1 4 2 5 8 3];y = x(1,3:6);`. Hence, `y = [2 5 8 3];`
`eps`	Floating-point relative accuracy. To avoid NA response in case there exists a zero-over-zero expression at a point, as in the `sinc` function, we, for example, write: `>>n=-4:4;x=sin(n*pi*.1)./((n*pi+eps);`
`i,j`	Imaginary unit
`pi`	Ratio of a circle's circumference to its diameter

A1.3.2 Matrix Manipulation

`diag`	Diagonal matrices and diagonals of a matrix. `>>x=[1 3 5;2 6 9;4 7 0];` `y=diag(x);` will give a column vector $\mathbf{y} = [1 \quad 6 \quad 0]^T$. `>>y1=diag(x,1);` will give a column vector $\mathbf{y}1 = [3 \quad 9]^T$ which is the diagonal above the main diagonal `>>y2=diag(x,-1);` will give the column vector $\mathbf{y}2 = [2 \quad 7]$ which is the diagonal just below the main diagonal `>>y3=diag(diag(x));` will give a 3×3 matrix with the diagonal 1, 6, 0 and the rest of the elements zeros

(*Continued*)

`fliplr`	Flips vectors and matrices left–right
`flipud`	Flip vectors and matrices up–down
`tril`	Lower triangular part of a matrix including the main diagonal and the rest are zero. If `x=[1 3 5;2 6 9; 4 7 0]`, then `y=tril(x)` is the matrix [1 0 0;3 6 0;4 7 0].
`triu`	Upper triangular part of a matrix
`toeplitz`	Produces a Toeplitz matrix given a vector. `>>x=[1 5 2];y= ...` `>>toeplitz(x)` produces the matrix $y = [1\ \ 5\ \ 2;5\ \ 1\ \ 5;2\ \ 5\ \ 1]$.

A1.4 ELEMENTARY MATHEMATICAL FUNCTIONS

A1.4.1 ELEMENTARY FUNCTIONS

abs	Absolute value of a number and the magnitude of a complex number
acos,acosh	Inverse cosine and inverse hyperbolic cosine
acot,acoth	Inverse cotangent and inverse hyperbolic cotangent
acsc,acsch	Inverse cosecant and inverse hyperbolic cosecant
angle	Phase angle of a complex number, for example, angle(1 + j) = 0.7854
asec,asech	Inverse secant and inverse hyperbolic secant
asin,asinh	Inverse sine and inverse hyperbolic sine
atan,atanh	Inverse tangent and inverse hyperbolic tangent
ceil	Round toward infinity, for example, ceil(4.22) = 5
conj	Complex conjugate, for example, conj(2 + j * 3) = 2 – j * 3
cos,cosh	Cosine and hyperbolic cosine
cot,coth	Cotangent and hyperbolic cotagent
csc,csch	Cosecant and hyperbolic cosecant
exp	Exponential, for example, exp(–1) = 1/e = 0.3679
fix	Rounds toward zero, for example, fix(–3.22) = –3
floor	Round toward minus infinity, for example, floor(–3.34) = –4, and floor(3.65) = 3
imag	Imaginary part of a complex number, for example, imag(2 + j * 5) = 5
log	Natural logarithm, for example, log(10) = 2.3026
log2	Based 2 logarithm, for example, log2(10) = 3.3219
log10	Common (base 10) logarithm, for example, log10(10) = 1
mod	Modulus (signed remainder after division), for example, mod(10,3) = 1, mod(10,4) = 2. In general, mod(x,y) = x – n * y
real	Real part of complex number
rem	Remainder after division, for example, rem(10,3) = 1, rem(10,5) = 0, rem(10,4) = 2
round	Round to the nearest integer, for example, round(3.22) = 3, round(3.66) = 4
sec,sech	Secant and hyperbolic secant
sign	Signum function, for example, sign(x) = 0 for $x = 0$, sign(x) = 1 for $x > 0$, and sign(x) = –1 for $x < 1$
sin,sinh	Sine and hyperbolic sine
sqrt	Square root, for example, sqrt(4) = 2
tan,tanh	Tangent and hyperbolic tangent
erf,erfc	Error and co-error function
gamma	Gamma function, for example, gamma(6) = 120 or 1 * 2 * 3 * 4 * (6 – 1) = 120

A1.5 NUMERICAL LINEAR ALGEBRA

A1.5.1 Matrix Analysis

det	Matrix determinant, for example, `>>a=[1 2; 3 4];det(a);` gives `-2`
norm	The norm of a vector, for example, `norm(v)=sum(abs(v).^2)^(1/2)`
rank	Rank of a matrix. `rank(A)` provides the number of independent columns or rows of matrix ***A***
trace	Sum of the diagonal elements, for example, `trace([1 3; 4 12]) = 13`
eig	Eigenvalues and eigenvectors, for example, `>>[v,d]=eig([1 3;5 8]);`. Therefore, $v = \begin{bmatrix} -0.8675 & -0.3253 \\ 0.4974 & -0.9456 \end{bmatrix}, \quad d = \begin{bmatrix} -0.7202 & 0 \\ 0 & 9.7202 \end{bmatrix}$
inv	Matrix inversion, for example, `>>A=[1 3;5 8];B=inv(A);`. Therefore, $B = \begin{bmatrix} -1.1429 & 0.4286 \\ 0.7143 & -0.1429 \end{bmatrix}, \quad A*B = \begin{bmatrix} 1.0000 & 0 \\ 0 & 1.0000 \end{bmatrix}$

A1.6 DATA ANALYSIS

A1.6.1 Basic Operations

`max`	Maximum element of an array, for example, `>>v=[1 3 5 2 1 7];x=max(v);` Therefore, $x = 7$
`mean`	Average or mean value of an array, for example, `>>v=mean([1 3 5 2 8]);` Therefore `v=3.8`
`median`	Median value of an array, for example, `>>a=median([1 3 5 2 8]);` Therefore `a=3.`
min	Minimum element of an array
sort	Sorts elements in ascending order, for example, `>>v=sort([1 3 5 2 8]);` `v=[1 2 3 5 8]`
std	Standard deviation
sum	Sum of an array elements, for example, `>>a=sum([1 3 5 2 8]);` Therefore `a=19`

A1.6.2 Filtering and Convolution

`conv`	Convolution and polynomial multiplication, for example, `conv([1 1 1]) = [1 2 3 2 1]`. If we have to multiply these two polynomials $(x^2 + 2x + 1)*(x + 2)$, we convolve their coefficients `conv([1 2 1],[1 2]) = [1 4 5 2]`. Therefore, we write the polynomial: $x^3 + 4x^2 + 5x + 2$
`conv2`	Two-dimensional convolution
`filter`	Filter data with infinite impulse response (IIR) or finite impulse response (FIR) filter. Let the FIR filter be given by $y(n) = 0.8x(n) + 0.2x(n-1) - 0.05x(n-2)$. Let the input data are $x = [0.5 \quad -0.2 \quad 0.6 \quad 0.1]$. Hence, $a = [1]$, $b = [0.5 \quad 0.2 \quad -0.05]$, and the output is given by `y = filter(a,b,x)`. The result is $y = [0.6250 \quad -0.4063 \quad 0.8906 \quad -0.1230]$

A1.6.3 Fourier Transforms

`abs`	Absolute value and complex magnitude, for example, `abs(4+j*3) = 5`, `abs([-0.2 3.2]) = [0.2 3.2]`
`angle`	Phase angle, e.g., `angle(4+j*3) = 0.6435` in radians
`fft`	One-dimensional fast Fourier transform. For example, `>>x = [1 1 1 0]; y = fft(x);`. Hence, `y = [3 0-1.0000i 1.0000 0+1.0000i]`. If we had written `z = fft(x,8)`, we would have obtained `z = [3 1.7071-1.7071i 0-1.0000i 0.2929+0.2929i 1 0.2929-0.2929i 0+1.0000i 1.7071+1.7071i]`
`fft2`	Two-dimensional fast Fourier transform
`fftshift`	Shift DC component of fast Fourier transform to the center of spectrum. For example, we write in the Command window: `>>x = [1 1 1 1 1 0]; y = fft(x,256);`. Then the command `plot(abs(fftshift(y)))` will center the spectrum. We can also write: `plot(abs(fftshift(fft(x,256))))`
`ifft`	Inverse one-dimensional fast Fourier transform
`ifft2`	Inverse two-dimensional fast Fourier transform

A1.7 2D PLOTTING

A1.7.1 2D Plots

`plot`	Linear plot. If we have three vectors of equal length such as `x` with numbers of equal distance, `y` and `z`, we can create the following simple plots: `plot(y)` will plot the values of `y` at numbers 1, 2, … in the `x`-direction, `plot(x,y)` will plot the `y` values versus the equal-distance values of the vector `x` in the `x`-direction, `plot(x,y,x,z)` will plot both vectors `y` and `z` on the same graph, we can plot the two vectors by writing `>>plot(x,y); hold on;plot(x,z);`. If we would like the second graph to have different color, we write `plot(x,z,'g')` for getting green color.
`loglog`	Log–log scale plot. For example, `loglog(y)` will produce the plot.
`semilogx`	Semilog scale plot. The log scale will be on the *x*-axis and the linear scale on the *y*-axis. The plot is accomplished by writing `semilogx(y)`.
`semilogy`	Semilog scale plot. The log scale will be on the *y*-axis and the linear scale on the *x*-axis. The plot is accomplished by writing `semilogy(y)`.
`axis`	Controls axis scaling. For example, if we want the axes to have specific ranges, we write after we create a plot using the MATLAB default: `axis([minx maxx miny max])`.
`grid`	Grid lines. After we create the plot, we write `grid on`.
`subplot`	Create axes in tiled positions. For example, when we write `subplot(3,1,1)`, we expect 3×1 plots in one page starting from plot one. Next, we write `subplot(3,1,2)` and then proceed to plot the second plot and so on. If we write `subplot(3,2,1)`, we expect 3×2 = 6 plots on the page. After we write `subplot(3,2,1)`, we proceed to plot the first of the 3×2 matrix-format plots. For example, if we write `subplot(3,2,2)` and proceed to plot the figure, we create a plot at line two and the second plot.

(*Continued*)

`legend`	Graph legend. For example, if we have two lines on the plot, one red and one green, and write `legend("one," "two")`, then a rectangle frame will appear on the graph with a red line and the word *one* and with a green line and the word *two.*
`title`	Graph title. For example, if we write `title ("This is a graph")`, then the script in parentheses will appear on the top of the graph.
`xlabel`	x-axis label. For example, if we write `xlabel ("n time")`, the *n time* will appear under the x-axis.
`gtext`	Place the text with mouse. After we have created a plot, if we write in the Command window `gtext("this is the 1st graph")`, at the return a crosshair will appear on the graph and at the click the phrase in parentheses will appear on the graph.

Appendix 2
Matrix Analysis

A2.1 DEFINITIONS*

Let A be an $m \times n$ matrix with elements a_{ij}, $i = 1, 2, \ldots, m$; $j = 1, 2, \ldots, n$. A shorthand description of A is

$$[A]_{ij} = a_{ij} \tag{A2.1}$$

The **transpose** of A, denoted by A^T, is defined as the $n \times m$ matrix with elements a_{ji} or

$$[A^T]_{ij} = a_{ji} \tag{A2.2}$$

Example A2.1.1

$$A = \begin{bmatrix} 1 & 2 \\ 4 & 9 \\ 3 & 1 \end{bmatrix}; \quad A^T = \begin{bmatrix} 1 & 4 & 3 \\ 2 & 9 & 1 \end{bmatrix}$$

■

A **square** matrix is a matrix in which $m = n$. A square matrix is **symmetric** if $A^T = A$.

The **rank** of a matrix is the number of linearly independent rows or columns, whichever is less. The **inverse** of a square $n \times n$ matrix A^{-1} in which

$$A^{-1}A = AA^{-1} = I \tag{A2.3}$$

where:

I is the identity matrix

$$I = \begin{bmatrix} 1 & 0 & \cdots & 0 \\ 0 & 1 & \cdots & 0 \\ \vdots & \vdots & & \vdots \\ 0 & 0 & \cdots & 1 \end{bmatrix} \tag{A2.4}$$

* In this appendix, capital letters represent matrices and lowercase letters without subscripts, excluding identifiers, indicate vectors. If lowercase letters have subscripts and indicate vectors, they should be written in a boldfaced format.

A **matrix** $\boldsymbol{A}$ is singular if its inverse does not exist.

The determinant of a square $n \times n$ matrix is denoted by $\det\{\boldsymbol{A}\}$, and it is computed as

$$\det\{\boldsymbol{A}\} = \sum_{j=1}^{n} a_{ij} C_{ij} \tag{A2.5}$$

where:

$$C_{ij} = (-1)^{i+j} M_{ij} \tag{A2.6}$$

and M_{ij} is the determinant of the submatrix $\boldsymbol{A}$ obtained by deleting the ith row and jth column and is called the **minor** of a_{ij}. C_{ij} is the **cofactor** of a_{ij}.

Example A2.1.2

$$\boldsymbol{A} = \begin{bmatrix} 1 & 2 & 4 \\ 4 & -3 & 9 \\ -1 & -1 & 6 \end{bmatrix},\ \det\{\boldsymbol{A}\} = (-1)^{1+1} \begin{vmatrix} -3 & 9 \\ -1 & 6 \end{vmatrix} + (-1)^{1+2}\, 2 \begin{vmatrix} 4 & 9 \\ -1 & 6 \end{vmatrix} + (-1)^{1+3}\, 4 \begin{bmatrix} 4 & -3 \\ -1 & -1 \end{bmatrix}$$

$$= C_{11} + C_{12} + C_{13}$$

$$= (18 + 9) + [-2(24 + 9)] + [4(-4 - 3)]$$

■

Any choice of i will yield the same value for $\det\{\boldsymbol{A}\}$.

A **quadratic form** Q is defined as

$$Q = \sum_{n=1}^{n} \sum_{j=1}^{n} a_{ij} x_i x_j \tag{A2.7}$$

In defining the quadratic form, it is assumed that $a_{ji} = a_{ij}$. This entails no loss in generality since any quadratic functions may be expressed in this manner. Q may also be expressed as

$$Q = \boldsymbol{x}^T \boldsymbol{A} \boldsymbol{x} \tag{A2.8}$$

where:

$\boldsymbol{x} = [x_1\ x_2\ x_n]^T$

$\boldsymbol{A}$ is a square $n \times n$ matrix with $a_{ji} = a_{ij}$ (symmetric matrix)

Example A2.1.3

$$Q = \begin{bmatrix} x_1 & x_2 \end{bmatrix} \begin{bmatrix} a_{11} & a_{12} \\ a_{21} & a_{22} \end{bmatrix} \begin{bmatrix} x_1 \\ x_2 \end{bmatrix}$$

$$= [a_{11}x_1 + a_{21}x_2 \quad a_{12}x_1 + a_{22}x_2] \begin{bmatrix} x_1 \\ x_2 \end{bmatrix}$$

$$= a_{11}x_1^2 + a_{21}x_1x_2 + a_{12}x_1x_2 + a_{22}x_2^2$$

■

A square $n \times n$ matrix $\boldsymbol{A}$ is **positive semidefinite** if $\boldsymbol{A}$ is symmetric and

$$\boldsymbol{x}^T \boldsymbol{A} \boldsymbol{x} \geq 0 \tag{A2.9}$$

for all $\boldsymbol{x} \neq 0$. If the quadratic form is strictly positive, the matrix $\boldsymbol{A}$ is called **positive definite**. If a matrix is positive definite or positive semidefinite, it is automatically assumed that the matrix is symmetric.

The **trace** of a **square** matrix is the sum of the diagonal elements or

$$\mathrm{tr}\{\boldsymbol{A}\} = \sum_{i=1}^{n} a_{ii} \tag{A2.10}$$

A partitioned $m \times n$ matrix $\boldsymbol{A}$ is a matrix that is expressed in terms of its submatrices. An example is the 2 × 2 partitioned matrix:

$$\boldsymbol{A} = \begin{bmatrix} A_{11} & A_{12} \\ A_{21} & A_{22} \end{bmatrix}, \begin{bmatrix} k \times l & k \times (n-l) \\ (m-k) \times l & (m-k) \times (n-l) \end{bmatrix} \tag{A2.11}$$

The MATLAB functions are as follows:

```
>>B=A';%B is the transpose of A
>>B=inv(A);%B is the inverse of A
>>a=det(A);% a is the determinant of A
>>I=eye(n);%I is an nxn identity matrix
>>a=trace(A);%a is the trace of A
```

A2.2 SPECIAL MATRICES

A **diagonal** matrix is a square $n \times n$ matrix with $a_{ij} = 0$ for $i \neq j$. A diagonal matrix has all the elements off the principal diagonal equal to zero. Hence,

$$\boldsymbol{A} = \begin{bmatrix} a_{11} & 0 & \cdots & 0 \\ 0 & a_{22} & \cdots & 0 \\ \vdots & \vdots & & \vdots \\ 0 & 0 & \cdots & a_{nn} \end{bmatrix} \tag{A2.12}$$

$$\boldsymbol{A}^{-1} = \begin{bmatrix} a_{11}^{-1} & 0 & \cdots & 0 \\ 0 & a_{22}^{-1} & \cdots & 0 \\ \vdots & \vdots & & \vdots \\ 0 & 0 & \cdots & a_{nn}^{-1} \end{bmatrix} \tag{A2.13}$$

A generalization of the diagonal matrix is the square $n \times n$ **block diagonal** matrix:

$$\boldsymbol{A} = \begin{bmatrix} \boldsymbol{A}_{11} & 0 & \cdots & 0 \\ 0 & \boldsymbol{A}_{22} & \cdots & 0 \\ \vdots & \vdots & & \vdots \\ 0 & 0 & \cdots & \boldsymbol{A}_{kk} \end{bmatrix} \tag{A2.14}$$

in which all $\boldsymbol{A}_{ii}$ matrices are square and the submatrices are identically zero. The submatrices may not have the same dimensions. For example, if $k = 2$, A_{11} may be a 2×2 matrix and A_{22} might be a scalar. If all A_{ii} are nonsingular, then

$$\boldsymbol{A}^{-1} = \begin{bmatrix} \boldsymbol{A}_{11}^{-1} & 0 & \cdots & 0 \\ 0 & \boldsymbol{A}_{22}^{-1} & \cdots & 0 \\ \vdots & \vdots & & \vdots \\ 0 & 0 & \cdots & \boldsymbol{A}_{kk}^{-1} \end{bmatrix} \tag{A2.15}$$

and

$$\det\{\boldsymbol{A}\} = \prod_{i=1}^{n} \det\{\boldsymbol{A}_{ii}\} \tag{A2.16}$$

A square $n \times n$ matrix is **orthogonal** if

$$\boldsymbol{A}^{-1} = \boldsymbol{A}^{T} \tag{A2.17}$$

Example A2.2.1

$$\boldsymbol{A} = \begin{bmatrix} \frac{2}{\sqrt{5}} & \frac{1}{\sqrt{5}} \\ -\frac{1}{\sqrt{5}} & \frac{2}{\sqrt{5}} \end{bmatrix}, \quad \boldsymbol{A}^{-1} = \frac{1}{\det\{\boldsymbol{A}\}} \begin{bmatrix} \frac{2}{\sqrt{5}} & \frac{1}{\sqrt{5}} \\ -\frac{1}{\sqrt{5}} & \frac{2}{\sqrt{5}} \end{bmatrix}^{T} = \begin{bmatrix} \frac{2}{\sqrt{5}} & -\frac{1}{\sqrt{5}} \\ \frac{1}{\sqrt{5}} & \frac{2}{\sqrt{5}} \end{bmatrix} = \boldsymbol{A}^{T}$$

■

A matrix is **orthogonal** if its columns (and rows) are orthonormal. Therefore, we must have

$$A = \begin{bmatrix} a_1 & a_2 & \cdots & a_n \end{bmatrix}$$

$$a_i^T a_j = \begin{cases} 0 & \text{for } i \neq j \\ 1 & \text{for } i = j \end{cases} \tag{A2.18}$$

An idempotent matrix is a square $n \times n$ matrix that satisfies the following relations:

$$\begin{aligned} A^2 &= A \\ A^m &= A \end{aligned} \tag{A2.19}$$

Example A2.2.2

The **projection** matrix $A = H(H^T H)^{-1} H^T$ becomes $A^2 = H(H^T H)^{-1} H^T H$

$$\begin{aligned} (H^T H)^{-1} H^T &= H(H^{-1} H^{-T} H^T H (H^T H)^{-1}) H^T \\ &= H(H^{-1} I H (H^T H)^{-1}) H^T = H(H^T H)^{-1} H^T \end{aligned}$$

Hence, it is an idempotent matrix.

■

A **Toeplitz** square matrix is defined as

$$[A]_{ij} = a_{i-j} \tag{A2.20}$$

$$A = \begin{bmatrix} a_0 & a_{-1} & a_{-2} & \cdots & a_{-(n-1)} \\ a_1 & a_0 & a_{-1} & \cdots & a_{-(n-2)} \\ \vdots & \vdots & \vdots & & \vdots \\ a_{n-1} & a_{n-2} & a_{n-3} & \cdots & a_0 \end{bmatrix} \tag{A2.21}$$

Each element along the northwest-to-southeast diagonals is the same. If in addition $a_{-k} = a_k$, then A is **symmetric Toeplitz**.

The MATLAB functions are as follows:

```
>>A=diag(x);%creates a diagonal matrix A with its diagonal
     %the vector x;
>>A=toeplitz(x);%A is a symmetric Toeplitz matrix;
>>A=toeplitz(x,y);%x and y must be of the same length, the
     %main diagonal will be the first element of
     %x, the first element of y is not used;
```

A2.3 MATRIX OPERATION AND FORMULAS

Addition and Subtraction

$$\boldsymbol{A}+\boldsymbol{B}=\begin{bmatrix} a_{11} & a_{12} & \cdots & a_{1n} \\ a_{21} & a_{22} & \cdots & a_{2n} \\ & \vdots & & \\ a_{m1} & a_{m2} & \cdots & a_{mn} \end{bmatrix}+\begin{bmatrix} b_{11} & b_{12} & \cdots & b_{1n} \\ b_{21} & b_{22} & \cdots & b_{2n} \\ & \vdots & & \\ b_{m1} & b_{m2} & \cdots & b_{mn} \end{bmatrix}$$
$$=\begin{bmatrix} a_{11}+b_{11} & a_{12}+b_{12} & \cdots & a_{1n}+b_{1n} \\ a_{21}+b_{21} & a_{22}+b_{22} & \cdots & a_{2n}+b_{2n} \\ & \vdots & & \\ a_{m1}+b_{m1} & a_{m2}+b_{m2} & \cdots & a_{mn}+b_{mn} \end{bmatrix} \tag{A2.22}$$

Both matrices must have the same dimension.

Multiplication

$$\boldsymbol{AB}\ (m\times n\times n\times k)=\boldsymbol{C}\,(m\times k)$$
$$c_{ij}=\sum_{j=1}^{n} a_{ij}b_{ji} \tag{A2.23}$$

Example A2.3.1

$$\boldsymbol{AB}=\begin{bmatrix} a_{11} & a_{12} \\ a_{21} & a_{22} \\ a_{31} & a_{32} \end{bmatrix}\begin{bmatrix} b_{11} & b_{12} \\ b_{21} & b_{22} \end{bmatrix}$$
$$=\begin{bmatrix} a_{11}b_{11}+a_{12}b_{21} & a_{11}b_{12}+a_{12}b_{22} \\ a_{21}b_{11}+a_{22}b_{21} & a_{21}b_{12}+a_{22}b_{22} \\ a_{31}b_{11}+a_{32}b_{21} & a_{31}b_{12}+a_{32}b_{22} \end{bmatrix};\ 3\times 2\times 2\times 2=3\times 2 \tag{A2.24}$$

■

Transposition

$$(\boldsymbol{AB})^T=\boldsymbol{B}^T\boldsymbol{A}^T \tag{A2.25}$$

Inversion

$$(\boldsymbol{A}^T)^{-1}=(\boldsymbol{A}^{-1})^T \tag{A2.26}$$

$$(\boldsymbol{AB})^{-1}=\boldsymbol{B}^{-1}\boldsymbol{A}^{-1} \tag{A2.27}$$

$$A^{-1} = \frac{C^T}{\det\{A\}} \qquad (A \equiv n \times n \text{ matrix}) \tag{A2.28}$$

$$c_{ij} = (-1)^{i+j} M_{ij} \tag{A2.29}$$

where:

M_{ij} is the minor of a_{ij} obtained by deleting the ith row and jth column of $\boldsymbol{A}$

Example A2.3.2

$$A^{-1} = \begin{bmatrix} 2 & 4 \\ -1 & 5 \end{bmatrix}^{-1} = \frac{1}{10+4}\begin{bmatrix} 5 & 1 \\ -4 & 2 \end{bmatrix}^T = \frac{1}{14}\begin{bmatrix} 5 & -4 \\ 1 & 2 \end{bmatrix}$$

$$AA^{-1} = \begin{bmatrix} 2 & 4 \\ -1 & 5 \end{bmatrix}\frac{1}{14}\begin{bmatrix} 5 & -4 \\ 2 & 2 \end{bmatrix} = \frac{1}{14}\begin{bmatrix} 14 & -8+8 \\ -5+5 & 4+10 \end{bmatrix} = \begin{bmatrix} 1 & 0 \\ 0 & 1 \end{bmatrix} = I$$

■

Determinant [see (A2.5)]

$$A = n \times n \text{ matrix}; \ B = n \times n \text{ matrix}$$

$$\det\{A^T\} = \det\{A\} \tag{A2.30}$$

$$\det\{cA\} = c^n \det\{A\} \tag{A2.31}$$

$$\det\{AB\} = \det\{A\}\det\{B\} \tag{A2.32}$$

$$\det\{A^{-1}\} = \frac{1}{\det\{A\}} \tag{A2.33}$$

Trace [see (A2.10)]

$$A = n \times n \quad \text{matrix}; \quad B = n \times n \quad \text{matrix}$$

$$\mathrm{tr}\{AB\} = \mathrm{tr}\{BA\} \tag{A2.34}$$

$$\mathrm{tr}\{A^T B\} = \sum_{i=1}^{n}\sum_{j=1}^{n} a_{ij} b_{ij} \tag{A2.35}$$

$$\mathrm{tr}\{xy^T\} = y^T x; \quad x, y = \text{vectors} \tag{A2.36}$$

Matrix Inversion Formula

$$A = n \times n; \ B = n \times m; \ C = m \times m; \ D = m \times n$$

$$(A + BCD)^{-1} = A^{-1} - A^{-1}B(DA^{-1}B + C^{-1})^{-1}DA^{-1} \tag{A2.37}$$

$$(A + xx^T)^{-1} = A^{-1} - \frac{A^{-1}xx^T A^{-1}}{1 + x^T A^{-1} x}, \quad x = n \times 1 \text{ vector} \tag{A2.38}$$

Partition Matrices

Examples of 2×2 partition matrices are given as follows:

$$\boldsymbol{AB}=\begin{bmatrix}\boldsymbol{A}_{11} & \boldsymbol{A}_{12}\\ \boldsymbol{A}_{21} & \boldsymbol{A}_{22}\end{bmatrix}\begin{bmatrix}\boldsymbol{B}_{11} & \boldsymbol{B}_{12}\\ \boldsymbol{B}_{21} & \boldsymbol{B}_{22}\end{bmatrix}=\begin{bmatrix}\boldsymbol{A}_{11}\boldsymbol{B}_{11}+\boldsymbol{A}_{12}\boldsymbol{B}_{21} & \boldsymbol{A}_{11}\boldsymbol{B}_{12}+\boldsymbol{A}_{12}\boldsymbol{B}_{22}\\ \boldsymbol{A}_{21}\boldsymbol{B}_{11}+\boldsymbol{A}_{22}\boldsymbol{B}_{21} & \boldsymbol{A}_{21}\boldsymbol{B}_{12}+\boldsymbol{A}_{22}\boldsymbol{B}_{22}\end{bmatrix} \quad \text{(A2.39)}$$

$$\begin{bmatrix}\boldsymbol{A}_{11} & \boldsymbol{A}_{12}\\ \boldsymbol{A}_{21} & \boldsymbol{A}_{22}\end{bmatrix}^T=\begin{bmatrix}\boldsymbol{A}_{11}^T & \boldsymbol{A}_{12}^T\\ \boldsymbol{A}_{21}^T & \boldsymbol{A}_{22}^T\end{bmatrix} \quad \text{(A2.40)}$$

$$\boldsymbol{A}=\begin{bmatrix}\boldsymbol{A}_{11} & \boldsymbol{A}_{12}\\ \boldsymbol{A}_{21} & \boldsymbol{A}_{22}\end{bmatrix}=\begin{bmatrix}k\times k & k\times(n-k)\\ (n-k)\times k & (n-k)\times(n-k)\end{bmatrix}$$

$$\boldsymbol{A}^{-1}=\begin{bmatrix}(\boldsymbol{A}_{11}-\boldsymbol{A}_{12}\boldsymbol{A}_{22}^{-1}\boldsymbol{A}_{21})^{-1} & -(\boldsymbol{A}_{11}-\boldsymbol{A}_{12}\boldsymbol{A}_{22}^{-1}\boldsymbol{A}_{21})^{-1}\boldsymbol{A}_{12}\boldsymbol{A}_{22}^{-1}\\ -(\boldsymbol{A}_{22}-\boldsymbol{A}_{21}\boldsymbol{A}_{11}^{-1}\boldsymbol{A}_{12})^{-1}\boldsymbol{A}_{21}\boldsymbol{A}_{11}^{-1} & (\boldsymbol{A}_{22}-\boldsymbol{A}_{21}\boldsymbol{A}_{11}^{-1}\boldsymbol{A}_{12})^{-1}\end{bmatrix} \quad \text{(A2.41)}$$

$$\det\{\boldsymbol{A}\}=\det\{\boldsymbol{A}_{12}\}\det\{\boldsymbol{A}_{11}-\boldsymbol{A}_{12}\boldsymbol{A}_{22}^{-1}\boldsymbol{A}_{21}\}=\det\{\boldsymbol{A}_{11}\}\det\{\boldsymbol{A}_{22}-\boldsymbol{A}_{21}\boldsymbol{A}_{11}^{-1}\boldsymbol{A}_{12}\} \quad \text{(A2.42)}$$

Some important theorems are as follows:

1. A square matrix $\boldsymbol{A}$ is singular (invertible) if and only if its columns (or rows) are linearly independent or, equivalently, if its $\det\{A\}\neq 0$. If this is true, $\boldsymbol{A}$ is of **full rank**. Otherwise, it is singular.
2. A square matrix $\boldsymbol{A}$ is positive definite, if and only if
   a.
   $$\boldsymbol{A}=\boldsymbol{CC}^T \quad \text{(A2.43)}$$
   where:
   $\boldsymbol{C}$ is a square matrix of the same dimension as $\boldsymbol{A}$ and it is of full rank (invertible)
   b. The principal minors are all positive. (The ith principal minor is the determinant of the submatrix formed by deleting all rows and columns with an index greater than i.) If $\boldsymbol{A}$ can be written as in (A2.43), but $\boldsymbol{C}$ is not of full rank or the principal minors are only nonnegative, then $\boldsymbol{A}$ is positive definite.
3. If $\boldsymbol{A}$ is positive definite, then
   $$\boldsymbol{A}^{-1}=(\boldsymbol{C}^{-1})^T\boldsymbol{C}^{-1} \quad \text{(A2.44)}$$
4. If $\boldsymbol{A}$ is positive definite and $B(m\times n)$ is of full rank ($m\leq n$), then $\mathbf{B},\mathbf{A},\mathbf{B}^T$ is positive definite.
5. If $\boldsymbol{A}$ is positive definite (or positive semidefinite), then the diagonal elements are positive (nonnegative).

A2.4 EIGENDECOMPOSITION OF MATRICES

Let λ denotes an **eigenvalue** of the matrix $\boldsymbol{A}$ $(n \times n)$, then

$$\boldsymbol{A}v = \lambda v \tag{A2.45}$$

where:

v is the **eigenvector** corresponding to the eigenvalue λ

If $\boldsymbol{A}$ is symmetric, then

$$\boldsymbol{A}v_i = \lambda_i v_i, \boldsymbol{A}v_j = \lambda_j v_j (\lambda_i \neq \lambda_j)$$

and

$$v_j^T \boldsymbol{A} v_i = \lambda_i v_j^T v_i \quad \text{(a)}$$

$$v_i^T \boldsymbol{A} v_j = \lambda_j v_i^T v_j \ \text{ or } \ v_j^T \boldsymbol{A} v_i = \lambda_j v_j^T v_i \quad \text{(b)}$$

Subtracting (a) from (b), we obtain $(\lambda_i - \lambda_j) v_j^T v_i = 0$. But $\lambda_i \neq \lambda_j$, and hence $v_j^T v_i = 0$, which implies that the eigenvectors of a symmetric matrix are orthogonal. We can proceed and normalize them producing orthonormal eigenvectors.

From (A2.45), we write

$$\boldsymbol{A}\begin{bmatrix} v_1 & v_2 & \cdots & v_n \end{bmatrix} = \begin{bmatrix} \lambda_1 v_1 & \lambda_2 v_2 & \cdots & \lambda_n v_n \end{bmatrix}$$

or (A2.46)

$$\boldsymbol{AV} = \boldsymbol{V\Lambda}$$

where:

$$\Lambda = \begin{bmatrix} \lambda_1 & 0 & 0 & \cdots & 0 \\ 0 & \lambda_2 & 0 & \cdots & 0 \\ & \vdots & & & \\ 0 & 0 & 0 & \cdots & \lambda_n \end{bmatrix} \tag{A2.47}$$

Because v_i are mutually orthogonal, $v_i^T v_j = \delta_{ij}$ makes $\boldsymbol{V}$ a **unitary** matrix, $\boldsymbol{V}^T\boldsymbol{V} = \boldsymbol{I} = \boldsymbol{V}\boldsymbol{V}^T$. Postmultiplying (A2.46) by $\boldsymbol{V}^T$, we obtain

$$\boldsymbol{A} = \boldsymbol{V\Lambda V}^T \sum_{i=1}^{n} \lambda_i v_i v_i^T \tag{A2.48}$$

which is known as **unitary decomposition** of $\boldsymbol{A}$. We also say that $\boldsymbol{A}$ is **unitary similar** to the diagonal Λ, because a unitary matrix $\boldsymbol{V}$ takes $\boldsymbol{A}$ to diagonal form: $\boldsymbol{V}^T\boldsymbol{AV} = \Lambda$.

If $\Lambda = \boldsymbol{I}$, then from (A2.48) $\boldsymbol{A} = \boldsymbol{V}\boldsymbol{V}^T = \boldsymbol{I}$, and hence,

$$\boldsymbol{I} = \boldsymbol{V}\boldsymbol{V}^T = \sum_{i=1}^{n} v_i v_i^T \tag{A2.49}$$

Each of the terms in the summation is of rank 1 projection matrix:

$$\boldsymbol{P}_i^2 = v_i v_i^T v_i v_i^T = v_i v_i^T = \boldsymbol{P}_i \qquad (v_i^T v_i = 1) \tag{A2.50}$$

$$\boldsymbol{P}_i^T = v_i v_i^T = \boldsymbol{P}_i \tag{A2.51}$$

Hence, we write [see (A2.48) and (A2.49)]

$$\boldsymbol{A} = \sum_{i=1}^{n} \lambda_i \boldsymbol{P}_i \tag{A2.52}$$

$$\boldsymbol{I} = \sum_{i=1}^{n} \boldsymbol{P}_i \tag{A2.53}$$

Inverse

Because $\boldsymbol{V}$ is unitary matrix, $\boldsymbol{V}\boldsymbol{V}^T = \boldsymbol{I}$ or $\boldsymbol{V}^T = \boldsymbol{V}^{-1}$ or $\boldsymbol{V} = (\boldsymbol{V}^T)^{-1}$, and therefore,

$$\boldsymbol{A}^{-1} = (\boldsymbol{V}^T)^{-1}\Lambda^{-1}\boldsymbol{V}^{-1} = \boldsymbol{V}\Lambda^{-1}\boldsymbol{V}^T = \sum_{i=1}^{n} \frac{1}{\lambda_i} v_i v_i^T \tag{A2.54}$$

Determinant

$$\det\{\boldsymbol{A}\} = \det\{\boldsymbol{V}\}\det\{\Lambda\}\det\{\boldsymbol{V}^{-1}\} = \det\{\Lambda\} = \prod_{i=1}^{n} \lambda_i \tag{A2.55}$$

A2.5 MATRIX EXPECTATIONS

$$\boldsymbol{E}\{\boldsymbol{x}\} = \boldsymbol{m}_x \tag{A2.56}$$

$$\boldsymbol{E}\{(\boldsymbol{x} - \boldsymbol{m}_x)(\boldsymbol{x} - \boldsymbol{m}_x)^T\} = \boldsymbol{R}_{xx} \tag{A2.57}$$

$$\boldsymbol{E}\{\text{tr}\{\boldsymbol{A}\}\} = \text{tr}\{\boldsymbol{E}\{\boldsymbol{A}\}\} \tag{A2.58}$$

$$\boldsymbol{E}\{\boldsymbol{A}\boldsymbol{x} + \boldsymbol{b}\} = \boldsymbol{A}\boldsymbol{m}_x + \boldsymbol{b} \tag{A2.59}$$

$$\boldsymbol{E}\{\boldsymbol{x}\boldsymbol{x}^T\} = \boldsymbol{R}_{xx} + \boldsymbol{m}_x\boldsymbol{m}_x^T \tag{A2.60}$$

$$\boldsymbol{E}\{\boldsymbol{x}\boldsymbol{a}^T\boldsymbol{x}\} = (\boldsymbol{R}_{xx} + \boldsymbol{m}_x\boldsymbol{m}_x^T)\,\boldsymbol{a} \tag{A2.61}$$

$$\boldsymbol{E}\{(\boldsymbol{x} + \boldsymbol{a})(\boldsymbol{x} + \boldsymbol{a})^T\} = \boldsymbol{R}_{xx} + (\boldsymbol{m}_x + \boldsymbol{a})(\boldsymbol{m}_x + \boldsymbol{a})^T \tag{A2.62}$$

$$E\{x^T x\} = \text{tr}\{R_{xx}\} + m_x^T m_x = \text{tr}\{R_{xx} + m_x m_x^T\} \tag{A2.63}$$

$$E\{x^T a x^T\} = a^T \{R_{xx} + m_x m_x^T\} \tag{A2.64}$$

$$E\{x^T A x\} = \text{tr}\{A R_{xx}\} + m_x^T A m_x = \text{tr}\{A(R_{xx} + m_x m_x^T)\} \tag{A2.65}$$

A2.6 DIFFERENTIATION OF A SCALAR FUNCTION WITH RESPECT TO A VECTOR

$$x = [x_1 \ \ x_2 \ \cdots \ x_n]^T; \quad \frac{\partial}{\partial x} = \left[\frac{\partial}{\partial x_1} \quad \frac{\partial}{\partial x_2} \quad \cdots \quad \frac{\partial}{\partial x_n}\right]^T$$

$$\frac{\partial}{\partial x}(y^T x) = \frac{\partial}{\partial x}(x^T y) = y \tag{A2.66}$$

$$\frac{\partial}{\partial x}(x^T A) = A \tag{A2.67}$$

$$\frac{\partial}{\partial x}(x^T) = I \tag{A2.68}$$

$$\frac{\partial}{\partial x}(x^T x) = 2x \tag{A2.69}$$

$$\frac{\partial}{\partial x}(x^T A y) = A y \tag{A2.70}$$

$$\frac{\partial}{\partial x}(y^T A x) = A^T y \tag{A2.71}$$

$$\frac{\partial}{\partial x}(x^T A x) = (A + A^T) x \tag{A2.72}$$

$$\frac{\partial}{\partial x}(x^T A x) = 2 A x \quad \text{if } A \text{ is symmetric} \tag{A2.73}$$

$$\frac{\partial}{\partial x}(a^T A x x^T) = (A + A^T) x x^T + x^T A x I \tag{A2.74}$$

Appendix 3
Mathematical Formulas

A3.1 TRIGONOMETRIC IDENTITIES

$$\cos(-a) = \cos a$$

$$\sin(-a) = -\sin a$$

$$\cos\left(a \pm \frac{\pi}{2}\right) = \mp \sin a$$

$$\sin\left(a \pm \frac{\pi}{2}\right) = \pm \cos a$$

$$\cos(a \pm \pi) = -\cos a$$

$$\sin(a \pm \pi) = -\sin a$$

$$\cos^2 a + \sin^2 a = 1$$

$$\cos^2 a - \sin^2 a = \cos 2a$$

$$\cos(a \pm b) = \cos a \cos b \mp \sin a \sin b$$

$$\sin(a \pm b) = \sin a \cos b \pm \cos a \sin b$$

$$\cos a \cos b = \frac{1}{2}[\cos(a-b) + \cos(a+b)]$$

$$\sin a \sin b = \frac{1}{2}[\cos(a-b) - \cos(a+b)]$$

$$\sin a \cos b = \frac{1}{2}[\sin(a-b) + \sin(a+b)]$$

$$c \cos a + d \sin a = \sqrt{a^2 - b^2} \cos\left[a - \tan^{-1}\left(\frac{d}{c}\right)\right]$$

$$\cos^2 a = \frac{1}{2}(1 + \cos 2a)$$

$$\cos^3 a = \frac{1}{4}(3\cos a + \cos 3a)$$

$$\cos^4 a = \frac{1}{8}(3 + 4\cos 2a + \cos 4a)$$

$$\sin^2 a = \frac{1}{2}(1 - \cos 2a)$$

$$\sin^3 a = \frac{1}{4}(3\sin a - \sin 3a)$$

$$\sin^4 a = \frac{1}{8}(3 - 4\cos 2a + \cos 4a)$$

$$\mathrm{e}^{\pm ja} = \cos a \pm j \sin a$$

$$\cosh a = \cos ja$$

$$\cos a = \frac{1}{2}(\mathrm{e}^{ja} + \mathrm{e}^{-ja})$$

$$\sin a = \frac{j}{2}(\mathrm{e}^{-ja} - \mathrm{e}^{ja})$$

$$\sinh a = -j \sin ja$$

$$\tanh a = -j \tan ja$$

A3.2 ORTHOGONALITY

$$\sum_{n=0}^{N-1} \cos\left(\frac{2\pi k}{N}\right)n \cos\left(\frac{2\pi l}{N}\right)n = 0 \qquad 1 \le k,l \le N-1,\ k \ne l$$

$$\sum_{n=0}^{N-1} \sin\left(\frac{2\pi k}{N}\right)n \sin\left(\frac{2\pi l}{N}\right)n = 0 \qquad 1 \le k,l \le N-1,\ k \ne l$$

$$\sum_{n=0}^{N-1} \sin\left(\frac{2\pi k}{N}\right)n \cos\left(\frac{2\pi l}{N}\right)n = 0 \qquad 1 \le k,l \le N-1,\ k \ne l$$

$$\sum_{n=0}^{N-1} \cos^2\left(\frac{2\pi k}{N}\right)n = \begin{cases} N/2 & 1 \le k,l \le N-1,\ k \ne l \\ N & k = 0, N/2 \end{cases}$$

$$\sum_{n=0}^{N-1} \sin^2\left(\frac{2\pi k}{N}\right)n = \begin{cases} N/2 & 1 \le k,l \le N-1,\ k \ne l \\ 0 & k = 0, N/2 \end{cases}$$

The above formulas are correct if all k, l, and n are replaced by $k \bmod N$ and $l \bmod N$.

A3.3 SUMMATION OF TRIGONOMETRIC FORMS

$$\sum_{n-0}^{N-1} \cos\left(\frac{2\pi k}{N}\right) n = \begin{cases} 0 & 1 \le k \le N-1 \\ N & k = 0, N \end{cases}$$

$$\sum_{n-0}^{N-1} \sin\left(\frac{2\pi k}{N}\right) n = \begin{cases} 0 & 1 \le k \le N-1 \\ N & k = 0, N \end{cases}$$

where:

k, l, and n are integers

A3.4 SUMMATION FORMULAS

A3.4.1 Finite Summation Formulas

$$\sum_{k=0}^{n} a^k = \frac{1-a^{n+1}}{1-a} \qquad a \ne 1$$

$$\sum_{k=1}^{n} ka^k = \frac{[a1-(n+1)a^n + na^{n+1}]}{(1-a)^2} \qquad a \ne 1$$

$$\sum_{k=1}^{n} k^2 a^k = \frac{a[(1+a)-(n+1)^2 a^n + (2n^2+2n-1)a^{n+1} - n^2 a^{n+2}]}{(1-a)^3} \qquad a \ne 1$$

$$\sum_{k=1}^{n} k = \frac{n(n+1)}{2}$$

$$\sum_{k=1}^{n} k^2 = \frac{n(n+1)(2n+1)}{6}$$

$$\sum_{k=1}^{n} k^3 = \frac{n^2(n+1)^2}{4}$$

$$\sum_{k=0}^{2n-1} (2k+1) = n^2$$

A3.4.2 Infinite Summation Formulas

$$\sum_{k=0}^{\infty} a^k = \frac{1}{1-a} \qquad |a| < 1$$

$$\sum_{k=0}^{\infty} k a^k = \frac{a}{(1-a)^2} \qquad |a| < 1$$

$$\sum_{k=0}^{\infty} k^2 a^k = \frac{a^2 + a}{(1-a)^3} \qquad |a| < 1$$

A3.5 SERIES EXPANSIONS

$$e^a = 1 + a + \frac{a^2}{2!} + \frac{a^3}{3!} + \cdots$$

$$\ln(1+a) = a - \frac{a^2}{2} + \frac{a^3}{3} - \frac{a^4}{4} + \cdots \qquad |a| < 1$$

$$\sin a = a - \frac{a^3}{3!} + \frac{a^5}{5!} - \frac{a^7}{7!} + \cdots$$

$$\cos a = 1 - \frac{a^2}{2!} + \frac{a^4}{4!} - \frac{a^6}{6!} + \cdots$$

$$\tan a = a + \frac{a^3}{3} + \frac{2a^5}{15} + \frac{17a^7}{315} + \cdots \qquad |a| < \frac{\pi}{2}$$

$$\sinh a = a + \frac{a^3}{3!} + \frac{a^5}{5!} + \frac{a^7}{7!} + \cdots$$

$$\cosh a = 1 + \frac{a^2}{2!} + \frac{a^4}{4!} + \frac{a^6}{6!} + \cdots$$

$$\tanh a = a - \frac{a^3}{3} + \frac{2a^5}{15} - \frac{17a^7}{315} + \cdots \qquad |a| < \frac{\pi}{2}$$

$$(1+a)^n = 1 + na + \frac{n(n-1)}{2!} a^2 + \frac{n(n-1)(n-2)}{3!} a^3 + \cdots \qquad |a| < 1$$

A3.6 LOGARITHMS

$$\log_b N = \log_a N \log_b a = \frac{\log_a N}{\log_a b}$$

A3.7 SOME DEFINITE INTEGRALS

$$\int_0^{\infty} x^2 e^{-ax} dx = \frac{2}{a^3}$$

$$\int_0^\infty x^n e^{-ax}\,dx = \frac{n!}{a^{n+1}} \qquad a > 0$$

$$\int_0^\infty e^{-a^2x^2}\,dx = \frac{\sqrt{\pi}}{2a} \qquad a > 0$$

$$\int_0^\infty xe^{-a^2x^2}\,dx = \frac{1}{2a^2} \qquad a > 0$$

$$\int_0^\infty \frac{e^{-ax}}{x}\sin(mx)\,dx = \tan^{-1}\left(\frac{m}{a}\right) \qquad a > 0$$

$$\int_0^\infty \frac{\sin mx}{x}\,dx = \frac{\pi}{2}$$

Appendix 4
Lagrange Multiplier Method

To solve a constrained optimization problem, the Lagrange method is used. The normalized least mean-square (NLMS) recursion, for example, can be obtained as a solution to the following problem:

$$\text{minimize } \min_{w} \left\| \boldsymbol{w}(n) - \boldsymbol{w}(n-1) \right\|^2 \text{ subject to constraint } d(n) = \boldsymbol{w}^T(n+1)\boldsymbol{x}(n)$$

The first step in the solution is to write the cost function as follows:

$$J(n) = [\boldsymbol{w}(n+1) - \boldsymbol{w}(n)]^T [\boldsymbol{w}(n+1) - \boldsymbol{w}(n)] + \lambda[d(n) - \boldsymbol{w}^T(n+1)\boldsymbol{x}(n)] \quad \text{(A4.1)}$$

Differentiating the cost function above with respect to $\boldsymbol{w}(n+1)$, we obtain

$$\frac{J(n)}{\partial \boldsymbol{w}(n+1)} = 2[\boldsymbol{w}(n+1) - \boldsymbol{w}(n)] - \lambda \boldsymbol{x}(n) \quad \text{(A4.2)}$$

Setting the results to 0, we obtain

$$\boldsymbol{w}(n+1) = \boldsymbol{w}(n) + \frac{1}{2}\lambda \boldsymbol{x}(n) \quad \text{(A4.3)}$$

Substituting this result into the constraint $d(n) = \boldsymbol{w}^T(n+1)\boldsymbol{x}(n)$, we obtain

$$d(n) = \left[\boldsymbol{w}(n) + \frac{1}{2}\lambda \boldsymbol{x}(n) \right]^T \boldsymbol{x}(n) = \boldsymbol{w}^T(n)\boldsymbol{x}(n) + \frac{1}{2}\lambda \left\| \boldsymbol{x}(n) \right\|^2 \quad \text{(A4.4)}$$

Since $e(n) = d(n) - \boldsymbol{w}^T(n)\boldsymbol{x}(n)$, solving (A4.4) for λ leads to

$$\lambda = \frac{2e(n)}{\left\| \boldsymbol{x}(n) \right\|^2} \quad \text{(A4.5)}$$

Substituting (A4.5) in (A4.3), we find

$$\boldsymbol{w}(n+1) = \boldsymbol{w}(n) + \frac{1}{\left\| \boldsymbol{x}(n) \right\|^2} e(n)\boldsymbol{x}(n) \quad \text{(A4.6)}$$

Finally, introducing a factor μ in (A4.6) to control the change in the weight vector, we obtain the conventional NLMS algorithm:

$$\mathbf{w}(n+1) = \mathbf{w}(n) + \frac{\mu}{\|\mathbf{x}(n)\|^2} e(n)\mathbf{x}(n) \quad \text{(A4.7)}$$

Bibliography

Brockwell, R. J. and Davis, R. A., *Time Series—Theory and Methods*, New York, Springer-Verlag, 1991.

Clarkson, P. M., *Optimal and Adaptive Signal Processing*, Boca Raton, FL, CRC Press, 1993.

Diniz, P. S. R., *Adaptive Filtering*, Fourth Edition, New York, Springer-Verlag, 2013.

Farhang-Boroujeny, B., *Adaptive Filters: Theory and Applications*, Second Edition, New York, John Wiley & Sons, 2013.

Hayes, M. H., *Statistical Digital Signal Processing and Modeling*, New York, John Wiley & Sons, 1996.

Haykin, S., *Adaptive Filter Theory*, Fourth Edition, Upper Saddle River, NJ, Prentice Hall, 2003.

Haykin, S. and Widrow, B., *Least-Mean-Square Adaptive Filters*, Hoboken, NJ, Wiley Interscience, 2003.

Sayed, A. H., *Fundamentals of Adaptive Filtering*, Hoboken, NJ, Wiley Interscience, 2004.

Sayed, A. H., *Adaptive Filters*, Hoboken, NJ, Wiley Interscience, 2008.

Widrow, B. and Sterns, S. D., *Adaptive Signal Processing*, Englewood Cliffs, NJ, Prentice Hall, 1985.

Index

Note: Locators followed by "*f*" and "*t*" denote figures and tables in the text

M

For Product Safety Concerns and Information please contact our EU representative GPSR@taylorandfrancis.com
Taylor & Francis Verlag GmbH, Kaufingerstraße 24, 80331 München, Germany

www.ingramcontent.com/pod-product-compliance
Lightning Source LLC
LaVergne TN
LVHW011110300926
849160LV00008B/328

* 9 7 8 1 4 8 2 2 5 3 3 5 1 *